THE KUBERNETES BIBLE

The definitive guide to deploying and managing
Kubernetes across major cloud platforms

Nassim Kebbani

Piotr Tylenda

Russ McKendrick

Packt>

BIRMINGHAM—MUMBAI

THE KUBERNETES BIBLE

Copyright © 2022 Packt Publishing

Group Product Manager: Rahul Nair

Publishing Product Manager: Shrilekha Malpani

Senior Editor: Athikho Sapuni Rishana

Content Development Editor: Nihar Kapadia

Technical Editor: Shruthi Shetty

Copy Editor: Safis Editing

Project Coordinator: Neil D'mello

Proofreader: Safis Editing

Indexer: Tejal Daruwale Soni

Production Designer: Alishon Mendonca

Marketing Coordinator: Sanjana Gupta

First published: February 2022

Production reference: 1130122

Published by Packt Publishing Ltd.

Livery Place

35 Livery Street

Birmingham

B3 2PB, UK.

ISBN 978-1-83882-769-4

www.packt.com

Contributors

About the authors

Nassim Kebbani is an experienced software engineer with in-depth expertise in Kubernetes and the Amazon Web Services cloud provider. He has an extensive background both in software development and operations teams, having implemented the entire spectrum of a DevOps life cycle chain, from application code to pipelines, and carried out monitoring in various industries such as e-commerce, media, and financial services.

He has implemented numerous cloud-native architectures and containerized applications on Docker and AWS and holds both Kubernetes CKA and CKAD certifications.

Piotr Tylenda is an experienced DevOps and software engineer with a passion for Kubernetes and Azure technologies. In his projects, he has focused on the adoption of microservices architecture for monolithic applications, developing big data pipelines for e-commerce, and architecting solutions for scalable log and telemetry analytics for hardware. His most notable contribution to Kubernetes' open source ecosystem is the development of Ansible automation for provisioning and deploying hybrid Windows/ Linux Kubernetes clusters. Currently, he works at Microsoft Development Center Copenhagen in Denmark as part of a team developing a Microsoft Dynamics 365 Business Central SaaS offering.

Russ McKendrick is an experienced DevOps practitioner and system administrator with a passion for automation and containers. He has been working in IT and related industries for the better part of 27 years. During his career, he has had varied responsibilities in many different sectors, including first-line, second-line, and senior support in both client-facing and internal teams for small and large organizations. He works almost exclusively with Linux, using open source systems and tools across both dedicated hardware and virtual machines hosted in public and private clouds at N4Stack, which is a Node4 company, where he holds the title of practice manager (SRE and DevOps). He also buys way too many records!

About the reviewer

Dushyant Nathalal Dubaria pursued a Master of Science degree in network and telecommunication engineering from Southern Methodist University, USA, in December 2019. He is passionate about DevOps, the cloud, networking, virtualization, and automation, and has a keen interest in technologies such as Docker and Kubernetes. He has published articles in three international journal publications in IEEE, held at Columbia University, New York, and at the University of Nevada, Las Vegas. He was a teaching assistant and lecturer at Southern Methodist University and loves teaching DevOps for networking, network automation, and programmability to graduate students. He likes contributing to open source projects and was a speaker at the Linux Foundation Event: Open Networking Summit, San Jose, April 2019, and at the Open Source Summit, San Diego, August 2019.

Reviewing a book is harder than I thought and more rewarding than I could have ever imagined. I would like to thank my friend, Jay Ashok Shah, who introduced me to the world of DevOps and Kubernetes.

To my caring parents for always encouraging me to accomplish my goals.

To my loving wife, Monika, for her continued support with everything I do.

Finally, my sincere appreciation to the Packt committee for the learning opportunities they have afforded me, especially Kajol Pawar and Neil D'mello.

Table of Contents

2

Kubernetes Architecture – From Docker Images to Running Pods

3

Installing Your First Kubernetes Cluster

Section 2: Diving into Kubernetes Core Concepts

4

Running Your Docker Containers

5

Using Multi-Container Pods and Design Patterns

6

Configuring Your Pods Using ConfigMaps and Secrets

7

Exposing Your Pods with Services

8
Managing Namespaces in Kubernetes

9

Persistent Storage in Kubernetes

Section 3: Using Managed Pods with Controllers

10

Running Production-Grade Kubernetes Workloads

11

Deployment – Deploying Stateless Applications

12

StatefulSet – Deploying Stateful Applications

15

Launching a Kubernetes Cluster on Amazon Web Services with Amazon Elastic Kubernetes Service

16

Kubernetes Clusters on Microsoft Azure with Azure Kubernetes Service

Section 5: Advanced Kubernetes

17
Working with Helm Charts

18
Authentication and Authorization on Kubernetes

19
Advanced Techniques for Scheduling Pods

20

Autoscaling Kubernetes Pods and Nodes

21

Advanced Traffic Routing with Ingress

Index

Other Books You May Enjoy

Preface

Containers have allowed a real leap forward since their massive adoption in the world of virtualization because they have allowed greater flexibility, especially these days, when buzzwords such as **cloud**, **agile**, and **DevOps** are on everyone's lips.

Today, almost no one questions the use of containers and they're basically everywhere, especially since the success of Docker.

Containers have brought tremendous flexibility to organizations, but they have remained questionable for a very long time when organizations went to face the challenge of deploying them in production. For years, companies were using containers for proof-of-concept projects, local development, and suchlike, but the use of containers for real production workloads was inconceivable for many organizations.

Container orchestrators were the game-changer, with Kubernetes in the lead.

Originally built by Google, Kubernetes is today the leading container orchestrator that is providing you with all the features you need in order to deploy containers in production at scale. Kubernetes is popular, but it is also complex. This tool is so versatile that getting started with it and progressing to advanced usages is not an easy task: it is not an easy tool to learn and operate.

As an orchestrator, Kubernetes has its own concepts independent of those of a container engine, such as Docker. But when both are used together, you get a very strong platform ready to deploy your cloud-native applications in production. As engineers working with Kubernetes daily, we were convinced, like many, that it was a technology to master and we decided to share our knowledge in order to make Kubernetes accessible by covering most of this orchestrator.

This book is entirely dedicated to Kubernetes and is the result of our work: it provides a broad view of Kubernetes and covers a lot of aspects of the orchestrators, from pure container Pod creation to deploying the orchestrator on the public cloud. We didn't want this book to be a *Getting started* guide.

We hope this book will teach you everything you want to learn about Kubernetes!

Who this book is for

This book is for people who intend to use Kubernetes with Docker. Although Kubernetes can be used together with a lot of different container engines and is not tied to Docker, the combination between the two remains the most frequent use case of Kubernetes.

This book is very technical. It mainly focuses on Kubernetes and Docker from an engineering perspective, and thus, it is dedicated to engineers, whether they come from a developer or a system background, and not to project managers. It is a Kubernetes bible for people who are going to use Kubernetes daily, or for people who wish to discover this tool. You shouldn't be afraid of typing some commands on a terminal.

Being a total beginner to Kubernetes or having an intermediate level is not a problem, but you must already have some technical ability with Docker to follow this book. Containers should be familiar to you. This book can also serve as a guide if you are in the process of migrating an existing application to Kubernetes.

The book incorporates content that will allow readers to deploy Kubernetes on public cloud offerings such as Amazon EKS or Google GKE. Cloud users who wish to add Kubernetes to their stack on the cloud will appreciate this book.

What this book covers

Chapter 1, Kubernetes Fundamentals, is an introduction to Kubernetes. We're going to explain what Kubernetes is, why it was created, who created it, who is making this project alive, and when and why you should use it as part of your stack.

Chapter 2, Kubernetes Architecture – from Docker Images to Running Pods, covers how Kubernetes is built as a distributed software, and is technically not a single monolith binary but built as a set of microservices interacting with each other. We're going to explain this architecture and how Kubernetes proceeds to translate your instructions into running Docker containers.

Chapter 3, Installing Your First Kubernetes Cluster, explains that Kubernetes is really difficult to install due to its distributed nature, so to make the process easier, it is possible to install by using one of its distributions. Kind and Minikube are two options we're going to discover in this chapter to have a Kubernetes cluster working on your machine.

Chapter 4, Running Your Docker Containers, is an introduction to the concept of Pods.

Chapter 5, Using Multi-Container Pods and Design Patterns, introduces multi-container Pods and the design patterns, such as a proxy or sidecar that you can build when running several containers as part of the same Pod.

Chapter 6, Configuring Your Pods Using ConfigMaps and Secrets, explains how, in Kubernetes, we separate Kubernetes applications from their configurations. Both applications and configurations have their own life cycle thanks to the ConfigMap and Secret resources. This chapter will be dedicated to these two objects and how to mount data in ConfigMap and Secret as environment variables or volumes mounted on your Pod.

Chapter 7, Exposing Your Pods with Services, teaches you the notion of services in Kubernetes. Each Pod in Kubernetes gets assigned its own IP address dynamically. Services are extremely useful if you want to provide a consistent one to expose Pods within your cluster to other Pods or to the outside world, with a single static DNS name. You'll learn here that there are three main service types, called ClusterIp, NodePort, and LoadBalancer, which are all dedicated to a single use case in terms of Pod exposition.

Chapter 8, Managing Namespaces in Kubernetes, explains how using namespaces is a key aspect of cluster management and forcibly, you'll have to deal with namespaces during your journey with Kubernetes. Though it's a simple notion, it is a key one, and you'll have to master namespaces perfectly in order to be successful with Kubernetes.

Chapter 9, Persistent Storage in Kubernetes, covers how, by default, Pods are not persistent. As they're just managing raw Docker containers in the end, destroying them will result in the loss of your data. The solution to that is the usage of persistent storage thanks to the PersistentVolume and PersistentVolumeClaim resource kinds. This chapter is dedicated to these two objects and the StorageClass object: it will teach you that Kubernetes is extremely versatile in terms of storage and that your Pods can be interfaced with a lot of different storage technologies.

Chapter 10, Running Production-Grade Kubernetes Workloads, takes a deep dive into high availability and fault tolerance in Kubernetes using ReplicationController and ReplicaSet.

Chapter 11, Deployment – Deploying Stateless Applications, is a continuation of the previous chapter and explains how to manage multiple versions of ReplicaSets using the Deployment object. This is the basic building block for stateless applications running on Kubernetes.

Chapter 12, StatefulSet – Deploying Stateful Applications, takes a look at the next important Kubernetes object: StatefulSet. This object is the backbone of running stateful applications on Kubernetes. We explain the most important differences between running stateless and stateful applications using Kubernetes.

Chapter 13, DaemonSet – Maintaining Pod Singletons on Nodes, covers DaemonSet, which is a special Kubernetes object that can be used for running operational or supporting workloads on Kubernetes clusters. Whenever you need to run precisely one container Pod on a single Kubernetes node, DaemonSet is what you need.

Chapter 14, Kubernetes Clusters on Google Kubernetes Engine, looks at how we can move our Kubernetes workload to Google Cloud using both the native command-line client and the Google Cloud console.

Chapter 15, Launching a Kubernetes Cluster on Amazon Web Services with Amazon Elastic Kubernetes Service, looks at moving the workload we launched in the previous chapter to Amazon's Kubernetes offering.

Chapter 16, Kubernetes Clusters on Microsoft Azure with Azure Kubernetes Service, looks at launching a cluster in Microsoft Azure.

Chapter 17, Working with Helm Charts, covers Helm Charts, which is a dedicated packaging and redistribution tool for Kubernetes applications. Armed with knowledge from this chapter, you will be able to quickly set up your Kubernetes development environment or even plan for the redistribution of your Kubernetes application as a dedicated Helm Chart.

Chapter 18, Authentication and Authorization on Kubernetes, covers authorization using built-in role-based access control and authorization schemes together with user management.

Chapter 19, Advanced Techniques for Scheduling Pods, takes a deeper look at Node affinity, Node taints and tolerations, and advanced scheduling policies in general.

Chapter 20, Autoscaling Kubernetes Pods and Nodes, introduces the principles behind autoscaling in Kubernetes and explains how to use Vertical Pod Autoscaler, Horizontal Pod Autoscaler, and Cluster Autoscaler.

Chapter 21, Advanced Traffic Routing with Ingress, covers Ingress objects and IngressController in Kubernetes. We explain how to use nginx as an implementation of IngressController and how you can use Azure Application Gateway as a native IngressController in Azure environments.

To get the most out of this book

It is necessary to have some prior knowledge to get the most out of this book. Indeed, this book is dedicated to Kubernetes, and although this orchestrator can be used with many container engines, this book will be about using Kubernetes in combination with Docker. It is therefore necessary to know Docker as much as possible. You don't have to be an expert, but you should be able to launch and manage applications on Docker before reading this book.

While it is possible to run Windows containers with Kubernetes, most of the topics covered in this book will be Linux-based. Having a good knowledge of Linux will be helpful, but not required. Again, you don't have to be an expert: knowing how to use a terminal session and basic Bash scripting should be enough.

Lastly, having some general knowledge of software architecture such as REST APIs will be beneficial.

Software/hardware covered in the book	OS requirements
Kubernetes >= 1.17	Windows, macOS X, and Linux (any)
Kubectl >= 1.17	Windows, macOS X, and Linux (any)

We strongly advise you to not attempt to install Kubernetes or Kubectl on your machine for now. Kubernetes is not a single binary but is a distributed software composed of several components and as such, it is really complex to install a complete Kubernetes cluster from scratch. Instead, we recommend that you follow the third chapter of this book, which is dedicated to the setup of Kubernetes.

If you are using the digital version of this book, we advise you to type the code yourself or access the code via the GitHub repository (link available in the next section). Doing so will help you avoid any potential errors related to the copying and pasting of code.

Please note that Kubernetes and Kubectl are the two tools we're going to use most frequently in this book, but there is a huge ecosystem around Kubernetes and we might install additional software not mentioned in this section. This book is also about using Kubernetes in the cloud, and we're going to discover how to provision Kubernetes clusters on public cloud platforms such as Amazon Web Services and Google Cloud Platform. As part of this setup, we might install additional software dedicated to these platforms that are not strictly bound to Kubernetes, but also to other services provided by these platforms.

Download the example code files

You can download the example code files for this book from GitHub at `https://github.com/PacktPublishing/The-Kubernetes-Bible`. In case there's an update to the code, it will be updated on the existing GitHub repository.

We also have other code bundles from our rich catalog of books and videos available at `https://github.com/PacktPublishing/`. Check them out!

Download the color images

We also provide a PDF file that has color images of the screenshots/diagrams used in this book. You can download it here: `https://static.packt-cdn.com/downloads/9781838827694_ColorImages.pdf`.

Conventions used

There are a number of text conventions used throughout this book.

`Code in text`: Indicates code words in the text, database table names, folder names, filenames, file extensions, pathnames, dummy URLs, user input, and Twitter handles. Here is an example: "Now, we need to create a `kubeconfig` file for our local Kubectl CLI."

A block of code is set as follows:

```
apiVersion: v1
kind: Pod
metadata:
  name: nginx-Pod
```

When we wish to draw your attention to a particular part of a code block, the relevant lines or items are set in bold:

```
apiVersion: v1
kind: ReplicationController
metadata:
  name: nginx-replicationcontroller-example
```

Any command-line input or output is written as follows:

```
$ kubectl get nodes
```

Bold: Indicates a new term, an important word, or words that you see on screen. For example, words in menus or dialog boxes appear in the text like this. Here is an example: "On this screen, you should see an **Enable Billing** button."

> **Tips or Important Notes**
> Appear like this.

Get in touch

Feedback from our readers is always welcome.

General feedback: If you have questions about any aspect of this book, mention the book title in the subject of your message and email us at customercare@packtpub.com.

Errata: Although we have taken every care to ensure the accuracy of our content, mistakes do happen. If you have found a mistake in this book, we would be grateful if you would report this to us. Please visit www.packtpub.com/support/errata, selecting your book, clicking on the Errata Submission Form link, and entering the details.

Piracy: If you come across any illegal copies of our works in any form on the internet, we would be grateful if you would provide us with the location address or website name. Please contact us at copyright@packt.com with a link to the material.

If you are interested in becoming an author: If there is a topic that you have expertise in and you are interested in either writing or contributing to a book, please visit authors.packtpub.com.

Share Your Thoughts

Once you've read *The Kubernetes Bible*, we'd love to hear your thoughts! Scan the QR code below to go straight to the Amazon review page for this book and share your feedback.

https://packt.link/r/1838827692

Your review is important to us and the tech community and will help us make sure we're delivering excellent quality content.

Section 1: Introducing Kubernetes

Kubernetes is a fantastic container orchestrator tool that can manage Docker containers at a large scale. Let's discover what Kubernetes is exactly, how it started as an internal project at Google to become a leading solution, and how it can help you today to manage Docker containers in production.

This part of the book comprises the following chapters:

- *Chapter 1, Kubernetes Fundamentals*
- *Chapter 2, Kubernetes Architecture – From Docker Images to Running Pods*
- *Chapter 3, Installing Your First Kubernetes Cluster*

1
Kubernetes Fundamentals

Welcome to *The Kubernetes Bible*. This is the first chapter of this book, and I'm happy to accompany you on your journey with Kubernetes. If you are working in the software development industry, you have probably heard about Kubernetes. This is normal because the popularity of Kubernetes has grown a lot in recent years.

Built by Google, Kubernetes is the leading container orchestrator solution in terms of popularity and adoption: it's the tool you need if you are looking for a solution to manage containerized applications in production at scale, whether it's on-premises or on a public cloud. Be focused on the word. Deploying and managing containers at scale is extremely difficult because, by default, container engines such as Docker do not provide any way on their own to maintain the availability and scalability of containers at scale.

Kubernetes first emerged as a Google project, and they put a lot of effort into building a solution to deploy a huge number of containers on their massively distributed infrastructure. By adopting Kubernetes as part of your stack, you'll get an open source platform that was built by one of the biggest companies on the internet, with the most critical needs in terms of stability.

Although Kubernetes can be used with a lot of different container runtimes, this book is going to focus on the Kubernetes + Docker combination.

Perhaps you are already using Docker on a daily basis, but the world of container orchestration might be completely unknown to you. It is even possible that you do not even see the benefits of using such technology because everything looks fine to you with just raw Docker. That's why, in this first chapter, we're not going to look at Kubernetes in detail. Instead, we will focus on explaining what Kubernetes is and how it can help you to manage your Docker containers in production. It will be easier for you to learn a new technology if you already understand why it was built.

In this chapter, we're going to cover the following main topics:

- Understanding monoliths and microservices
- Understanding containers and Docker
- What is Kubernetes?
- How can Kubernetes help you to manage Docker containers?
- What problem does Kubernetes solve?
- Understanding the story of Kubernetes

Understanding monoliths and microservices

Let's put Kubernetes and Docker to one side for the moment, and instead, let's talk a little bit about how internet and software development evolved together over the past 20 years. This will help you to gain a better understanding of where Kubernetes sits and what problem it solves.

Understanding the growth of the internet since the late 1990s

Since the late 1990s, the popularity of the internet has grown rapidly. Back in the 1990s, and even in the early 2000s, the internet was only used by a few hundred thousand people in the world. Today, almost 2 billion people are using the internet, whether for email, web browsing, video games, or more.

There are now a lot of people on the internet, and we're using it to answer tons of different needs, and these needs are adressed by dozens of applications deployed on dozens of devices.

Additionally, the number of connected devices has increased, as each person can now have several devices of a different nature connected to the internet: laptops, computers, smartphones, TVs, tablets, and more.

Today, we can use the internet to shop, to work, to entertain, to read, or to do whatever. It has entered almost every part of our society and has led to a profound paradigm shift for the last 20 years. All of this has given the utmost importance to software development.

Understanding the need for more frequent software releases

To cope with this ever-increasing number of users who are always demanding more in terms of features, the software development industry had to evolve in order to make new software releases faster and more frequent.

Indeed, back in the 1990s, you could build an application, deploy it to production, and simply update it once or twice a year. Today, companies must be able to update their software in production, sometimes several times a day, whether to deploy a new feature, to integrate with a social media platform, to support the resolution of the latest fashionable smartphone, or even to release a patch to a security breach identified the day before. Everything is far more complex today, and you must go faster than before.

We constantly need to update our software, and in the end, the survival of many companies directly depends on how often they are able to offer releases to their users. But how do we accelerate software developments life cycles so that we can deliver new versions of our software to our users more frequently?

IT departments of companies had to evolve, both in an organizational sense and a technical sense. Organizationally, they changed the way they managed projects and teams in order to shift to agile methodologies, and technically, technologies such as cloud computing platforms, containers, virtualization were adopted widely and helped a lot to align technical agility with organizational agility. All of this to ensure more frequent software releases! So, let's focus on this evolution next.

Understanding the organizational shift to agile methodologies

From a purely organizational point of view, agile methodologies such as Scrum, Kanban, and DevOps became the standard way to organize IT teams.

Typical IT departments that do not apply agile methodologies are often made of three different teams, each of them having a single responsibility toward the development and release process life cycle.

Before the adoption of agile methodologies, there was very strong opposition between them:

- **The business team**: These teams are in charge of explaining the need for a new feature to other teams, especially the developers. Their job is hard because they need to translate business needs into concrete technical features that can be understood by the developers.

- **The development team**: These teams are in charge of writing the code. First, they take the specs from the business team, and then they implement the software and features. If they do not understand the need, the development of new features can go back and forth between them and the business team, which can lead to a massive loss of time. Even worse, back in the old days, these guys had no clear vision of the type of environment their code would ultimately run on because it was kept at the sole discretion of the operation team.

- **The operation team**: These teams are in charge of deploying the software to the production servers and operating it. Often, they are not happy when they hear that a new version of a piece of software, which includes new features, has to be deployed because the management judges them on their ability to provide stability to the app. In general, they are here to deploy something that was developed by another team without having a clear vision of what it contains and how it is configured since they did not participate in its development.

These are what we call silos. The roles are clearly defined, people do not work together that much, and when something goes wrong, everyone loses time in an attempt to find the right information from the proper person.

This kind of siloed organization has led to major issues:

- A significantly longer development time
- Greater risk in the deployment of a release that might not work at all in production

And that's essentially what agile methodologies and DevOps broke. The change agile methodologies wrought was to make people work together by creating multidisciplinary teams.

An agile team consists of a product owner describing concrete features by writing them as user stories that are readable by the developers who are working in the same team as them. Developers should have visibility over the production environment and the ability to deploy on top of it, preferably using a **continuous integration and continuous deployment (CI/CD)** approach. Testers should also be part of agile teams in order to write tests.

Simply put, by adopting agile methodologies and DevOps, these silos were broken and multidisciplinary teams capable of formalizing a need, implementing it, testing it, releasing it, and maintaining it in the production environment were created.

> **Important Note**
> Rest assured, even though we are currently discussing agile methodologies and the whole internet in a lot of detail, this book is really about Kubernetes! We just need to explain some of the problems that we have faced before introducing Kubernetes for real!

Agile development teams are complete operational units that are capable of handling all development steps on their own. An agile team should understand the business value brought by a new feature. They should have a minimal view of the software architecture, understand how to build it, how to test it, and the production environment it will run on.

That's the purpose of the expression *You Build It, You Run It* that you'll see everywhere when reading about this subject: an agile team should be able to cover all aspects of an app's development, release, and maintenance life cycles.

You just have to bear in mind that before this, teams were siloed and each had its own scope and working process. So, we've covered the organizational transition brought by the adoption of the agile methodologies, now let's discuss the technical evolution that we've gone through over the past several years.

Understanding the shift from on-premises to the cloud

Having agile teams is very nice. But agility must also be applied to how the software is built and hosted.

With the aim to always achieve faster and more recurrent releases, agile software development teams had to revise two important aspects of software development and release:

- Hosting
- Software architecture

Today, apps are not just for a few hundred users but potentially for millions of users concurrently. Having more users on the internet also means having more computing power capable of handling them. And indeed, hosting an application became a very big challenge.

Back in the old days, there were two ways to get machines to host your apps. We call this on-premises hosting:

- Renting servers from established hosting providers
- Building your own data center, only for companies willing to invest a large amount of money in data centers

When your user base grows, the need to get more powerful machines to handle the load. The solution is to purchase a more powerful server and install your app on it from the start or to order and rack new hardware if you manage your data center. This is not very flexible. Today, a lot of companies are still using an on-premises solution, and often, it's not super flexible.

The game-changer was the adoption of the public cloud, which is the opposite of on-premises. The whole idea behind cloud computing is that big companies such as Amazon, Google, and Microsoft, which own a lot of data centers, decided to build virtualization on top of their massive infrastructure to ensure the creation and management of virtual machines was accessible by APIs. In other words, you can get virtual machines with just a few clicks or just a few commands.

Understanding why the cloud is well suited for scalability

Today, virtually anyone can get hundreds or thousands of servers, in just a few clicks, in the form of virtual machines or instances created on physical infrastructure maintained by cloud providers such as **Amazon Web Services**, **Google Cloud Platform**, and **Microsoft Azure**. A lot of companies decided to migrate their workload from on-premises to a cloud provider, and their adoption has been massive over these last years.

Thanks to that, now, computing power is one of the simplest things you can get.

Cloud computing providers are now typical hosting solutions that agile teams possess in their arsenal. The main reason for this is that the cloud is extremely well suited to modern development.

Virtual machine configurations, CPUs, OSes, network rules, and more are publicly displayed and fully configurable, so there are no secrets for your team in terms of what the production environment is made of. Because of the programmable nature of cloud providers, it is very easy to replicate a production environment in a development or testing environment, providing more flexibility to teams and helping them face their challenges when developing software.

That's a useful advantage for an agile development team built around the DevOps philosophy that needs to manage development, release, and application maintenance in production.

Cloud providers have brought many benefits, as follows:

- Offering elasticity and scalability
- Helping to break up silos and enforcing agile methodologies
- Fitting well with agile methodologies and DevOps
- Offering low costs and flexible billing models
- Ensuring there is no need to manage physical servers
- Allowing virtual machines to be destroyed and recreated at will
- More flexible compared to renting a bare-metal machine monthly

Due to these benefits, the cloud is a wonderful asset in the arsenal of an agile development team. Essentially, you can build and replicate a production environment over and over without the hassle of managing the physical machine by yourself. The cloud enables you to scale your app based on the number of users using it or the computing resources they are consuming. You'll make your app highly available and fault-tolerant. The result is a better user experience for your end users.

> **Important Note**
> Please note that Kubernetes can run both on the cloud and on-premises. Kubernetes is very versatile, and you can even run it on a Raspberry Pi. However, you'll discover that it's better to run it on a cloud due to the benefits they provide. Kubernetes and the public cloud are a good match, but you are not required or forced to run it on the cloud.

Now that we have explained what the cloud brought, let's move on to software architecture, as over the years, a few things have also changed there.

Essentially, software architecture consists of design paradigms that you can choose when developing software. In the 2020s, we can name two architectures:

- Monolithic architecture
- Microservices architecture

Exploring the monolithic architecture

In the past, applications were mostly composed as monoliths. A typical monolith application consists of a simple process, a single binary, or a single package.

This unique component is responsible for the entire implementation of the business logic, to which the software must respond. Monoliths are a good choice if you want to develop fairly simple applications that might not necessarily be updated frequently in production. Why? Well, because monoliths have one major drawback. If your monolith becomes unstable or crashes for some reason, your entire application will become unavailable:

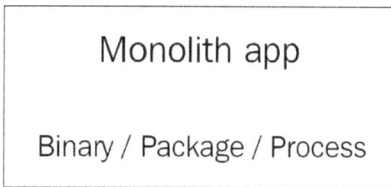

Figure 1.1 – A monolith application consists of one big component that contains all your software

The monolithic architecture can allow you to gain a lot of time during your development and that's perhaps the only benefit you'll find by choosing this architecture. However, it also has many disadvantages. Here are a few of them:

- A failed deployment to production can break your whole application.

- Scaling activities become difficult to achieve; if you fail to scale, all your applications might become unavailable.

- A failure of any kind on a monolith can lead to the complete outage of your app.

In the 2010s, these drawbacks started to cause real problems. With the increase in the frequency of deployments, it became necessary to think of a new architecture that would be capable of supporting frequent deployments and closer update cycles, while reducing the risk or general unavailability of the application. This is why the microservices architecture was designed.

Exploring the microservices architecture

The microservices architecture consists of developing your software application as a suite of independent micro-applications. Each of these applications, which is called a **microservice**, has its own versioning, life cycle, environment, and dependencies. Additionally, it can have its own deployment life cycle. Each of your microservices must only be responsible for a limited number of business rules, and all of your microservices, when used together, make up the application. Think of a microservice as real full-featured software on its own, with its own life cycle and versioning process.

Since microservices are only supposed to hold a subset of all the features that the entire application has, they have to be accessible to expose their functions. You have to get data from a microservice, but you might also want to push data into it. You can make your microservice accessible through widely supported protocols such as HTTP or AMQP, and they need to be able to communicate with each other if needed.

That's why microservices are generally built as web services that are accessible through HTTP REST APIs. This is something that greatly differs from the monolithic architecture:

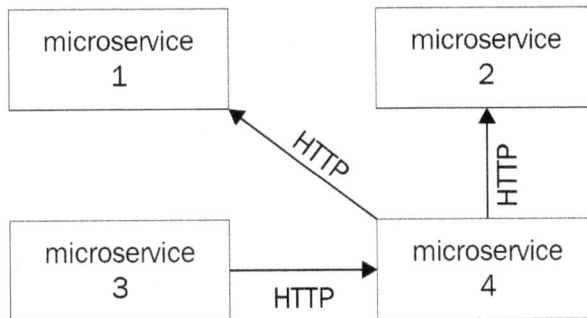

Figure 1.2 – A microservice architecture where different microservices communicate with the HTTP protocol

Another key aspect of the microservice architecture is that microservices need to be decoupled: if a microservice becomes unavailable or unstable, it must not affect the other microservices nor the entire application's stability. You must be able to provision, scale, start, update, or stop each microservice independently without affecting anything else. If your microservices need to work with a database engine, bear in mind that even the database must be decoupled. Each microservice should have its own SQL database and so on. So, if the database of **microservice A** crashes, it won't affect **microservice B**:

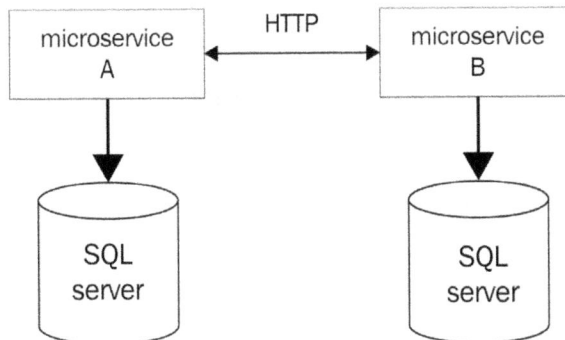

Figure 1.3 – A microservice architecture where different microservices communicate with the HTTP protocol and also with a dedicated SQL server; this way, the microservices are isolated and have no common dependencies

The key rule is to decouple as much as possible so that your microservices are fully independent. Because they are meant to be independent, microservices can also have completely different technical environments and be implemented in different languages. You can have one microservice implemented in Go, another one in Java, and another one in PHP, and all together they form one application. In the context of a microservice architecture, this is not a problem. Because HTTP is a standard, they will be able to communicate with each other even if their underlying technologies are different.

Microservices must be decoupled from other microservices, but they must also be decoupled from the operating system running them. Microservices should not operate at the host system level but at the upper level. You should be able to provision them, at will, on different machines without needing to rely on a strong dependency with the host system; that's why microservice architectures and containers are a good combination.

If you need to release a new feature in production, you simply deploy the microservices that are impacted by the new feature version. The others can remain the same.

As you can imagine, the microservice architecture has tremendous advantages in the context of modern application development:

- It is easier to enforce recurring production deliveries with minimal impact on the stability of the whole application.
- You can only upgrade to a specific microservice each time, not the whole application.
- Scaling activities are smoother since you might only need to scale specific services.

However, on the other hand, the microservice architecture has a few disadvantages, too:

- The architecture requires more planning and is considered to be hard to develop.
- There are problems in managing each microservice's dependencies.

Indeed, microservice applications are considered hard to develop, and it is easy to just do it incorrectly. This approach might be hard to understand, especially for junior developers. On the other hand, dependency management also becomes complex since all microservices can potentially have different dependencies.

Choosing between monolithic and microservices architectures

Presented in this way, you might think that microservices are the better of the two architectures. However, this is not always the case.

Although the monolithic architecture is older than microservice architecture, monolithic applications are not dead yet, and they can still be a good choice in certain situations. Microservices are not necessarily the ideal answer to all projects. If your application is simple, if there are only a few developers on your team working on your project, or if you can tolerate outages when you deploy a new version in production, then you can still opt for an application architecture that is a monolith.

On the other hand, if your application is more complex, if there are many developers with different skills on your team, or if you have a high level of requirements in terms of operational quality in production, scalability, and availability, then you should opt for a microservice architecture.

The problem is that microservices are slightly more complex to develop and manage in production since managing microservices essentially consists of managing multiple applications that each have their own dependencies and life cycles. Thankfully, the rise of Docker has enabled a lot of developers to adopt the microservice architecture.

Understanding containers and Docker

Following this comparison between monolithic and microservice architectures, you should have understood that the architecture that best combines with agility and DevOps is the microservice architecture. It is this architecture that we will discuss throughout the book because this is the architecture that Kubernetes manages well.

Now, we will move on to discuss how Docker, which is a container engine for Linux, is a good option in which to manage microservices. If you already know a lot about Docker, you can skip this section. Otherwise, I suggest that you read through it carefully.

Understanding why Docker is good for microservices

Recall the two important aspects of the microservice architecture:

- Each microservice can have its own technical environment and dependency.
- At the same time, it must be decoupled from the operating system it's running on.

Let's put the latter point aside for the moment and discuss the first one: two microservices of the same app can be developed in two different languages or be written in the same language but as two different versions. Now, let's say that you want to deploy these two microservices inside the same Linux machine. That would be a nightmare.

The reason for this is that you'll have to install all the multiple versions of the different runtimes, as well as the dependencies, and there might also be different versions or overlaps between the two microservices. Additionally, all of this will be on the same host operating system. Now, let's imagine you want to remove one of these two microservices from the machine to deploy it on another server and clean the former machine of all the dependencies used by that microservice. Of course, if you are a talented Linux engineer, you'll succeed in doing this. However, for most people, the risk of conflict between the dependencies is huge, and in the end, you might just make your app unavailable while running such a nightmarish infrastructure.

There is a solution to this: you could build a machine image for each microservice and then put each microservice on a dedicated virtual machine. In other words, you refrain from deploying multiple microservices on the same machine. However, in this example, you will need as many machines as you have microservices. Of course, with the help of AWS or GCP, it's going to be easy to bootstrap tons of servers, each of them tasked to run one and only one microservice, but it would be a huge waste of money to not mutualize the computing power offered by the host.

That's why the second requirement exists: microservices should be decoupled from the microservice they are running on. To achieve this, we use Docker containers.

Understanding the benefit of Docker container isolation

Docker allows you to manage containers that are, in fact, isolated Linux namespaces. Docker's job is to expose a user-friendly API to manage containers, which are like small virtual machines that run on top of the Linux kernel, not at the hypervisor level. By installing Docker on top of your Linux system, you, therefore, add an additional layer of virtualization on top of your host machine. Your microservices are going to be launched on top of this layer, not directly on the host system, whose sole role will be to run Docker.

Since containers are isolated, you can run as many containers as you want and have them run applications written in different languages without any conflict. Microservice relocation becomes as easy as stopping a running container and launching another one from the same image on another machine.

The usage of Docker with microservices offers three main benefits:

- It reduces the footprint on the host system.
- It mutualizes the host system without the conflict between different microservices.
- It removes coupling between the microservice and the host system.

Once a microservice has been containerized, you can eliminate its coupling with the host operating system. The microservice will only depend on the container in which it will operate. Since a container is much lighter than a real full-featured Linux operating system, it will be easy to share and deploy on many different machines. Therefore, the container and your microservice will work on any machine that is running Docker.

The following diagram shows a microservice architecture where each microservice is actually wrapped by a Docker container:

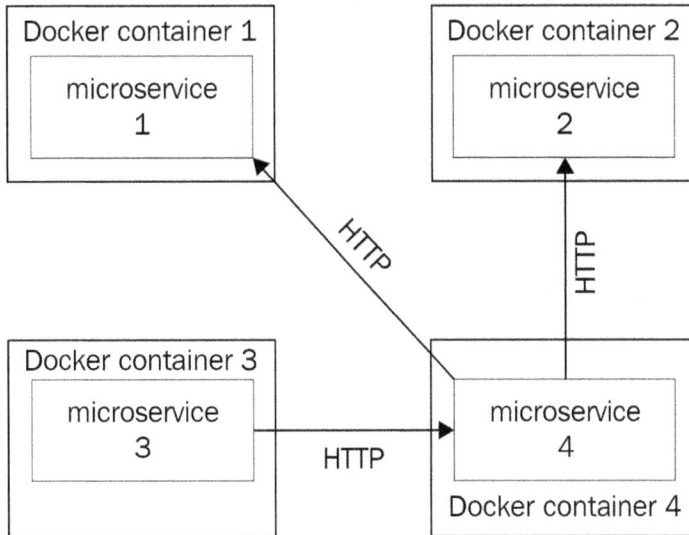

Figure 1.4 – A microservice application where all microservices are wrapped by a Docker container; the life cycle of the app becomes tied to the container, and it is easy to deploy it on any machine that is running Docker

Docker fits well with the DevOps methodology, too. By developing locally in a Docker container, which would be later be built and deployed in production, you ensure you develop in the same environment as the one that will eventually run the application.

Docker is not only capable of managing the life cycle of a container, it is actually an entire ecosystem around containers. It can manage networks, the intercommunication between different containers, and all of these features respond particularly well to the properties of the microservice architecture that we mentioned earlier.

By using the cloud and Docker together, you can build a very strong infrastructure to host your microservice. The cloud will give you as many machines as you want. You simply need to install Docker on each of them, and you'll be able to deploy multiple containerized microservices on each of these machines.

Docker is a very nice tool on its own. However, you'll discover that it's hard to run it in production alone, just as it is. The reason is that Docker was built in order to be an ecosystem around Linux containers, not a production platform. When it comes to production, everything is particular, because it is the concrete environment where everything happens for real. This environment deserves special treatment, and deploying Docker on it is risky. This is because Docker cannot alone address the particular needs that are related to production.

There are a number of questions, such as *how to relaunch a container that failed automatically* and *how to autoscale my container based on its CPU utilization*, that Docker alone cannot answer. This is the reason why some people were afraid to run Docker-based workloads in production a few years ago.

To answer these questions, we will need a container orchestrator, such as the one discussed in this book: Kubernetes.

How can Kubernetes help you to manage your Docker containers?

Now, we will focus a little bit more on Kubernetes, which is the purpose of this book. Here, we're going to discover that Kubernetes was meant to use container runtimes in production, by answering operational needs mandatory for production.

Understanding that Kubernetes is meant to use Docker in production

If you open the official Kubernetes website (at `https://kubernetes.io`), the title you will see is **Production-Grade Container Orchestration**:

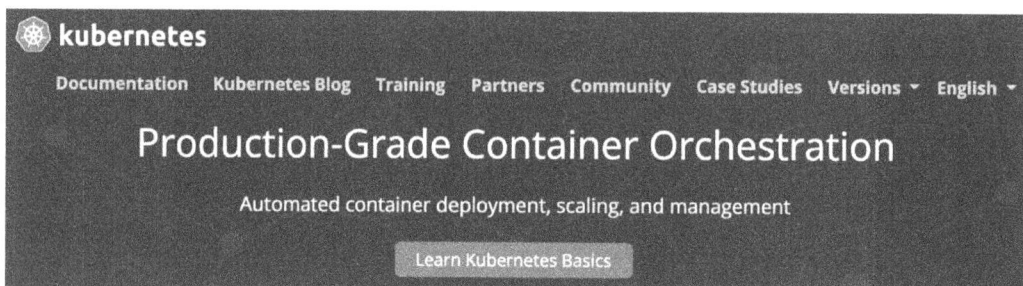

Figure 1.5 – The Kubernetes home page showing the header and introducing Kubernetes as a production container orchestration platform

These four words perfectly sum up what Kubernetes is: it is a container orchestration platform for production. Kubernetes does not aim to replace Docker nor any of the features of Docker; rather, it aims to help us to manage clusters of machines running Docker. When working with Kubernetes, you use both Kubernetes and the full-featured standard installations of Docker.

The title refers to **production**. Indeed, the concept of production is absolutely central to Kubernetes: it was thought and designed to answer modern production needs. Managing production workloads is different today compared to what it was in the 2000s. Back in the 2000s, your production workload would consist of just a few bare metal servers, if not a single one on-premises. These servers mostly ran monoliths directly installed on the host Linux system. However, today, thanks to public cloud platforms such as **Amazon Web Services** (**AWS**) or **Google Cloud Platform** (**GCP**), anyone can now get hundreds or even thousands of machines in the form of instances or virtual machines with just a few clicks. Even better, we no longer deploy our applications on the host system but as containerized microservices on top of the Docker engine instead, thereby reducing the footprint of the host system.

A problem will arise when you have to manage Docker installations on each of these virtual machines on the cloud. Let's imagine that you have 10 (or 100 or 1,000) machines launched on your preferred cloud and you want to achieve a very simple task: deploy a containerized Docker app on each of these machines.

You could do this by running the `docker run` command on each of your machines. It would work, but of course, there is a better way to do it. And that's by using a **container orchestrator** such as **Kubernetes**. To give you an extremely simplified vision of Kubernetes, it is actually a **REST API** that keeps a registry of your machines executing a Docker daemon.

Again, this is an extremely simplified definition of Kubernetes. In fact, it's not made of a single centralized REST API, because as you might have gathered, Kubernetes itself was built as a suite of microservices.

Exploring the problems that Kubernetes solves

You can imagine that launching containers on your local machine or a development environment is not going to require the same level of planning as launching these same containers on remote machines, which could face millions of users. Problems specific to production will arise, and Kubernetes is a top solution with which to address these problems when using containers in production:

- Ensuring high availability
- Handling release management and container deployments
- Autoscaling containers

Ensuring high availability

High availability is the central principle of production. This means that your application should always remain accessible and should never be down. Of course, it's utopian. Even the biggest companies such as Google or Amazon are experiencing outages. However, you should always bear in mind that this is your goal. Microservice architecture is a way to mitigate the risk of a total outage in the event of a failure. Using microservices, the failure of a single microservice will not affect the overall stability of the application. Kubernetes includes a whole battery of functionality to make your Docker containers highly available by replicating them on several host machines and monitoring their health on a regular and frequent basis.

When you deploy Docker containers, the accessibility of your application will directly depend on the health of your containers. Let's imagine that for some reason, a container containing one of your microservice becomes inaccessible; how can you automatically guarantee that the container is terminated and recreated using only Docker without Kubernetes? This is impossible because, by default, Docker cannot do it alone. With Kubernetes, it becomes possible. Kubernetes will help you design applications that can automatically repair themselves by performing automating tasks such as health checking and container replacement.

If one machine in your cluster were to fail, all of the containers running on it would disappear. Kubernetes would immediately notice that and reschedule all of the containers on another machine. In this way, your applications will become highly available and fault-tolerant as well.

Release management and container deployment

Deployment management is another of these production-specific problems that Kubernetes answers. The process of deployment consists of updating your application in production in order to replace an old version of a given microservice with a new version.

Deployments in production are always complex because you have to update the containers that are responding to requests from end users. If you miss them, the consequences can be great for your application because it could become unstable or inaccessible, which is why you should always be able to quickly revert to the previous version of your application by running a rollback. The challenge of deployment is that it needs to be performed in the least visible way to the end user, with as little friction as possible.

When using Docker, each release is preceded by a build process. Indeed, before releasing a new container, you have to build a new Docker image containing the new version. A Docker image is a kind of template used by Docker to launch containers. A container can be considered a running instance of a Docker image.

> **Important Note**
> The Docker build process has absolutely nothing to do with Kubernetes: it's pure Docker. Kubernetes will come into play later when you'll have to deploy new containers based on a newly built image.

Triggering a build is straightforward. Perform the following steps:

1. You just need to run the `docker build` command:

   ```
   $ docker build .
   ```

2. Docker reads build instructions from the `Dockerfile` file inside the `.` directory and starts the build process.

3. The build completes.

4. The resulting image is stored on the local machine where the build ran.

5. Then, you push the new image to a Docker repository with a specific tag to identify the software version included in the new image.

Once the push has been completed, another process starts, that is, the deployment. To deploy a containerized Docker app, you simply need to pull the image from the machine where you want to run it and then run a `docker run` command.

This is what you'll need to do to release a new version of your containerized software, and this is exactly where things can become hard if you don't use an orchestrator such as Kubernetes.

The next step to achieve the release is to delete the existing container and replace it with new containers created from this new image.

Without Kubernetes, you'll have to run a `docker run` command on the machine where you want to deploy a new version of the container and destroy the container containing the old version of the application. Then, you will have to repeat this operation on each server that runs a copy of the container. It should work, but it is extremely tedious since it is not automated. And guess what? Kubernetes can automate this for you.

Kubernetes has features that allow it to manage deployments and rollbacks of Docker containers, and this will make your life a lot easier when responding to this problem. With a single command, you can ask Kubernetes to update your containers on all of your machines. Here is the command, which we'll learn later, that allows you to do that:

```
$ kubectl set image deploy/myapp myapp_container=myapp:1.0.0
# Meaning of the command
# kubectl set image <deployment_name> <container_name>=<docker_
image>:<docker_tag>
```

On a real Kubernetes cluster, this command will update the container called `myapp_container`, which is running as part of the application called `myapp`, on every single machine where `myapp_container` runs to the `1.0.0` tag.

Whether it has to update one container running on one machine or millions over multiple data centers, this command works the same. Even better, it ensures high availability.

Remember that the goal is always to meet the requirement of high availability; a deployment should not cause your application to crash or cause a service disruption. Kubernetes is natively capable of managing deployment strategies such as rolling updates aimed at avoiding service interruptions.

Additionally, Kubernetes keeps in memory all the revisions of a specific deployment and allows you to revert to a previous version with just one command. It's an incredibly powerful tool that allows you to update a cluster of Docker containers with just one command.

Autoscaling containers

Scaling is another production-specific problem that has been widely democratized through the use of public clouds such as **Amazon Web Services (AWS)** and **Google Cloud Platform (GCP)**. Scaling is the ability to adapt your computing power to the load you are facing – again to meet the requirement of high availability. Never forget that the goal is to avoid outages and downtime.

When your production machines are facing a traffic spike and one of your containers is no longer able to cope with the load, you need to find a way in which to identify the failing container. Decide whether you wish to scale it vertically or horizontally; otherwise, if you don't act and the load doesn't decrease, your container or even the host machine will eventually fail, and your application might become inaccessible:

- **Vertical scaling**: This allows your container to use more computing power offered by the host machine.
- **Horizontal scaling**: You can duplicate your container to another machine, and you can load balance the traffic between the two containers.

Again, Docker is not able to respond to this problem alone; however, when you manage your Docker with Kubernetes, it becomes possible. Kubernetes is capable of managing both vertical and horizontal scaling automatically. It does this by letting your containers consume more computing power from the host or by creating additional containers that can be deployed on another node on the cluster. And if your Kubernetes cluster is not capable of handling more containers because all your nodes are full, Kubernetes will even be able to launch new virtual machines by interfacing with your cloud provider in a fully automated and transparent manner by using a component called a Cluster Autoscaler.

> **Important Note**
> The Cluster Autoscaler only works if the Kubernetes cluster is deployed on a cloud provider.

These goals cannot be achieved without using a container orchestrator. The reason for this is simple. You can't afford to do these tasks; you need to think about DevOps' culture and agility and seek to automate these tasks so that your applications can repair themselves, be fault-tolerant, and be highly available.

Contrary to scaling out your containers or cluster, you must also be able to decrease the number of containers if the load starts to decrease in order to adapt your resources to the load, whether it is rising or falling. Again, Kubernetes can do this, too.

When and where is Kubernetes not the solution?

Kubernetes has undeniable benefits; however, it is not always advisable to use it as a solution. Here, we have listed several cases where another solution might be more appropriate:

- **Container-less architecture**: If you do not use a container at all, Kubernetes won't be of any use to you.

- **Monolithic architecture**: While you can use Kubernetes to deploy containerized monoliths, Kubernetes shows all of its potential when it has to manage a high number of containers. A monolithic application, when containerized, often consists of a very small number of containers. Kubernetes won't have much to manage, and you'll find a better solution for your use case.

- **A very small number of microservices or applications**: Kubernetes stands out when it has to manage a large number of containers. If your app consists of two to three microservices, a simpler orchestrator might be a better fit.

- **No cluster**: Are you only running one machine and only one Docker installation? Kubernetes is good at managing a cluster of computers that executes a Docker daemon. If you do not plan to manage a real cluster, then Kubernetes is not for you.

Understanding the history of Kubernetes

To finish this chapter, let's discuss the history of the Kubernetes project. It will be really useful for you to understand the context in which the Kubernetes project started and the people who are keeping this project alive.

Understanding how and where Kubernetes started

Kubernetes started as an internal project at Google. Since its founding in 1998, Google gained huge experience in managing high-demanding workloads at scale, especially container-based workloads. Today, in addition to Google, Amazon and Microsoft are also releasing a lot of open source and commercial software to allow smaller companies to benefit from their experience of managing cloud-native applications. Kubernetes is one example of this open source software that has been released by Google.

At Google, everything has been developed as Linux containers since the mid-2000s. The company understood the benefit of using containers long before Docker made them simple to use for the general public. Essentially, everything at Google runs as a container. And they are undoubtedly the first to have felt the need to develop an orchestrator that would allow them to manage their container-based resources along with the machines that launch them. This project is called Borg, and you can consider it to be the ancestor of Kubernetes. Another container orchestrator project, called Omega, was then started by Google in order to improve the architecture of Borg to make it easier to extend and become more robust. Many of the improvements brought by Omega were later merged into Borg.

> **Important Note**
>
> Borg is actually not the ancestor of Kubernetes because the project is not dead and is still in use at Google. It would be more appropriate to say that a lot of ideas from Borg were actually reused to make Kubernetes. Bear in mind that Kubernetes is not Borg nor Omega. Borg was built in C++ and Kubernetes in Go. In fact, they are two entirely different projects, but one is heavily inspired by the other. This is important to understand: Borg and Omega are two internal Google projects. They were not built for the public.

As the interest in containers became greater during the early 2010s, Google decided to develop and release a third container orchestrator. This time, it was meant to be an open source one that was built for the public. Therefore, Kubernetes was born and would eventually be released in 2014.

Kubernetes was developed with the experience gained by Google to manage containers in production. Most importantly, it inherited Borg and Omega's ideas, concepts, and architectures. Here is a brief list of ideas and concepts taken from Borg and Omega, which have now been implemented in Kubernetes:

- The concept of pods to manage your containers: Kubernetes uses a logical object, called a pod, to create, update, and delete your containers.

- Each pod has its own IP address in the cluster.

- There are distributed components that all watch the central Kubernetes API in order to retrieve the cluster state.

- There is internal load balancing between pods and services.

- Labels and selectors are two metadata used together to build interaction between Kubernetes

That's why Kubernetes is so powerful when it comes to managing containers in production at scale: in fact, the concepts you'll learn in Kubernetes are older than Kubernetes itself. They have existed for more than a decade, running Google's entire infrastructure as part of Borg and Omega. So, although Kubernetes is a young project, it was built on solid foundations.

Who manages Kubernetes today?

Kubernetes is no longer maintained by Google. They gave Kubernetes to an organization called **Cloud Native Computing Foundation** (**CNCF**), which is a big consortium whose goal is to promote the usage of container technologies. This happened in 2018.

Google is a founding member of CNCF along with companies such as Cisco, Red Hat, and Intel. The Kubernetes source code itself is hosted on GitHub and is an extremely active project on the platform. The code is under License Apache version 2.0, which is a permissive open source license. You won't have to pay in order to use Kubernetes, as the software is available for free, and if you are good at coding with Go, you can even contribute to the code.

Where is Kubernetes today?

Kubernetes has a lot of competitors, and some of them are open source, too. Others are bound to a specific cloud provider. We can name a few, as follows:

- Apache Mesos
- Hashicorp Nomad
- Docker Swarm
- Amazon ECS

These container orchestrators all have their pros and cons, but it's fair to say that Kubernetes is, by far, the most popular of them all.

Kubernetes has won the fight of popularity and adoption and is really about to become the de facto standard way of deploying container-based workloads in production. As its immense growth made it one of the hottest topics in IT industry, it has become crucial for cloud providers to come up with a Kubernetes offering as part of their services. Therefore, Kubernetes is supported almost everywhere now.

The following Kubernetes-based services can help you to get a Kubernetes cluster up and running with just a few clicks:

- Google GKE
- Amazon EKS
- Microsoft Azure AKS
- Alibaba ACK

It's not just about the cloud offerings. It's also about the Platform-as-a-Service market. Recently, Red Hat OpenShift decided to rewrite their entire platform to rebuild it on Kubernetes. Now they are offering a complete set of enterprise tools to build, deploy, and manage Docker containers entirely on top of Kubernetes. In addition to this, other projects such as Rancher were built as *Kubernetes distributions* to offer a complete set of tools around the Kubernetes orchestrator, whereas projects such as Knative offers to manage serverless workloads with the Kubernetes orchestrator.

> **Important Note**
>
> AWS is an exception because it has two container orchestrator services. The first one is Amazon ECS, which is entirely made by AWS and is a competitor to Kubernetes. The second one is Amazon EKS, which was released later than the first one and is a complete Kubernetes offering on AWS. These services are not the same, so do not be misguided by their similar names.

Learning Kubernetes today is one of the smartest decisions you can take if you are into managing cloud-native applications in production. Kubernetes is evolving rapidly, and there is no reason to think why its growth would stop.

By mastering this wonderful tool, you'll get one of the hottest skills being searched for in the IT industry today. I hope you are now convinced!

Summary

This first chapter gave us room for a big introduction. We covered a lot of subjects, such as monoliths, microservices, Docker containers, cloud computing, and Kubernetes. We also discussed how this project came to life. You should now have a global vision of how Kubernetes can be used to manage your containers in production.

In the next chapter, we will discuss the process Kubernetes follows to launch a Docker container. You will discover that you can issue commands to Kubernetes, and these commands will be interpreted by Kubernetes as instructions to run containers. We will list and explain each component of Kubernetes and its role in the whole cluster. There are a lot of components that make up a Kubernetes cluster, and we will discover all of them. We will explain how Kubernetes was technically built with a focus on the distinction between master nodes, worker nodes, and control plane components.

2

Kubernetes Architecture – From Docker Images to Running Pods

In the previous chapter, we laid the groundwork regarding what Kubernetes is from a functional point of view. You should now have a better idea of how Kubernetes can help you to manage clusters of machines running containerized microservices. Now, let's go a little deeper into the technical details. In this chapter, we will examine how Kubernetes enables you to manage containers that are distributed on different machines. Following this chapter, you should have a better understanding of the anatomy of a Kubernetes cluster; in particular, you will have a better understanding of Kubernetes components and know the responsibility of each of them in the execution of your containers.

Kubernetes is made up of several distributed components, each of which plays a specific role in the execution of Docker containers. To understand the role of each Kubernetes component, we will follow the life cycle of a Docker container as it is created and managed by Kubernetes: that is, from the moment you execute the command to create the container to the point when it is actually executed on a machine that is part of your Kubernetes cluster.

In this chapter, we're going to cover the following main topics:

- Understanding the difference between the master and worker nodes
- The `kube-apiserver` component
- The `kubectl` command-line tool and YAML syntax
- The `Etcd` datastore
- The kubelet and worker node components
- The `kube-scheduler` component
- The `kube-controller-manager` component
- How to make Kubernetes highly available

Understanding the difference between the master and worker nodes

To run Kubernetes, you will require Linux machines, which are called **nodes** in Kubernetes. A node could be a physical machine or a virtual machine on a cloud provider, such as an EC2 instance. There are two types of nodes in Kubernetes:

- Master nodes
- Worker nodes

Master nodes are responsible for maintaining the state of the Kubernetes cluster, whereas worker nodes are responsible for executing your Docker containers.

While using Linux, you will have probably used commands such as `apt-get install` or `yum install` to get a new, fully functional software preconfigured that just works out of the box. With Kubernetes, things are a slightly more complex.

The good news is that you can also use Windows-based nodes to launch Windows-based containers in your Kubernetes cluster. The thing to know is that you can mix Linux and Windows machines on your cluster and it will work the same, but you cannot launch a Windows container on a Linux worker node and vice versa.

By nature, Kubernetes is a distributed application. What we call *Kubernetes* is not a single monolithic app released as a single build that you would install on a dedicated machine. What we mean by Kubernetes is a collection of small projects. Each project is written in **Go** and forms part of the overall project that is Kubernetes.

To get a fully functional Kubernetes cluster, you need to set up each of these components by installing and configuring them separately and have them communicate with each other. When these two requirements are met, you can start running your containers using the Kubernetes orchestrator.

For development or local testing, it is fine to install all of the Kubernetes components on the same machine; however, in production, these components should be spread across different hosts. This will help you to make your Kubernetes cluster highly available. By spreading the different components across multiple machines, you gain two benefits:

- You make your cluster highly available and fault-tolerant.
- You make your cluster a lot more scalable. Components have their own lifecycle, they can be scaled without impacting others.

In this way, having one of your servers down will not break the entire cluster but just a small part of it, and adding more machines to your servers becomes easy.

Each Kubernetes component has its own clearly defined responsibility. It is important for you to understand each component's responsibility and how it articulates with the other components to understand how Kubernetes works overall.

Depending on its role, a component will have to be deployed on a master node or a worker node. While some components are responsible for maintaining the state of a whole cluster and operating the cluster itself, others are responsible for running our application containers by interacting with Docker daemons directly. Therefore, the components of Kubernetes can be grouped into two families:

- Components belonging to the **Control Plane**:

  These components are responsible for maintaining the state of the cluster. They should be installed on a master node. These are the components that will keep the list of containers executed by your Kubernetes cluster or the number of machines that are part of the cluster. As an administrator, when you interact with Kubernetes, you actually interact with the control plane components.

- • Components belonging to the **Worker Nodes**:

  These components are responsible for interacting with the Docker daemon in order to launch containers according to the instructions they receive from the control plane components. Worker node components must be installed on a Linux machine running a Docker daemon. You are not supposed to interact with these components directly. It's possible to have hundreds or thousands of worker nodes in a Kubernetes cluster.

Clustering technologies use this architecture a lot. They define two types of nodes: masters and workers. The master(s) nodes are responsible for the management of the cluster and all operational tasks according to the instructions received from the administrator. The worker(s) nodes are responsible for the execution of the actual workload based on instructions received from the master(s).

Kubernetes works in relatively the same way. You are not supposed to launch your Docker containers by yourself, and therefore, you do not interact directly with the worker nodes. Instead, you send your instructions to the control plane. Then, it will delegate the actual container creation and maintenance to the worker node on your behalf. You never run a `docker` command directly:

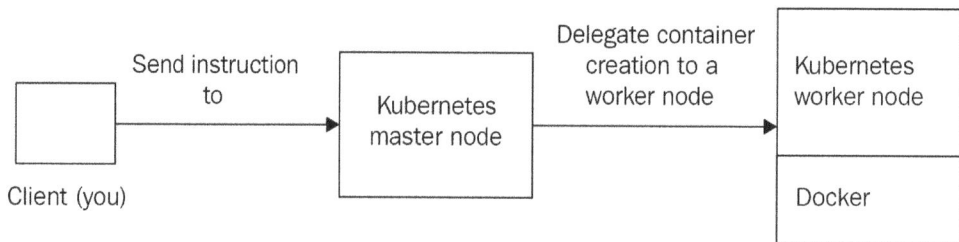

Figure 2.1 – A typical Kubernetes workflow. The client interacts with the master node/control plane components, which delegate container creation to a worker node. There is no communication between the client and the worker node

When using Kubernetes you'll notice here and there the concepts of *control plane* and the *master node*. They're almost the same: both expressions are meant to designate the Kubernetes components responsible of cluster administration, and by extension, the machines (or nodes) on which these components have been installed. In Kubernetes, we generally try to avoid talking about *master nodes*. Instead, we talk about the *control plane*.

The reason is because saying *"master node"* supposes the components allowing the management of the cluster are installed on the same machine and have a strong coupling with the machine that is running them. However, due to the distributed nature of Kubernetes, its *master node* components can actually be spread across multiple machines. This is quite tricky, but there are, in fact, two ways in which to set up the control plane components:

- You run all of them on the same machine, and you have a master node.
- You run them on different machines, and you no longer have a master node.

Master node terminology tends to imply that all control plane components are running on the same machine, but not the case. To achieve maximum fault tolerance, it's a good idea to spread them across different machines. Kubernetes is so distributed that even its *master node* can be broken into multiple machines, where each of them has the responsibility to execute a single component that allows the management of the cluster.

The idea is that control plane components must be able to communicate with each other, and this can be achieved by installing them onto different hosts. In fact, later, you'll discover that the control plane components can even be launched as Docker containers on worker node machines. This is a very advanced topic, but it's possible. That's why the master node terminology is not that accurate to Kubernetes, and we prefer the *control plane* terminology instead.

However, for the sake of simplicity, I'll consider, in this chapter, and also for a huge part of this book, that we are using dedicated *master node* machines to execute all of the control plane components in the same place. Throughout the different examples listed here, control plane components will be tightly coupled with the machine that is executing them. This will help you to understand the role of each component. Later, we will explore the advanced techniques related to the management of control plane components, such as launching them as Docker containers and more.

That being said, things are simpler when it comes to worker nodes: you start from a standard machine running Docker, and you install the worker node components next to the Docker runtime. These components will interface with the local container engine that is installed on the said machine and execute containers based on the instructions you send to the control plane components. You can control all the aspects of Docker from Kubernetes thanks to this worker nodes mechanics: container creation, network management, scaling containers, and so on. Adding more computing power to your cluster is easy; you just need to add more worker nodes and have them join the cluster to make room for more containers.

> **Important Note**
>
> By splitting the control plane and worker node components of different machines, you are making your cluster highly available and scalable. Kubernetes was built with all of the cloud-native concerns in mind; its components are stateless, easy to scale, and built to be distributed across different hosts. The whole idea is to avoid having a single point of failure by grouping all of the components onto the same host.

Here is a simplified diagram of a full-featured Kubernetes cluster with all the components listed. In this chapter, we're going to explain all of the components listed on this diagram, their roles, and their responsibilities. Here, all of the control plane components are installed on a single master node machine:

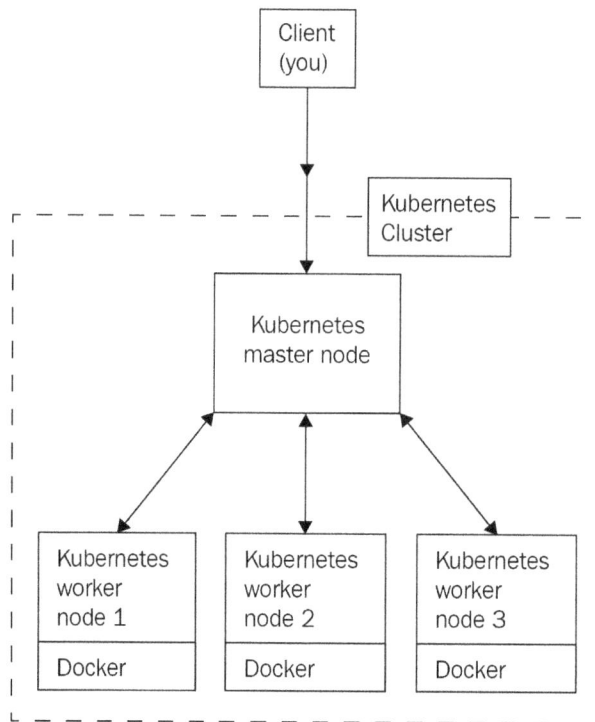

Figure 2.2 – A full-featured Kubernetes cluster with one master node and three worker nodes

The preceding diagram displays a 4-node Kubernetes cluster with all of the necessary components. As you can see, there are quite a few components.

The following is a table that lists all of the Kubernetes components that we will discuss in this chapter. In addition to all of these components, we will also examine the kubectl client that will allow you to interact with your Kubernetes clusters:

Component name	Control plane, worker node, or client
Kube-apiserver	Control plane (master node)
Etcd	Control plane (master node)
Kube-scheduler	Control plane (master node)
Kube-controller-manager	Control plane (master node)
Kubelet	Worker node
Kube-proxy	Worker node
Container Engine (Docker...)	Worker node
kubectl	Client

Bear in mind that Kubernetes is modified and, therefore, can be modified to fit a given environment. When Kubernetes is deployed and used as part of a distribution such as Amazon EKS or Red Hat Openshift, additional components could be present, or the behavior of the default ones might differ. In this book, for the most part, we will discuss bare Kubernetes. The components discussed in this chapter, and which are listed here, are the default ones. You will find them everywhere; they are the backbone of Kubernetes.

You might have noticed that the majority of these components have a name starting with Kube: these are the components that are part of the Kubernetes project. Additionally, you might have noticed that there are two components with a name that does not start with Kube. The other two (Etcd and Container Engine) components are two external dependencies that are not strictly part of the Kubernetes project, but which Kubernetes needs in order to work:

- Etcd is a third-party database used by the Kubernetes project. Don't worry; you won't have to master it in order to use Kubernetes.

- The container runtime is also a third-party engine. For us, it's going to be Docker because Kubernetes forcibly needs to manage something.

Rest assured, you will not have to install and configure these components all by yourself. Almost no one bothers with managing the components by themselves, and, in fact, it's super easy to get a working Kubernetes without having to install the components.

For development purposes, you can use Minikube, which will install all of the Kubernetes components on your local machine and run them as a single virtual machine. This is absolutely NOT recommended for production, but it is incredibly useful for development and tests.

For production purposes, you can use a cloud offering such as Amazon EKS or Google GKE. This will give you a production-grade Kubernetes cluster that is fully integrated with your preferred cloud provider along with all of the scaling mechanics that you are already familiar with. If you don't have access to such platforms, you can use Kubeadm. This is a Kubernetes installation utility that is capable of installing and configuring all Kubernetes components with just one command.

For educational purposes, you can install them, one by one, from scratch. A very famous tutorial, called *Kubernetes the Hard Way*, is available on the internet. It teaches you how to install and configure the components of Kubernetes from scratch. From **public key infrastructure** (**PKI**) management to networking and computing provisioning, this tutorial teaches you how to install Kubernetes on bare Linux machines in Google Cloud. Do not try to follow this tutorial if you are still new to Kubernetes as it's quite advanced.

You will observe many references to this tutorial on the internet because it's very famous. Bear in mind that installing, securing, configuring, and managing a production-grade Kubernetes cluster is time-consuming and not easy at all. You should avoid using the result of the *Kubernetes the Hard Way* tutorial in production.

The kube-apiserver component

Kubernetes' most important component is a REST API called `kube-apiserver`, Essentially, this is an API that exposes all Kubernetes features. You interact with Kubernetes by calling this REST API through the kubectl command-line tool.

The role of kube-apiserver

This is a component that is part of the control plane, meaning it's something to install and run on a master node. The `kube-apiserver` component is so important in Kubernetes that, sometimes, people believe it is Kubernetes itself. However, it's not. `kube-apiserver` is simply one component of the orchestrator. It's the most important, yes, but it's just one of them.

It's implemented in Go. Its code is open source, under the Apache 2.0 license, and you can find it hosted on GitHub.

When working with Kubernetes, the workflow is simple. When you want to give an instruction to Kubernetes, you will always have to send an HTTP request to kube-apiserver. When you want to tell Kubernetes to create, delete, or update a container, you should always do so by calling the correct kube-apiserver endpoint with the correct HTTP verb. This is how we will work all of the time with Kubernetes; kube-apiserver is the single entry point for all operations issued to the orchestrator. This best practice consists of never having to interact with one of the Docker daemons by yourself. You do run containers by sending instructions to the kube-apiserver component through the HTTP(S) protocol, and you let the Kubernetes components update the state of the Docker daemons for you.

Let's update the previous architecture diagram to make it a little more accurate with what happens in the real world:

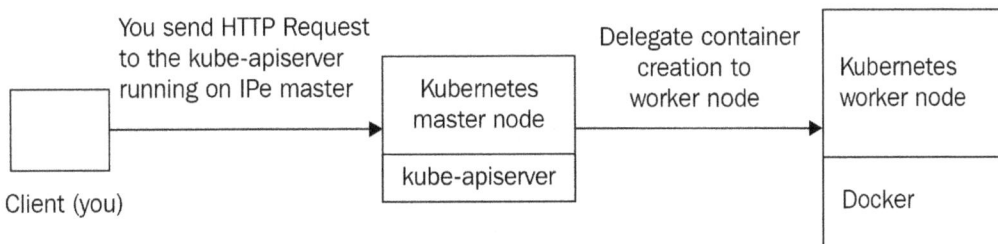

Figure 2.3 – When interacting with Kubernetes, you are actually issuing HTTP requests to a kube-apiserver component running on a master node; that's the only component you will interact with directly

The kube-apiserver component is built according to the REST standard. REST is very effective at exposing features via HTTP endpoints that you can request through the various methods of the HTTP protocol (for example, GET, POST, PUT, PATCH, and DELETE). To illustrate this, consider a dummy API that is managing users. You would have a path called /users. By sending HTTP requests to this /users path, you can command different operations against the users resource:

- The following request will retrieve a list of all the users on an API running on localhost:

```
GET https://127.0.0.1/users
```

- This request would delete the users with an ID of 1 on an API running on my.api.com:

```
DELETE https://my.api.com/users/1
POST https://127.0.0.1:8080/users
```

As you might have gathered, the REST standard relies entirely on the HTTP protocol. By combining HTTP methods and paths, you can run different operations defined by the method, against resources defined by the path.

Additionally, the REST standard brings a lot of flexibility. Because adding new resources means adding new paths, any REST API can be extended. To bring things to life, REST APIs generally use a datastore to maintain the state of the objects or resources they are managing. With our dummy API, it could, for example, use a MySQL database and a `users` table to keep the users created by calling the API in the long term.

There are two ways in which such an API can enable data retention:

- The REST API keeps its data in its own memory:

  This could work; however, in this case, the API is stateful and impossible to scale.

- By using a full-featured database engine such as MariaDB or PostgreSQL:

  This is the go-to way; delegating the storage to a third-party engine on another host makes the API stateless. In this scenario, the API is scalable horizontally; you can add a lot of instances of your API as long as they can retrieve the state in the common database.

Any REST API can be easily upgraded or extended to do more than its initial intent. To sum up, here are the essential properties of a REST API:

- It relies on the HTTP protocol.

- It defines a set of resources identified by URL paths.

- It defines a set of actions identified by HTTP methods.

- It can run actions against resources based on a properly forged HTTP request.

- It keeps the state of their resources on a datastore.

Just like this dummy `users` API, `kube-apiserver` is nothing more than a REST API, which is at the heart of any Kubernetes cluster you will set up, no matter if it's local, on the cloud, or on-premises. It is also stateless; that is, it keeps the state of the resources by relying on a database engine called **Etcd**. This means you can horizontally scale the `kube-apiserver` component by deploying it onto multiple machines and load balance request issues to it using a layer 7 load balancer without losing data.

> **Important Note**
> By default, HTTP servers listen to port 80, whereas HTTPS services listen to
> port 443. This is not the case with kube-apiserver. By default, the port
> it listens to is port 6443. However, this configuration data can be overridden.

Because HTTP is supported almost everywhere, it is very easy to communicate with and issue instructions to a Kubernetes cluster. However, most of the time, we interact with Kubernetes thanks to a command-line utility named kubectl, which is the HTTP client that is officially supported as part of the Kubernetes project. This book will also focus on how to communicate with the API server through the use of the kubectl command-line tool. In this chapter, we will explore this tool, too.

When you download kube-apiserver, you'll end up with a Go-compiled binary that is ready to be executed on any Linux machine. The Kubernetes developers defined a set of resources for us that are directly bundled within the binary. These resources are the ones that Kubernetes manages. Unlike the dummy users API mentioned earlier, Kubernetes does not describe a *Users* resource. This is simply because Kubernetes was not meant to manage users but containers. So, do expect that all of the resources you will find in kube-apiserver are related to container management, networking, and computing in general.

Let's name a few of these resources, as follows:

- Pod
- ReplicaSet
- PersistentVolume
- NetworkPolicy
- Deployment

Of course, this list of resources is not exhaustive. If you want a full list of the Kubernetes components, you can access it from the official Kubernetes documentation API reference page at https://kubernetes.io/docs/reference/generated/ kubernetes-api/v1.18/.

You might be wondering why there are no *Containers* resources here. As mentioned in *Chapter 1, Kubernetes Fundamentals*, Kubernetes makes use of a resource called a pod to manage the containers. For now, you can think of pods as though they were containers. We will learn a lot about them in the coming chapters. Each of these resources is associated with a dedicated URL path, and just as we witnessed earlier, changing the HTTP method when calling the URL path will have a different effect. All of these behaviors are defined in kube-apiserver; note that these behaviors are not something you have to develop, they are directly implemented as part of the kube-apiserver.

After the Kubernetes objects are stored on the `Etcd` database, other Kubernetes components will *convert* these objects into raw Docker instructions. This way, multiple Docker daemon can mirror the state of the cluster as it is stored in the `Etcd` datastore and described by `kube-apiserver`.

Another important point to bear in mind is that `kube-apiserver` is the single entry point and the single source of truth for the whole cluster. Everything in Kubernetes has been designed to revolve around `kube-apiserver`. You'll see that other Kubernetes components also have to read or change the state of the cluster: even they will do so by calling `kube-apiserver` through HTTP and never directly. As an administrator, aside from very rare occasions, you must never interact directly with the other components of the cluster. So, ensure you refrain from using SSH to sign in to a machine that is part of the cluster and start doing things manually.

The is because `kube-apiserver` does not just manage the state of the cluster, it also has tons of different mechanisms related to authentication, authorization, and HTTP response formatting. That's why doing things manually is really bad.

How do you install kube-apiserver?

In *Chapter 3, Installing Your First Kubernetes Cluster*, we will focus on how to install and configure a Kubernetes cluster locally. However, I would like you to know that all of the components of Kubernetes are available to download for free from Google's servers. You can download a specific version of `kube-apiserver`, along with any component of Kubernetes, using a simple `wget` command:

```
$ wget -q --show-progress --https-only –timestamping \
https://storage.googleapis.com/kubernetes-release/release/
v1.14.0/bin/linux/amd64/kube-apiserver
```

By calling this command, you will get the official build of `kube-apiserver` in version 1.14. However, running this binary won't be enough to have a fully working Kubernetes cluster. This is because, again, Kubernetes is not just the `kube-apiserver` component. You will require the other components of Kubernetes to make it work.

If you are wondering how to run `kube-apiserver`, essentially, there are two ways to proceed:

- From a Docker image on a worker node
- From a `systemd` unit file on a dedicated master node

Let's put aside the Docker method because it's a little bit more advanced. Instead, let's explain the simpler solution and run it as a `systemd` service. Essentially, `systemd` is a daemon management tool that is available by default on Linux and will be helpful to run `kube-apiserver`. I don't recommend you do that as it's going to be useless, but let me show you how easy it is to run a Kubernetes component on Linux thanks to `systemd`:

1. We need to create a file at the following path: `/etc/systemd/system/kube-apiserver.service`.

2. We need to put the following content inside the file. Essentially, this `systemd` unit file executes the `kube-apiserver` binary that was downloaded earlier with some arguments to configure it.

3. Once the unit file has been created, reload the `systemd` daemon, enable the new service, and start it using the following commands:

```
$ systemctl daemon-reload
$ systemctl enable kube-apiserver
$ systemctl start kube-apiserver
```

Ensure that the following file exists in the directory with the following content. As you might have gathered, it's just a basic `systemd` unit file, `/etc/systemd/system/kube-apiserver.service`:

```
[Unit]
Description=Kubernetes API Server
Documentation=https://github.com/kubernetes/kubernetes
[Service]
ExecStart=/usr/local/bin/kube-apiserver
  --advertise-address=${INTERNAL_IP}
  --allow-privileged=true
  --apiserver-count=3
  --audit-log-path=/var/log/audit.log
  --authorization-mode=Node,RBAC
  --bind-address=0.0.0.0
  --client-ca-file=/var/lib/kubernetes/ca.pem
  --enable-admission-plugins=NamespaceLifecycle,NodeRestriction
,LimitRanger,ServiceAccount,DefaultStorageClass,ResourceQuota
  --etcd-cafile=/var/lib/kubernetes/ca.pem
  --etcd-certfile=/var/lib/kubernetes/kubernetes.pem
  --etcd-keyfile=/var/lib/kubernetes/kubernetes-key.pem
```

```
    --etcd-servers=https://10.240.0.10:2379,https://10.240.0.11:2
379,https://10.240.0.12:2379
    --encryption-provider-config=/var/lib/kubernetes/encryption-
config.yaml
    --kubelet-certificate-authority=/var/lib/kubernetes/ca.pem
    --kubelet-client-certificate=/var/lib/kubernetes/kubernetes.
pem
    --kubelet-client-key=/var/lib/kubernetes/kubernetes-key.pem
    --kubelet-https=true \\
    --runtime-config=api/all \\
    --service-account-key-file=/var/lib/kubernetes/service-
account.pem \\
    --service-cluster-ip-range=10.32.0.0/24
    --service-node-port-range=30000-32767
    --tls-cert-file=/var/lib/kubernetes/kubernetes.pem
    --tls-private-key-file=/var/lib/kubernetes/kubernetes-key.pem
    --v=2
Restart=on-failure
RestartSec=5
[Install]
WantedBy=multi-user.target
```

Don't attempt to run this example by yourself, as there are some missing parts for now. In the next chapter, we are going to set up a local Kubernetes, but at least now you know that it is easy to launch Kubernetes components as `systemd`: in the end, they are plain old Linux-compiled binaries.

Where do you install kube-apiserver?

Cloud services such as Amazon EKS or Google GKE, or other similar cloud offerings, will install and configure all the components of Kubernetes properly and expose you to a Kubernetes endpoint (or, if you prefer, the `kube-apiserver` endpoint) without giving you too much information regarding the underlying machines or load balancers provisioned. The following is a screenshot of a Kubernetes cluster that was created on the Amazon EKS service:

Details

API server endpoint

https://C45D9685DDEBFE00FEC7634FB85EB589.yl4.eu-west-1.eks.amazonaws.com

OpenID Connect provider URL

https://oidc.eks.eu-west-1.amazonaws.com/id/C45D9685DDEBFE00FEC7634FB85EB589

Cluster ARN

arn:aws:eks:eu-west-1: ███████ :cluster/kluster

Creation time

May 13th 2020 at 9:34 AM

Certificate authority

LS0tLS1CRUdJTiBDRVJUSUZJQ0FURS0tLS0tCkTJSUN5RENDQW0yZ0F3SUJBZ0lCQURBTkJna3Foa2lHOXcwQkFRc0ZBREF2TVNNd0lRWUdRVRWUUVERXccmRXSmZSmVKY2KY201bGRIRB

Cluster IAM Role ARN

arn:aws:iam: ███████ :role/role

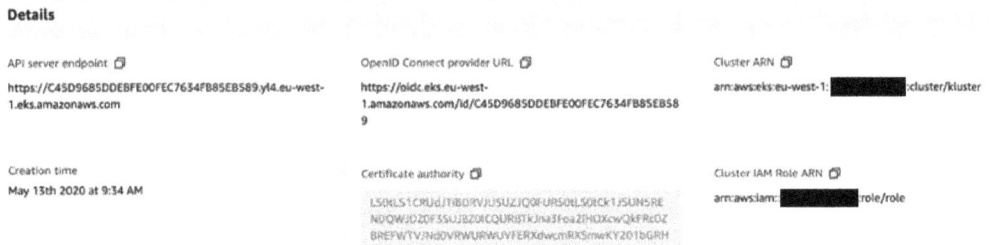

Figure 2.4 – The UI console showing details of a Kubernetes cluster provisioned on Amazon EKS

Note that the `kube-apiserver` component should be installed on a sufficiently powerful machine dedicated to its execution. Indeed, it is a very sensitive component. If your `kube-apiserver` component becomes inaccessible, your Docker containers will continue to run, but you won't be able to interact with them through Kubernetes. Instead, they become "orphan" containers running on isolated machines that are no longer managed by a Kubernetes cluster.

You want to avoid this situation; that's why you should pay special attention to your `kube-apiserver` instance(s).

In addition to this, be aware that the other Kubernetes components, as we'll discover a little later, constantly send HTTP requests to `kube-apiserver` in order to understand the state of the cluster or to update it. Because these components need to know the state of the cluster at any moment, they are running a lot of HTTP requests against the `kube-apiserver` component. And the more worker nodes you have, the more HTTP requests that will be issued against `kube-apiserver`. That's why `kube-apiserver` should be independently scaled as the cluster itself scales out.

As mentioned earlier, `kube-apiserver` is a stateless component that does not directly maintain the state of the Kubernetes cluster itself and relies on a third-party database to do so. You can scale it horizontally by hosting it on group of machines that are behind a load balancer such as an HTTP API. When using such a setup, you interact with `kube-apiserver` by calling your load balancer endpoint.

We have mentioned a lot of things about `kube-apiserver`. Now, let's take a look at the kubectl client. This is the official HTTP client that will allow you to interact with Kubernetes, or more exactly, with the `kube-apiserver` component of your Kubernetes cluster.

Exploring the kubectl command-line tool and YAML syntax

kubectl is the official command-line tool used by Kubernetes. You must install it on your local machine; we will learn how to install it properly in the next chapter. For now, we need to understand its role. This is an HTTP client that is fully optimized to interact with Kubernetes and allows you to issue commands to your Kubernetes cluster. You can install it right now since you are going to need it in the coming chapters:

- In Linux, use the following commands:

```
$ curl -LO https://storage.googleapis.com/kubernetes-
release/release/v1.14.10/bin/linux/amd64/kubectl
$ chmod +x ./kubectl
$ sudo mv ./kubectl /usr/local/bin/kubectl
$ kubectl version # Should output kubectl version
```

- On macOS, use the following commands:

```
$ curl -LO https://storage.googleapis.com/kubernetes-
release/release/v1.14.10/bin/darwin/amd64/kubectl
$ chmod +x ./kubectl
$ sudo mv ./kubectl /usr/local/bin/kubectl
$ kubectl version # Should output kubectl version
```

- In Windows, use the following commands

```
curl -LO https://storage.googleapis.com/kubernetes-
release/release/v1.14.10/bin/windows/amd64/kubectl.exe
  kubectl version # Should output kubectl version
```

This last example for Windows users requires you to add `kubectl.exe` to your `PATH` variable before being able to call `kubectl`.

The role of kubectl

Since `kube-apiserver` is nothing more than an HTTP API, any HTTP client will work to interact with a Kubernetes cluster. You can even use CURL to manage your Kubernetes cluster, but of course, there is a better way to do that.

The developers of Kubernetes have developed a client for the `kube-apiserver` component. This client comes in the form of a command-line tool called kubectl. So, why would you want to use such a client and not go directly with CURL calls? Well, the reason is simplicity. Indeed, `kube-apiserver` manages a lot of different resources and each of them has its own URL path.

Calling `kube-apiserver` constantly through CURL would be possible but extremely time-consuming. This is because remembering the path of each resource, and how to call it is not user-friendly. Essentially, CURL is not the way to go since `kubectl` also manages different aspects related to authentication against Kubernetes authentication layer, managing cluster contexts and more.

You would have to go to the documentation all of the time to remember which URL path, HTTP header, or query string. kubectl will do that for you by letting you call `kube-apiserver` through commands that are easy to remember, secure, and entirely dedicated to Kubernetes management.

When you call kubectl, it reads the parameters you pass to it, and based on them, will forge and issue HTTP requests to the `kube-apiserver` component of your Kubernetes cluster:

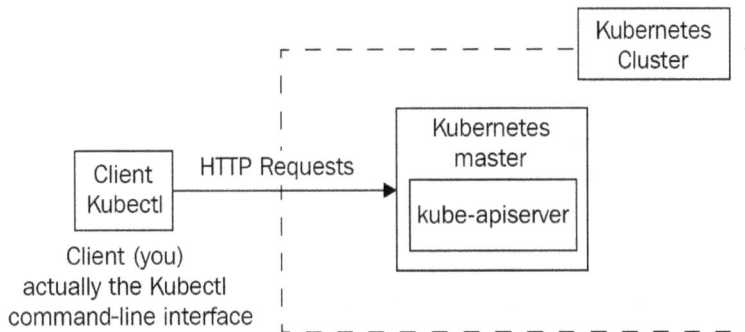

Figure 2.5 – The kubectl command line will call kube-apiserver with the HTTP protocol; you'll interact with your Kubernetes cluster through kubectl all of the time

This is the general workflow with Kubernetes: you will always interact with Kubernetes through kubectl. Once the `kube-apiserver` component receives a valid HTTP request coming from you, it will read or update the state of the cluster in `Etcd` based on the request you submitted. If it's a write operation, for example, to update the image of a running container, `kube-apiserver` will update the state of the cluster in `Etcd`. Then, the components running on the worker node where the said container is being hosted will issue the proper Docker commands in which to launch a new container based on the new image. This is so that the actual state of the Docker daemons always reflects what's in `Etcd`.

Given that you won't have to interact with Docker yourself, or with Etcd, we can say that the mastery of Kubernetes is largely based on your knowledge of the kubectl commands. To be effective with Kubernetes, you must master kubectl because it is the command that you will type constantly. You won't have to interact with any other components than kube-apiserver and the kubectl command-line tool that allows you to call it.

kubectl is available to download for free directly from Google's servers. This is an Apache 2.0 licensed project that is officially part of the Kubernetes project. Its source code is available on GitHub at https://github.com/kubernetes/kubectl.

> **Important Note**
>
> Since the kube-apiserver component is accessible through the HTTP(S) protocol, you could interact with any Kubernetes cluster with any HTTP-based library and even programmatically through your favorite programming language. Tons of kubectl alternatives exist. However, kubectl, being the official tool of the Kubernetes project, is the tool you'll systematically see in action in the documentation, and most of the examples you will find use kubectl.

How does kubectl work?

When you call the kubectl command, it will try to read a configuration file that must be created in $HOME/.kube/config. This configuration file is called kubeconfig.

All of the information, such as the kube-apiserver endpoint, its port, and the client certificate used to authenticate against kube-apiserver, must be written in this file. Its path can be overridden on your system by setting an environment variable, called KUBECONFIG, or by using the --kubeconfig parameter when calling kubectl:

```
$ export KUBECONFIG="/custom/path/.kube/config"
$ kubectl --kubeconfig="/custom/path/.kube/config"
```

Each time you run a kubectl command, the kubectl command-line tool will look for a kubeconfig file in which to load its configuration in the following order:

1. First, it checks whether the --kubeconfig parameter has been passed and loads the config file.

2. At that point, if no kubeconfig file is found, kubectl looks for the KUBECONFIG environment variable.

3. Ultimately, it falls back to the default one in $HOME/.kube/config.

To view the config file currently used by your local kubectl installation, you can run this command:

```
$ kubectl config view
```

The `kubeconfig` file contains different information about the Kubernetes cluster it communicates with. That information includes the URL of `kube-apiserver`, the configured user for authenticating against the said `kube-apiserver`, and more. The role of the kubectl tool is to forge an HTTP request based on the arguments you pass to the `kubectl` command and to execute it against the cluster specified on its loaded `kubeconfig` file. Then, the HTTP request is actually sent to `kube-apiserver`, which produces an HTTP response that kubectl will reformat in a human-readable format and output to your Terminal.

The following command is probably one that you'll type almost every day when working with Kubernetes:

```
$ kubectl get pods
```

This command lists the *Pods*. Essentially, it will issue a GET request to `kube-apiserver` to retrieve the list of containers (pods) on your cluster. Internally, kubectl associates the *pods* parameter passed to the command to the `/api/v1/pods` URL path, which is the path that `kube-apiserver` uses to expose the pod resource.

Here is another command:

```
$ kubectl run nginx --restart Never --image nginx
```

This one is slightly trickier because `run` is not an HTTP method. In fact, this command will issue a POST request against the `kube-apiserver` component, which will result in the creation of a container called `nginx`, based on the `nginx` image hosted on Docker Hub.

> **Important Note**
>
> In fact, this command won't create a container but a Pod. We will discuss the pod resource extensively in *Chapter 4, Running Your Docker Containers*. Let's try not to talk about containers anymore; instead, let's move on to pods and familiarize ourselves with Kubernetes concepts and wordings. From now on, if you come across the word *container*, it means a real container from a Docker perspective. Additionally, *pods* refer to the Kubernetes resource.

The YAML syntax

kubectl supports two kinds of syntaxes:

- The imperative syntax
- The declarative syntax

Almost every instruction that you send to `kube-apiserver` through kubectl can be written using one of these two forms.

Essentially, the imperative form runs `kubectl` commands on a shell session. By passing in correct sub-commands and arguments, you can forge instructions for the `kube-apiserver` component. Let me give you a few examples of the imperative syntax. All of the following commands, once executed, issue HTTP requests against a given `kube-apiserver` endpoint:

- This command creates a pod, called `my-pod`, based on the `busybox:latest` Docker image:

  ```
  $ kubectl run my-pod --restart Never --image
  busybox:latest
  ```

- This command lists all the `ReplicaSet` resources in the `my-namespace` namespace created on the Kubernetes cluster:

  ```
  $ kubectl get rs -n my-namespace
  ```

- This command deletes a pod, called `my-pod`, in the default namespace:

  ```
  $ kubectl delete pods my-pod
  ```

Imperative syntax has multiple benefits. If you already understand what kind of instructions to send to Kubernetes and the proper command to achieve this, you are going to be incredibly fast. The imperative syntax is easy to type, and you can do a lot with just a few commands. Some operations are only accessible with imperative syntax, too. For example, listing existing resources in the cluster is only possible with the imperative syntax.

However, the imperative syntax has a big problem. It is very complicated having to keep records of what you did previously in the cluster. If for some reason, you were to lose the state of your cluster and need to recreate it from scratch, it's going to be incredibly hard to remember all of the imperative commands that you typed in earlier to bring your cluster back to the state you want. You could read your `.bash_history` file but of course, there is a better way to do this.

So, let's move on to declarative syntax now. The declarative syntax is about writing a JSON or YAML file on disk first and then applying it against the cluster using the kubectl command line. Both JSON and YAML formats are supported; however, by convention, Kubernetes users prefer YAML syntax because of its simplicity. YAML is not a programming language. There is no real logic behind it. It's simply a kind of `key:value` configuration syntax that is used by a lot of projects nowadays, and Kubernetes is one of them.

Using kubectl in the declarative way requires you to create a file on your disk and to write some `key:value` pairs, in YAML format, inside of it. Each `key:value` pair represents the configuration data that you want to set to the Kubernetes resource you want to create.

The following is the imperative command that created the pod named `my-pod` using the `busybox:latest` Docker image we used earlier:

```
$ kubectl run my-pod --restart Never --image busybox:latest
```

We will now do the same but with the declarative syntax instead:

```
apiVersion: v1
kind: Pod
metadata:
  name: my-pod
spec:
  containers:
  -  name: busybox-container
     image: busybox:latest
```

Let's say this file is saved on disk with the name of `pod.yaml`. To create the actual pod, you'll need to run the following command:

```
$ kubectl create -f pod.yaml
```

This result will be the equivalent of the previous command.

Each YAML file that is created for Kubernetes must contain four mandatory keys:

- `apiVersion:`

  This field tells you in which API version the resource is declared. Each resource type has an `apiVersion` key that must be set in this field. The pod resource type is in API version `v1`.

- Kind:

  This Kind field indicated the resource type the YAML file will create. Here, it is a pod that is going to be created.

- Metadata:

  This field tells Kubernetes about the name of the actual resource. Here, the pod is named my-pod. This field describes the Kubernetes resource, not the Docker one. This metadata is for Kubernetes, not for Docker.

- Spec:

  This field tells Kubernetes what the object is made of. In the preceding example, the pod is made of one container that will be named busybox-container based on the busybox:latest image. These are the containers that are going to be created on Docker.

Another important aspect of the declarative syntax is that it enables you to declare multiple resources in the same file using three dashes as a separator between the resources. Here is a revised version of the YAML file, which will create two pods:

```
apiVersion: v1
kind: Pod
metadata:
  name: my-pod
spec:
  containers:
  -  name: busybox-container
     image: busybox:latest
---
apiVersion: v1
kind: Pod
metadata:
  name: my-second-pod
spec:
  containers:
  - name: nginx-container
    image: nginx:latest
```

You should be able to read this file by yourself and understand it; it just creates two pods. The first one uses the busybox image, and the second one uses the nginx image.

Of course, you don't have to memorize all of the syntaxes and what value to set for each key. It would be useless for you, so just use copy and paste from the documentation when needed.

The declarative syntax offers a lot of benefits, too. With it, you'll be slower because writing these YAML files is a lot more time-consuming than just issuing a command in the imperative way. However, it offers two major benefits:

- **Infrastructure as Code (IaC) management**:

  You'll be able to keep the configuration stored somewhere and use Git to version your Kubernetes resources, just as you would do with IaC. If you were to lose the state of your cluster, keeping the YAML files versioned in Git will enable you to recreate it in a clean and effective manner.

- **Create multiple resources at the same time**:

  Since you can declare multiple resources in the same YAML file, you can have entire applications and all of their dependencies in the same place. Additionally, you get to create and recreate complex applications with just one command. Later, you'll discover a tool called Helm that can achieve templating on top of the Kubernetes YAML files. Thanks to this tool, it's going to be easier to manage resources through YAML files.

There is no *better* way to use kubectl; these are just two ways to interact with it, and you need to master both. This is because some features are not available with the imperative syntax, while some others are not available with the declarative syntax. For example, you cannot create a multi-container pod from the imperative syntax, and you cannot list running pods from the declarative syntax. That's why you need to master both.

Remember that, in the end, both call the `kube-apiserver` component by using the HTTP protocol.

kubectl should be installed on any machine that needs to interact with the cluster

From a technical point of view, you must install and configure a kubectl runtime whenever and wherever you want to interact with a Kubernetes cluster.

Of course, it should be on your local machine, since this is where you are going to work with Kubernetes because you have to be able to communicate with your cluster from your own workstation. However, in larger projects, it's also a good idea to install kubectl in the agents/runner of your continuous integration platform.

Indeed, you will probably want to automate maintenance or deployment tasks to run against your Kubernetes cluster, and you will probably use a **continuous integration** (**CI**) platform such as GitLab CI or Jenkins to do that.

If you want to be able to run Kubernetes commands in a CI pipeline, you will need to install kubectl on your CI agents and have a properly configured `kubeconfig` file written on the CI agent filesystem. This way, your CI/CD pipelines will be able to issue commands against your Kubernetes cluster and update the state of your cluster, too.

Just to add, kubectl should not be seen as a Kubernetes client for *human* users only. It should be viewed as a generic tool to communicate with Kubernetes: install it wherever you want to communicate with your cluster.

Remember that the golden rule is very simple: you never do anything in Kubernetes without going through kubectl. It doesn't matter if it's made by you as a human administrator or as part of an automation script: make use of kubectl.

The Etcd datastore

We explained that `kube-apiserver` is a stateless API that can be scaled horizontally. However, it is necessary for `kube-apiserver` to store the state of the cluster somewhere, such as the number of containers created, on which machines, the names of the pods, which Docker images they use, and more. To achieve that, it uses the `Etcd` database.

The role of the Etcd datastore

`Etcd` is part of the control plane. The `kube-apiserver` component relies on a distributed NoSQL database called `Etcd`. Strictly speaking, `Etcd` is not a component of the Kubernetes project. As you might have gathered, `Etcd` is not named according to the same nomenclature as the other components (`kube*`). This is because `Etcd` is not actually a Kubernetes project but a project completely independent of Kubernetes.

Instead of using a full-featured relational database such as MySQL or PostgreSQL, Kubernetes relies on this NoSQL distributed datastore called `Etcd` to store its state persistently. `Etcd` can be used independently on any project, but Kubernetes cannot work without `Etcd`. This is something that you have to understand: Kubernetes has an external dependence. Instead of rewriting a database engine from scratch, Kubernetes developers decided to use `Etcd`. The good news is that `Etcd` is an open source project too, which is available on GitHub under license Apache 2.0, written in Go (just like Kubernetes). You can locate it at `https://github.com/etcd-io/etcd`. It's also a project incubated by the **Cloud Native Computing Foundation** (**CNCF**), which is the organization that maintains Kubernetes.

> **Important Note**
>
> Don't expect to be able to use another database engine with Kubernetes. It is not possible to replace `Etcd` with MySQL or another datastore. Kubernetes was really built to rely specifically on `Etcd`.

So, to sum up, `Etcd` is the database that Kubernetes uses to keep its state. When you call the `kube-apiserver`, each time you implement a read or write operation by calling the Kubernetes API, you will read or write data from or to `Etcd`. So, this is the main memory of your Kubernetes cluster, and the `kube-apiserver` can be seen as a proxy in front of `Etcd`.

Let's zoom into what is inside the master node:

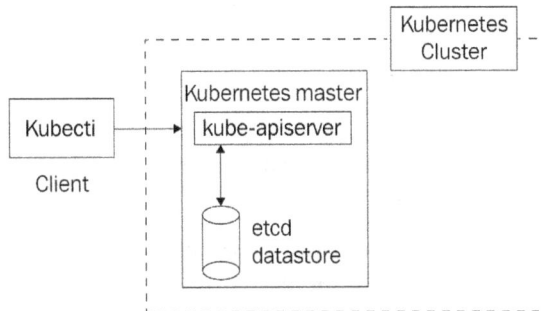

Figure 2.6 – The kube-apiserver component is in front of the Etcd datasore and acts as a proxy in front of it; kube-apiserver is the only component that can read or write from and to etcd

`Etcd` is the most sensitive point of your cluster. This is because if you were to lose the data inside it, your Kubernetes cluster would become completely unusable. It's even more sensitive than the `kube-apiserver` component. The reason is simple; after all, if `kube-apiserver` crashes, you can still relaunch it. However, if you lose the data in your `Etcd` datastore, or it somehow gets corrupted, and you don't have a backup to restore it, your Kubernetes cluster is dead.

Fortunately, you do not need to master Etcd in depth to use Kubernetes. It is even strongly recommended that you do not touch it at all if you do not know what you are doing. This is because a bad operation could corrupt the data stored in Etcd and, therefore, the state of your cluster.

Remember, the general rule in Kubernetes architecture says that every component has to go through the kube-apiserver to read or write in Etcd. This is because, from a technical point of view, the kubectl authenticates itself against the kube-apiserver through a TLS client certificate that only the kube-apiserver has. Therefore, it is the only component of Kubernetes that has the right to read or write in Etcd. This is a very important notion in the architecture of Kubernetes. All of the other components won't be able to read or write anything to or from Etcd without calling the kube-apiserver endpoints through HTTP.

> **Important Note**
> Please note that Etcd is also designed as a REST API. By default, it listens to port 2379.

Let's go back to the command that we mentioned earlier:

```
$ kubectl run nginx --restart Never --image nginx
```

When you execute the preceding command, the kubectl tool will forge an HTTP POST request that will be executed against the kube-apiserver specified in the kubeconfig file. The kube-apiserver will write a new entry in Etcd, which will be persistently stored on disk.

At that point, the state of the Kubernetes changes: it will then be the responsibility of the other Kubernetes component to reconcile the actual state of the cluster to the desired state of the cluster (that is, the one in Etcd).

Unlike Redis or Memcached, Etcd is not in-memory storage. If you reboot your machine, you do not lose the data because it is kept on disk.

Where do you install Etcd?

As with the kube-apiserver, cloud services such as Amazon EKS or Google GKE will install and configure a pool of servers running Etcd for you, without giving you much information about the machines. They offer you a working Kubernetes cluster without letting you know too much about the underlying resources.

If you were to run an `Etcd` datastore on-premises, or if you intend to manage it by yourself, you should know that `Etcd` is part of the Kubernetes control plane and should run on a master node.

It can be run as part of a `systemd` unit file, just as we saw with the `kube-apiserver`, but also as a Docker container, too. `Etcd` has the ability to natively include features, allowing you to deploy them in a distributed way. Therefore, it is also natively ready to be scaled horizontally since it's capable of spreading its dataset through multiple servers: `Etcd` is built as a clustering solution on its own.

In general, you can decide on the following:

- Install `Etcd` on a machine that is dedicated to it.
- Group `Etcd` on the same machines running other control plane components.

The first solution is better because you reduce the risk of the unavailability of your cluster in the event of an outage. Having a set of machines dedicated to running `Etcd` also reduces any security risks.

The problem is that it will always cost more than the first solution since it involves more machines, more planning, and also more maintenance. It's fine to run `Etcd` on the same machines as the `kube-apiserver`; however, if you can afford it, don't hesitate to use dedicated machines for `Etcd`.

The Kubelet and worker node components

So far, we have described key Kubernetes control plane components: the `kube-apiserver` component and the `Etcd` datastore. You also know that in order to communicate with the `kube-apiserver` component, you have to use the `kubectl` command-line utility to get data to and from the `Etcd` datastore through the help of `kube-apiserver`.

However, all of this is not telling us where and how these instructions result in running containers on worker nodes. We will dedicate this part of the chapter to explain the anatomy of a worker node by explaining the three components running on it:

- The container engine

The first component that should be installed on a worker node is the Docker daemon. Kubernetes is not limited to Docker; it can manage other container engines, such as `rkt`. However, in this book, we will be using Kubernetes with Docker, which is the most common setup.

Therefore, any Linux machine running Docker can be used as a base on which to build a Kubernetes worker node. Please note that a Kubernetes worker node can also run Docker containers that were not launched by Kubernetes. That is no problem.

The Docker daemon running on a Kubernetes worker node is a plain old Docker installation; it has nothing special compared to the one you might run on your local machine.

The Kubelet agent

As its name suggests (`Kube*`), the Kubelet is a component of Kubernetes, which is part of the worker node. In fact, it is the most important component of the worker node since it is the one that will interact with the local Docker daemon installed on the worker node.

The Kubelet is a daemon running on the system. It cannot run as a Docker container itself. Running it on the host system is mandatory; that's why we usually set up using `systemd`. The Kubelet differs from the other Kubernetes components because it is the only one that cannot run as a Docker container. The Kubelet strictly requires you to run on the host machine.

When the Kubelet gets started, by default, it reads a configuration file located at `/etc/kubernetes/kubelet.conf`.

This configuration specifies two values that are really important for the Kubelet to work:

- The endpoint of the `kube-apiserver` component
- The local Docker daemon UNIX socket

When the worker node launches, it will *join* the cluster by issuing an HTTP request against the `kube-apiserver` component to add a *Node* entry to the `Etcd` datastore. After that, running this `kubectl` command should list the new worker node:

```
$ kubectl get nodes
```

Note that `kube-apiserver` keeps a registry of all the worker nodes that are part of the Kubernetes cluster. You can retrieve a list of all your nodes by simply running this command.

Once the machine has joined the cluster, the Kubelet will act as a bridge between `kube-apiserver` and the local Docker daemon. Kubelet is constantly running HTTP requests against `kube-apiserver` to retrieve pods it has to launch. By default, every 20 seconds, the Kubelet runs a GET request against the `kube-apiserver` component to list the pods created on `Etcd` that are destined to it.

Once it receives a pod specification in the body of an HTTP response from `kube-apiserver`, it can convert this into Docker containers specification that will be executed against the specified UNIX socket. The result is the creation of your containers on your local Docker daemon.

> **Important note**
>
> Remember that Kubelet does not read directly from `Etcd`, rather it interacts with the `kube-apiserver` that exposes what is inside the `Etcd` data layer. Kubelet is not even aware that an `Etcd` server runs behind the `kube-apiserver` it polls.

These polling mechanisms, called **watch** mechanisms in Kubernetes terminology are precisely how Kubernetes proceeds to run and delete containers against your worker nodes at scale. Each Kubelet instance on each worker node is watching `kube-apiserver` to get noticed when a change occurs in the `Etcd` datastore. And once a change is noticed, Kubelet is capable of converting the changes into corresponding Docker instructions and communicating with the local Docker daemon using the Docker UNIX socket specified in the Kubelet configuration file. There are two things to pay attention to here:

- Kubelet and the `kube-apiserver` must be able to communicate with each other through HTTP. That's why HTTPS port `6443` must be opened between the worker and master nodes.

- Because they are both running on the same machine, Kubelet and the Docker daemon are interfaced through the usage of UNIX sockets.

Each worker node must have its own Kubelet, which means more HTTP polling against the `kube-apiserver` as you add more worker nodes to your cluster. If your Kubernetes cluster has hundreds of machines, it could result in a negative performance impact on the `kube-apiserver`. Ultimately, you can even DDoS your `kube-apiserver` by having too many worker nodes. That's why it's important to make the `kube-apiserver` and the other control plane component highly available by scaling them efficiently.

If you create Docker containers manually on a worker node, the Kubelet won't be able to manage it. You can completely bypass Kubernetes and create containers on your worker nodes without having to use the Kubelet. Running a plain old `docker run` command would work. Just bear in mind that the **Kubelet is only capable of managing the Docker containers it created**. The reason for this is that the containers won't be created as part of a pod in the `Etcd` datastore, and the sole job of Kubelet is that its local Docker daemon reflects the configuration that is stored in `Ectd`.

> **Important Note**
>
> Please note that the container engine running on the worker node has no clue that it is managed by Kubernetes through a local Kubelet agent. A worker node is nothing more than a Linux machine running a Docker daemon with a Kubelet agent installed next to it, executing raw Docker instructions. You can execute the same with the Docker client manually.

The kube-proxy component

This is the last of the three components running on a worker node. An important part of Kubernetes is networking. We will have to opportunity to dive into networking later; however, you need to understand that Kubernetes has tons of mechanics when it comes to exposing pods to the outside world or exposing pods to one another in the Kubernetes cluster.

These mechanics are implemented at the kube-proxy level; that is, each worker node requires an instance of a running kube-proxy so that the pods running on them are accessible. We will explore a Kubernetes feature called **Service**, which is implemented at the level of the kube-proxy component. Just like the Kubelet, the kube-proxy component also communicates with the `kube-apiserver` component.

There are also several other sub-components or extensions that are operating at the worker node level, such as cAdvisor or **Container Network Interface** (**CNI**). However, they are advanced topics that we will discuss later.

The kube-scheduler component

The `kube-scheduler` component is a control plane component. It should run on the master node.

This is a component that is responsible for electing a worker node out of those available to run a newly created pod.

The role of the kube-scheduler component

Similar to the Kubelet, `kube-scheduler` queries the `kube-apiserver` at regular intervals in order to list the pods that have not been *scheduled*. At creation, pods are not scheduled, which means that no worker node has been elected to run them. A pod that is not scheduled will be registered in `Etcd` but without any worker node assigned to it. Therefore, no running Kubelet will ever be aware that this pod needs to get launched, and ultimately, no container described in the pod specification will ever run.

Internally, the pod object, as it is stored in Etcd, has a property called nodeName. As the name suggests, this property should contain the name of the worker node that will host the pod. When this property is set, we say that the pod has been *scheduled*; otherwise, the pod is *pending* for schedule.

We need to find a way to fill this value, and that is the role of the kube-scheduler. To do this, the kube-scheduler poll continues the kube-apiserver at regular intervals. It searches for pod resources with an empty nodeName property. Once it finds such pods, it will execute an algorithm to elect a worker node. Then, it will update the nodeName property in the pod by issuing an HTTP request to the kube-apiserver component. While electing a worker node, the kube-scheduler component will take into account some configuration values that you can pass:

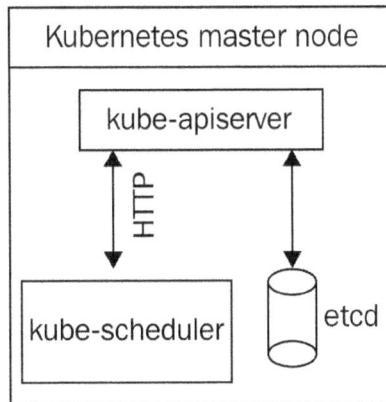

Figure 2.7 – The kube-scheduler component polls the kube-apiserver component to find unscheduled pods; it then schedules the pods by setting a nodeName property and updates the entry. Then, the pod gets launched on the proper worker node by the local Kubelet

The kube-scheduler component will take into account some configuration values that you can pass optionally. By using these configurations, you can precisely control how the kube-scheduler component will elect a worker node. Here are some of the features to bear in mind when scheduling pods on your preferred node:

- Node selector
- Node affinity and anti-affinity
- Taint and toleration

There are also advanced techniques for scheduling that will completely bypass the kube-scheduler component. We will examine these features later.

> **Important Note**
>
> The `kube-scheduler` component can be replaced by a custom one. You can implement your own `kube-scheduler` component with your own custom logic to select a node and use it on your cluster. It's one of the strengths of the distributed nature of the Kubernetes components.

Where do you install kube-scheduler?

Cloud providers such as Amazon EKS and Google GKE will create a highly available installation of the `kube-scheduler` component without exposing you to much information regarding the underlying machines that execute it.

Since the `kube-scheduler` component is part of the Kubernetes control plane, you should not install it on a worker node. Again, you can choose to install `kube-scheduler` on a dedicated machine or the same machine as `kube-apiserver`. It's a short process and won't consume many resources, but there are some things to pay attention to.

The `kube-scheduler` component should be highly available. That's why you should install it on at least two different machines. If your cluster does not have a working `kube-scheduler` component, new pods won't be scheduled, and the result will be a lot of pending pods. If no `kube-scheduler` component is present, it won't have an impact on the already scheduled pods. However, newly created pods won't result in new containers.

The kube-controller-manager component

This is the last Kubernetes component that we will review in this chapter. We will go through it very quickly because some notions might appear too abstract to you at that point.

The `kube-controller-manager` component is part of the Kubernetes control plane, too. It's a binary that runs what we call the reconciliation loop . It tries to maintain the actual state of the cluster with the one described in the Etcd so that there are no differences between the states.

The role of the kube-controller-manager component

Historically, the `kube-controller-manager` component is a big binary that implements a lot of things. Essentially, it embeds what is called a **Controller**. Kubernetes developers tend to break the different controllers being executed as part of the `kube-controller-manager` component into multiple smaller binaries. This is because, with time, `kube-controller-manager` holds multiple and diverse responsibilities that are not compliant with the microservice philosophy.

Sometimes, the *actual* state of the cluster differs from the *desired* state that is stored on `Etcd`. The reason for this could be because of a pod failure and more. Therefore, the role of the `kube-controller-manager` component is to reconcile the *actual* state with the *desired* state.

For example, let's take `ReplicationController`, which is one of the controllers running as part of the `kube-controller-manager` binary.

Later, you'll observe that it's possible to tell Kubernetes to create and maintain a specific number of pods across the different worker nodes. If for some reason, the actual number of pods differs from the number you asked for, `ReplicationController` will start running requests against the `kube-apiserver` component in order to recreate a new pod in `Etcd` so that the failed one is replaced on a worker node.

Here is a list of the few controllers that are part of it:

- `NodeController`
- `NamespaceController`
- `EndpointsController`
- `ServiceaccountController`

As you can gather, the `kube-controller-manager` component is quite big. But essentially, it's a binary that is responsible for reconciling the actual state of the cluster with the desired state of the cluster that is stored in `etcd`.

Where do you install kube-controller-manager?

The `kube-controller-manager` component is part of the Kubernetes control plane. Cloud providers such as Amazon EKS or Google GKE will create and configure a `kube-controller-manager` component for you.

Since it's a control plane component, it should run on a master node. The `kube-controller-manager` component can run as a Docker container or a `systemd` service similarly to `kube-apiserver`. Additionally, you can decide to install the `kube-controller-manager` component on a dedicated machine.

How to make Kubernetes highly available

As you've observed earlier, Kubernetes is a clustering solution. Its distributed nature allows it to run on multiple machines. By splitting the different components across different machines, you'll be able to make your Kubernetes cluster highly available. Next, we will have a brief discussion on the different Kubernetes setups.

The single-node cluster

Installing all Kubernetes components on the same machine is the worst possible idea if you want to deploy Kubernetes in production. However, it is perfectly fine for testing your development. The single-node way consists of grouping all of the different Kubernetes components on the same host or a virtual machine:

Kubernetes single - node cluster

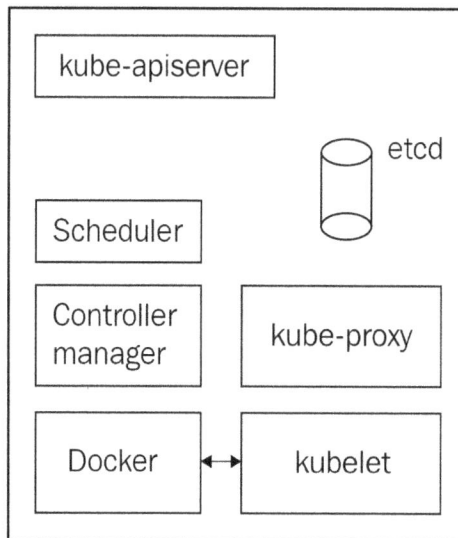

Figure 2.8 – All of the components are working on the same machine

In general, this setup is considered a good way to start your journey with Kubernetes using local testing. There is a project called Minikube that will help you set up a single-node Kubernetes on your local machine by running a virtual machine from a pre-built image with all of the components properly configured inside it. Minikube is good for testing locally, but of course, all of the features requiring multiple nodes won't be available with Minikube. Please note that it is, in fact, possible to run Minikube as a multi-node cluster but should absolutely not be used in production. However, still, it's a good way to start. Just remember to never deploy such clusters in production since it's a really terrible idea:

PROS	CONS
Good for testing	Impossible to scale
Easy to set up locally	Not highly available
Supported natively by Minikube	Not recommended for production

Overall, this setup is the best way to start experimenting with Kubernetes, although it prevents us from experimenting with multi-node scenarios. However, it's the worst process for production. We recommend that you do not ever deploy such a setup for a production workload.

The single-master cluster

This setup consists of having one node executing all of the control plane components with as many worker nodes as you want:

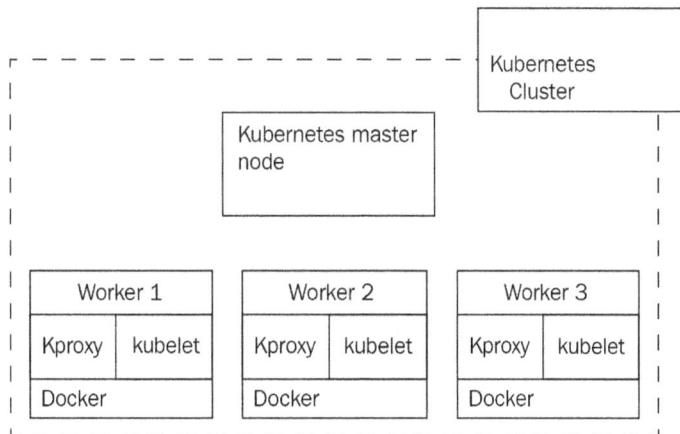

Figure 2.9 – A single master node rules all of the worker nodes (here, it is three)

This setup is quite good, and the fact that there are multiple worker nodes will enable high availability for your containerized application. However, there is still room for improvement.

There is a single point of failure since there is only one master node. If this single master node fails, you won't be able to manage your running Docker containers in a Kubernetes way anymore. Your containers will become orphans, and the only way to stop/update them would be to SSH on the worker node and run plain old Docker commands.

Also, there is a major problem here: by using a single Etcd instance, there is a huge risk that you lose your dataset if the master node gets corrupted. If this happens, your cluster will be impossible to recover.

Lastly, your cluster will encounter an issue if you start scaling your worker nodes. Each worker node brings its own Kubelet agent, and periodically, Kubelet polls the kube-apiserver every 20 seconds. If you start adding dozens of servers, you might DDoS your kube-apiserver, resulting in an outage of your control plane. Remember that your master node/control plane must be able to scale in parallel with your worker nodes:

PROS	CONS
It has high-availability worker nodes.	The master node is a single point of failure.
It supports multi-node features.	A single Etcd instance is running.
It is possible to run it locally with projects such as Kind, but it is not perfect.	It can't scale effectively.
	It is harder to run locally.

Overall, this setup will always be better than a single-node Kubernetes; however, it's still not highly available.

The multi-master multi-node cluster

This is the best way to achieve a highly available Kubernetes cluster. Both your running containers and your control plane are replicated to avoid a single point of failure.

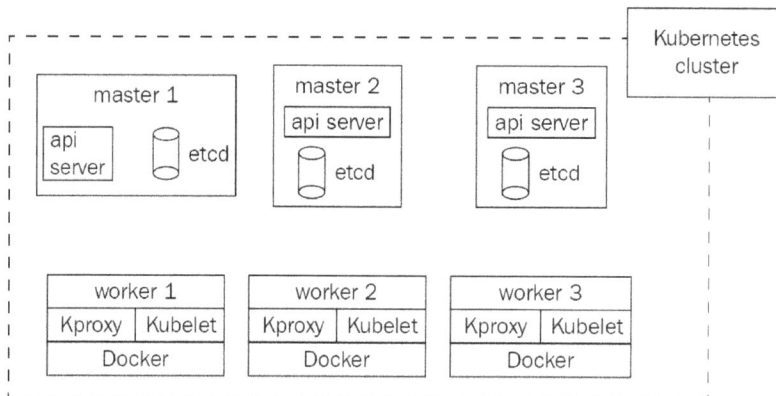

Figure 2.10 – This time, the cluster is highly available

By using such a cluster, you are eliminating all the risks because you are running multiple instances of your worker nodes and your master nodes. Both of the control plane components that are your running containers will be scalable, and since the kube-apiserver component is a stateless API, it's also ready to be scaled.

You will need a load balancer on top of your kube-apiserver instances in order to spread the load evenly between all of them, which will require a little bit more planning. Even though we don't have much information about this, it is almost certain that cloud providers such as Amazon EKS or Google GKE are provisioning Kubernetes clusters that are multi-master multi-worker clusters. If you wish to take it a step further, you can also split all of the different control plane components across a dedicated host. It's better but not mandatory, though. The cluster described in the preceding diagram is perfectly fine.

Before we end this chapter, I'd like to sum up all of the Kubernetes components. This table will help you to memorize all of their responsibilities:

Component name	Communicates with	Role
kube-apiserver	kubectl client(s), Etcd, kube-scheduler, kube-controller-manager, Kubelet, and kube-proxy	The HTTP REST API. It read and writes the state stored in Etcd. The only component that is able to communciate with Etcd directly.
Etcd	kube-apiserver	This stores the state of the Kubernetes cluster.
kube-scheduler	kube-apiserver	This reads the API every 20 seconds to list unscheduled pods (an empty nodeName property), elects a worker node, and updates the nodeName propery in the pod entry by calling kube-apiserver.
kube-controller-manager	kube-apiserver	This polls the API and runs the reconciliation loops.
kubelet	kube-apiserver and Docker (container runtime)	This reads the API every 20 seconds to get pods scheduled to the node it's running on, and translates the pod specs into running containers by calling the local Docker daemon.
kube-proxy	kube-apiserver	This implements the networking layer of Kubernetes.
Container Engine (Docker...)	Kubelet	This runs the containers by receiving instructions from the local Kubelet.

These components are the default ones and are officially supported as part of the Kubernetes project. Remember that other Kubernetes distributions might bring additional components, or they might change the behavior of these ones.

These components are the strict minimum that you need to have a working Kubernetes cluster.

Summary

This was quite a big chapter, but at least, now, you have a list of all of the Kubernetes components. Everything we will do later on will be related to these components: they are the core of Kubernetes. This chapter was full of technical details too, but it was still relatively theoretical. Don't worry if things are still not very clear to you. You will gain a better understanding through practice.

The good news is that you are now completely ready to install your first Kubernetes cluster locally and things are going to be a lot more practical from now on. It is the next step, and that's what we will do in the next chapter. After the next chapter, you'll have a running Kubernetes cluster locally on your workstation, and you will be ready to run your first pods using Kubernetes!

3
Installing Your First Kubernetes Cluster

In the previous chapter, we had the opportunity to explain what Kubernetes is, its distributed architecture, the anatomy of a working cluster, and how it can manage your Docker containers on multiple Linux machines. Now, we are going to get our hands dirty because it's time to install Kubernetes. The main objective of this chapter is to get you a working Kubernetes for the coming chapters. This is so that you have your own cluster to work on, practice with, and learn about while reading this book.

All Kubernetes installations require two steps. First, you need to install the Kubernetes cluster itself, and second, you need to configure your kubectl HTTP client so that it can perform API calls to the `kube-apiserver` component installed on your master node.

Installing Kubernetes means that you have to get the different components to work together. Of course, we won't do that the hard way with `systemd`; instead, we will use automated tools.These tools have the benefit of launching and configuring all of the components for us locally.

If you don't want to have a Kubernetes cluster on your local machine, we're also going to set up a minimalist yet full-featured production-ready Kubernetes cluster on **Google Kubernetes Engine** (**GKE**), Amazon Elastic Kubernetes Service (EKS), and Azure Kubernetes Service (AKS). These are cloud-based and production-ready solutions. In this way, you will be able to practice and learn on a real-world Kubernetes cluster hosted on the cloud.

Whether you want to go local or on the cloud, it is your choice. You'll have to choose the one that suits you best by considering each solution's benefits and drawbacks. In both cases, however, you'll require a working Kubectl installed on your local workstation to communicate with the resulting Kubernetes cluster. Installation instructions for Kubectl are available in the previous chapter, *Chapter 2, Kubernetes Architecture – From Docker Images to Running Pods*.

In this chapter, we're going to cover the following main topics:

- Installing a single-node cluster using Minikube
- Installing a multi-node local cluster using Kind
- Installing a full-featured Kubernetes cluster on GKE
- Installing a full-featured Kubernetes cluster on Amazon EKS
- Installing a full-featured Kubernetes cluster on AKS

Technical requirements

To follow along with the examples in this chapter, you will require the following:

- Kubectl installed on your local machine
- Reliable internet access
- A **Google Cloud Platform** (**GCP**) account with a valid payment method to follow the *Installing a full-featured Kubernetes cluster on Google GKE* section
- An AWS account with a valid payment method to follow the *Installing a full-featured Kubernetes cluster on Amazon EKS* section
- An Azure account with a valid payment method to follow the *Installing a full-featured Kubernetes cluster on Azure AKS* section

Installing a single-node cluster with Minikube

In this section, we are going to learn how to install a local Kubernetes cluster using **Minikube**. It's probably the easiest way to get a working Kubernetes installation locally. By the end of this section, you're going to have a working single-node Kubernetes installation on your local machine.

Minikube is easy to use and completely free. It's going to install all of the Kubernetes components on your local machine and configure all of them. Uninstalling all of the components through Minikube is easy too, so you won't be stuck with it if, one day, you want to destroy your local cluster.

Minikube has one big advantage: it's a super useful tool in which to test Kubernetes scenarios quickly. However, its major drawback is that it prevents you from running a multi-node Kubernetes cluster and, therefore, all of the multi-node scenarios we will discuss later won't be testable for you. That being said, if you do not wish to use Minkube, you can completely skip this section and choose another method described in this chapter.

Launching a single-node Kubernetes cluster using Minikube

The main purpose of Minikube is to launch the Kubernetes components on your local system and have them communicate with each other. To do this, Minikube can work with two different methods:

- A virtual machine
- A Docker container

The first method requires you to install a hypervisor on top of your local system. Then, Minikube will wrap all of the Kubernetes components onto a virtual machine that will be launched. This method is fine, but it requires you to install a hypervisor such as KVM on Linux or HyperKit on macOS to work.

The other method is simpler. Instead of using a virtual machine, Minikube uses a local Docker daemon to launch the Kubernetes components inside a big Docker container. This method is simpler since you just need to have Docker installed on your system. That's the solution we will use.

If you do not have Docker installed on your system, make sure that you install it following the instructions at `https://docs.docker.com/get-docker/`. The installation process is easy, but the steps are slightly different depending on your operating system.

After Docker has been installed, we need to install Minikube itself. Here, again, the process is slightly different depending on your operating system:

- Use the following commands for Linux:

```
$ curl -Lo minikube https://storage.googleapis.com/
minikube/releases/latest/minikube-linux-amd64
$ chmod +x minikube
$ mkdir -p /usr/local/bin
$ sudo install minikube /usr/local/bin
```

- Use the following commands for macOS:

```
$ curl -Lo minikube https://storage.googleapis.com/
minikube/releases/latest/minikube-linux-amd64
$ chmod +x minikube
$ mkdir -p /usr/local/bin
$ sudo install minikube /usr/local/bin
```

- You can also install it from Homebrew using the following command:

```
$ brew install minikube
```

- For Windows, once Minikube has been installed on your system, you can immediately run your first Kubernetes cluster using the following command:

```
$ minikube start --driver="docker"
minikube v1.8.2 on Darwin 10.14.5
Using the docker driver based on user configuration
```

Appending --driver might not be required. If you do not have any hypervisor such as VirtualBox or HyperKit, then Minikube should automatically fall back to Docker. However, if you do have multiple hypervisors on your system, you can set the Docker value explicitly, as shown in the previous example. The result of this command will be a working Kubernetes cluster on your localhost. It's actually as simple as that; you now have a Kubernetes cluster on your system with literally just one command.

Once the command is complete, you can run the following command to check the state of the Kubernetes components installed by Minikube:

```
$ minikube status
host: Running
kubelet: Running
```

```
apiserver: Running
kubeconfig: Configured
```

The preceding output shows that the cluster is up and running, and we can now start interacting with it.

Now, we need to create a `kubeconfig` file for our local Kubectl CLI to be able to communicate with this new Kubernetes installation. The good news is that Minikube also generated one on the fly for us when we launched the `minikube start` command. The `kubeconfig` file generated by Minikube is pointing to the local `kube-apiserver` endpoint, and your local Kubectl was configured to call this cluster by default. So, essentially, there is nothing to do: the `kubeconfig` file is already formatted and in the proper location.

Use the following command to display the current `kubeconfig` file. You should observe a cluster, named `minikube`, that points to a local IP address:

```
$ kubectl config view
```

Following this, you can run this command, which will show the Kubernetes cluster your Kubectl is pointing to right now:

```
$ kubectl config current-context
minikube
```

Now, let's try to issue a real Kubectl command to list the nodes that are part of our Minikube cluster. If everything is okay, this command should reach the `kube-apiserver` component launched by Minikube, which will return only one node since Minikube is a single-node solution. Let's list the nodes with the following command:

```
$ kubectl get nodes
NAME    STATUS    ROLES     AGE     VERSIO
m01     Ready     master    2m41    v1.17.3
```

If you don't view any errors when running this command, it means that your Minikube cluster is ready to be used and is fully working!

This is the very first real `kubectl` command you ran as part of this book. Here, a real `kube-apiserver` component received your API call and answered back with an HTTP response containing data coming from a real `Etcd` datastore. In our scenario, this is the list of the nodes in our cluster. Good job! You just set up your first ever Kubernetes cluster!

> **Important note**
>
> Since Minikube creates a single-node Kubernetes cluster, this command only outputs one node. This node is both a master node and a worker node at the same time. It will run both Kubelet and the control plane components. It's good for local testing, but do not deploy such a setup in production.

What we can do now is list the status of the control plane components so that you can start familiarizing yourself with `kubectl`. In fact, they are both the same; the second one is just an alias for the first one. There are a lot of aliases in `kubectl` along with more than one way to type the same command:

```
$ kubectl get componentstatuses
$ kubectl get cs # The exact same command, "cs" is an alias
NAME                   STATUS     MESSAGE              ERROR
scheduler              Healthy    ok
controller-manager     Healthy    ok
etcd-0                 Healthy    {"health":"true"}
```

This command should output the status of the control plane components. You should see the following:

- A running `Etcd` datastore
- A running `kube-scheduler` component
- A running `kube-controller-manager` component

Stopping and deleting the local Minikube cluster

You might want to stop or delete your local Minikube installation. To proceed, do not kill the Docker container directly, but rather, use the Minikube command-line utility. Here are the two commands to do so:

```
$ minikube stop
Stopping "minikube" in docker ...
Node "m01" stopped.
```

The preceding command will stop the cluster. However, it will continue to exist; its state will be kept, and you will be able to resume it later using the following `minikube start` command again. You can check it by calling the `minikube status` command again:

```
$ minikube status

host: Stopped
kubelet: Stopped
apiserver: Stopped
kubeconfig: Stopped
```

If you want to completely destroy the cluster, use the following command:

```
$ minikube delete
```

If you use this command, the cluster will be completely destroyed. Its state will be lost and impossible to recover.

Now that your Minikube cluster is operational, it's up to you to decide whether you want to use it to follow the next chapters or pick another solution. Minikube is fine, but you won't be able to practice when we get to multi-node scenarios. If you are fully aware of this and you still wish to continue, you can skip the next sections of this chapter. Otherwise, let's examine another tool in which to set up a local Kubernetes cluster, called **Kind**.

Launching a multi-node Kubernetes cluster with Kind

In this section, we are going to discuss another tool called Kind, which is far less known than Minikube, but which resembles it a lot. This tool is also designed to run a Kubernetes cluster locally just like Minikube. The main difference is that Kind is capable of launching multi-node Kubernetes clusters contrary to Minikube, which is a single-node solution.

The whole idea behind Kind is to use Docker containers such as Kubernetes worker nodes thanks to the **Docker-in-Docker (DIND)** model. By launching Docker containers, which themselves contain the Docker daemon and the Kubelet, you can manage to make them behave as Kubernetes worker nodes.

This is exactly the same as when you use the Docker driver for Minikube, except that there, it will not be done in a single container but in several. The result is a local multi-node cluster. Similar to Minikube, Kind is free, and you don't have to pay to use it.

> **Important Note**
>
> Similar to Minikube, Kind is a tool that is used for local development and testing. Please never use it in production because it is not designed for it.

Installing Kind onto your local system

Since Kind is a tool entirely built around Docker, you need to have the Docker daemon installed and working on your local system. If you do not have Docker installed on your system, make sure that you install it by following the instructions at `https://docs.docker.com/get-docker/`. The installation process is easy, but the steps are slightly different depending on your operating system.

After Docker has been installed, we need to install Kind itself. Again, the process will be different depending on your operating system:

- Use the following commands for Linux:

  ```
  $ curl -Lo ./kind https://kind.sigs.k8s.io/dl/v0.8.1/
  kind-$(uname)-amd64
  $ chmod +x ./kind
  $ mv ./kind /usr/local/bin/kind
  ```

- Use the following commands for macOS:

  ```
  $ curl -Lo ./kind https://kind.sigs.k8s.io/dl/v0.8.1/
  kind-$(uname)-amd64
  $ chmod +x ./kind
  $ mv ./kind /usr/local/bin/kind
  ```

- You can also install it with Homebrew:

  ```
  $ brew install kind
  ```

- Use the following commands for Windows:

  ```
  $ curl.exe -Llo kind-windows-amd64.exe https
  ```

- You can also install it with Chocolatey:

  ```
  $ choco install kind
  ```

Once Kind has been installed on your system, you can immediately proceed to launch a new Kubernetes cluster using the following command:

```
$ kind create cluster
Creating cluster "kind" ...
```

When you run this command, Kind will start to build a Kubernetes cluster locally by pulling a Docker image containing all of the control plane components. The result will be a single-node Kubernetes cluster with a Docker container acting as a *master node*. That being said, we do not want this setup since we can already achieve it with Minikube. What we want is a multi-node cluster. To do this, we must write a very small configuration file and tell Kind to use it as a template to build the local Kubernetes cluster. So, let's get rid of the single-node Kind cluster that we just built, and let's rebuild it as a multi-node cluster:

1. Run this command to delete the cluster:

    ```
    $ kind delete cluster
    Deleting cluster "kind" ...
    ```

2. Then, we need to create a config file that will serve as a template for Kind to build our cluster. Simply copy the following content to a local file in this directory, for example, ~/.kube/kind_cluster:

    ```
    Kind: Cluster
    apiVersion: kind.sigs.k8s.io/v1alpha3
    nodes:
    - role: control-plane
    - role: worker
    - role: worker
    - role: worker
    ```

    Please note that this file is in YAML format. Pay attention to the nodes array, which is the most important part of the file. This is where you tell Kind how many nodes you want in your cluster. The role key can take two values: control plane and worker.

    Depending on which role you chose, a different node will be created.

3. Let's relaunch the kind create command with this config file to build our multi-node cluster. For the given file, the result will be a one-master three-worker Kubernetes cluster:

    ```
    $ kind create cluster --config ~/.kube/kind_cluster
    Creating cluster "kind" ...
    ```

Following this, you should have four new Docker containers: one running as a master node and the other three as worker nodes of the same Kubernetes cluster.

Now, as always with Kubernetes, we need to write a `kubeconfig` file in order for our Kubectl utility to be able to interact with the new cluster. And guess what, Kind already generated the proper configuration and appended it to our `~/.kube/config` file, too. Additionally, Kind set the current context to our new cluster, so there is essentially nothing left to do. We can immediately start querying our new cluster. Let's list the node using the `kubectl get nodes` command. If everything is okay, we should view four nodes:

```
$ kubectl get nodes
```

Everything seems to be perfect. Your Kind cluster is working!

Just as we did with Minikube, you can also check for the component's statuses using the following command:

```
$ kubectl get cs
```

Stopping and deleting the local Kind cluster

You might want to stop or remove everything Kind created on your local system to clean the place after your practice. To do so, you can use the following command:

```
$ kind stop
```

This command will stop the Docker containers that Kind is managing. You will achieve the same result if you run the Docker `stop` command on your containers manually. Doing this will stop the containers but will keep the state of the cluster. That means your cluster won't be destroyed, and simply relaunching it using the following command will get the cluster back to its state before you stopped it.

If you want to completely remove the cluster from your system, use the following command. Running this command will result in removing the cluster and its state from your system. You won't be able to recover the cluster:

```
$ kind delete cluster
```

Now that your Kind cluster is operational, it's up to you to decide whether you want to use it to practice while reading the coming chapters. You can also decide whether to pick another solution described in the following sections of this chapter. Kind is particularly nice because it's free to use and allows you to install a multi-node cluster. However, it's not designed for production and remains a development and testing solution for a non-production environment. Kind makes use of Docker containers to create *Kubernetes nodes*, which, in the real world, are supposed to be Linux machines.

Installing a Kubernetes cluster using Google GKE

Google GKE is a cloud offering that is part of GCP, which is Google's cloud offering. If you do not want to install anything on your local system, this solution might be good for you. Indeed, Google GKE is a cloud-based solution, meaning that the resulting Kubernetes won't run on your own system.

Google GKE allows you to set up a full-featured Kubernetes cluster on the cloud that is built for production. The result of the service will be a Kubernetes cluster that runs on Google's machines on their cloud platform, not on your own local machine. Your local machine will only serve as a Kubernetes client thanks to Kubectl pointing to a remote endpoint exposed by Google.

The main advantage is that you will have a real production-ready and full-featured Kubernetes cluster to practice with, which is ideal in which to train and improve your skill. However, Google GKE is a commercial product, so you'll have to spend some dollars in order to use it, as well as having a valid payment method linked to your GCP account. Indeed, Google requires you to link your credit card to your GCP account prior to accessing any of their services.

Here, we assume you already have a GCP account with a valid payment method linked, and you are ready to set up a Kubernetes cluster on Google GKE.

Launching a multi-node Kubernetes cluster on Google GKE

There are different ways in which to set up a Kubernetes cluster on Google GKE. The easiest way is through the `gcloud` command-line utility. Indeed, GCP exposes all its cloud services as a command-line utility called `gcloud`. This tool can literally set up a Kubernetes cluster on the GCP cloud using just one command. So, let's proceed to install the `gcloud` utility onto your local system:

- Use the following commands for Linux:

  ```
  $ curl https://sdk.cloud.google.com | bash
  $ exec -l $SHELL
  $ gcloud init
  ```

- Use the following commands for macOS:

  ```
  $ curl https://sdk.cloud.google.com | bash
  $ exec -l $SHELL
  $ gcloud init
  ```

- On Windows, you need to download the ZIP file at `https://dl.google.com/dl/cloudsdk/channels/rapid/google-cloud-sdk.zip`. Then, unzip it, run the `google-cloud-sdk\install.bat` script, and follow the instructions. Restart your command line and run the following command:

  ```
  C:\> gcloud init
  ```

The `gcloud init` command is an interactive command. Simply follow the instructions it outputs and it should be fine. At some point, `gcloud` will require you to authenticate to your GCP account by opening your default web browser. Simply proceed, and `gcloud` should notice that you are successfully authenticated.

Then, `gcloud` will require you to create a new project. To explain a little bit about how GCP works, the platform is divided into projects that are logical units used to organize your workloads.

With GCP, the first thing to do is create a project that will contain your resources. So, let's create a project that will host our Kubernetes cluster on GKE. There is a little problem here: project IDs are meant to be unique across all GCP accounts. That means if someone in the world ever uses a project ID, you can't use it anymore. In order to find a free project ID, you can add some random values to it. I named mine `gke-cluster-0123456789`:

1. Use the following command to create your project:

   ```
   $ gcloud create projects gke-cluster-0123456789
   ```

   Please choose a unique name, and then try to append some random characters to it to find a free project ID. Use this command and replace my `gke-cluster-0123456789` ID with your own project ID.

2. Once the project has been created, select it from the drop-down menu on the GCP web console:

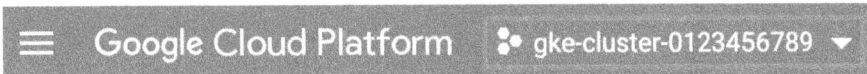

Figure 3.1 – The GCP web console

3. Now that the project is ready, let's define it as our current project at the `gcloud` level, too:

   ```
   $ gcloud config set project gke-cluster-0123456789
   Updated property [core/project].
   ```

4. The last thing you need to do is to enable billing on your GCP account. To do that, open the Google GKE console on the menu on the left-hand side and select **Clusters**:

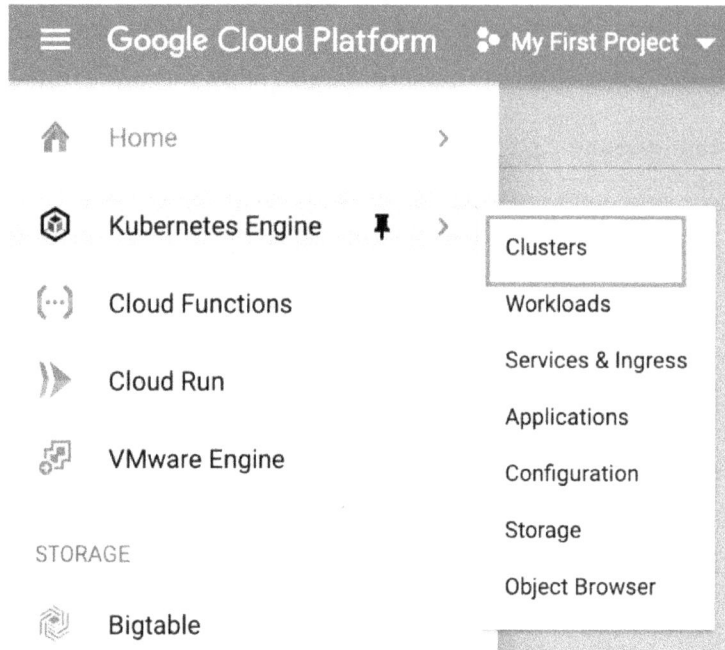

Figure 3.2 – The GCP web console GKE clusters menu

On this screen, you should see an **Enable Billing** button. Click on this button and enter your billing information. Once done, GCP will proceed in enabling the Kubernetes Engine API for your project. This can take a few minutes, but after that, you will be able to create your first cluster on GKE.

To create the cluster, there are two ways you can use, as follows:

- You can use the web console.

- You can use the gcloud command-line tool.

Let's use the `gcloud` command line. The good news is that `gcloud` can literally bootstrap a GKE cluster and configure Kubectl afterward in just one command:

1. Run the following command:

    ```
    $ gcloud container clusters create mygkecluster
    --num-nodes 3 --machine-type e2-medium --europe-west1
    ```

    This command will create a Kubernetes cluster on GKE, called `mygkecluster`, that will have three worker nodes. These worker nodes will be run as `e2-medium` instances, which are two CPU virtual machines.

    The cluster is going to be created in the `europe-west1` region of the GCP cloud. You should pick the nearest region to where you are in order to reduce network latency. In my case, `europe-west-1` is fine, but you should `pick us-west-1` if you live in the US, for example. The complete list of available regions for GCP is available at `https://cloud.google.com/compute/docs/regions-zones`.

2. After you have launched the command, the new cluster should appear on the Google GKE screen, as shown in *Figure 3.3*:

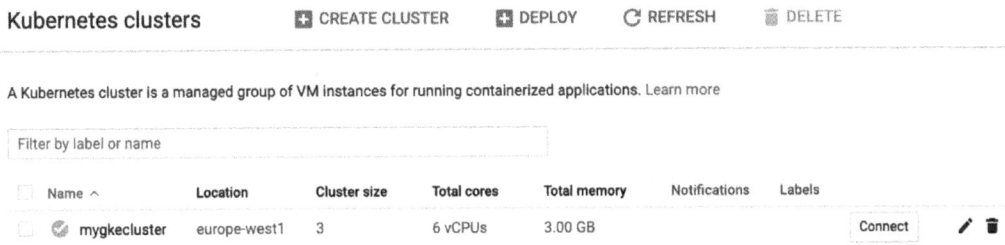

Figure 3.3 – The GCP web console

Good job! Wait a little bit for the cluster creation to complete; it can take a few minutes. Bear in mind that you will be billed the moment the cluster gets launched!

After a few minutes, the Kubernetes cluster is launched on Google GKE and is ready to use. Remember, the `gcloud` command we ran also generated a `kubeconfig` file for our local environment. So, we will not have to write the configuration file by ourselves.

3. Run the following command to make sure your Kubernetes cluster is up and running and the Kubectl command line is properly configured to query it:

    ```
    $ kubectl get nodes
    ```

This command should authenticate against the Kubernetes cluster running on Google GKE and retrieve the list of the nodes running as part of the cluster. Good job! You have successfully set up a Kubernetes cluster on GKE!

Stopping and deleting a Kubernetes cluster on Google GKE

Cloud providers bill you when you use their services. Every second a Kubernetes cluster is in the running state in GKE is a second that you will be charged for at the end of the month. So, it's a good idea to stop cloud services when they are not being used and resume them when needed.

Unfortunately, you can't stop a GKE cluster the way you can with Minikube or Kind. We are forced to completely destroy it and recreate it, meaning that we will lose the state of the cluster each time. It's not a big deal, though, as, for now, we're just looking for a practice cluster that we can break and recreate at will. To proceed with cluster removal, type in the following command:

```
$ gcloud container clusters delete mygkecluster --region
europe-west-1
```

You are all set to go! If you are happy with GKE, you can proceed to the next chapter and use your brand-new cluster to practice while reading this book. Otherwise, you can pick another solution. Bear in mind that GKE is not free; you are charged $0.1 per hour to use it, resulting in a $72/month if the cluster is running the whole month. Add to this price the cost of the worker nodes, which are instances that are billed independently based on their instance types.

Installing a Kubernetes cluster using Amazon EKS

If you do not wish to have a local Kubernetes cluster nor use Google GKE, you can also set up a remote cluster on the AWS cloud. This solution is ideal if AWS is your preferred cloud provider. This section will demonstrate how to get a Kubernetes cluster using Amazon EKS, which is the public cloud offering by AWS.

Using this solution is not free. Amazon EKS is a commercial product that is a competitor to Google GKE. Amazon EKS offers to manage a Kubernetes control plane just for you. That means the service will set up some nodes running all of the control plane components without exposing you to these machines. You will get an endpoint for a remote `kube-apiserver` component and that's all.

Once done, you'll have to set up a few Amazon EC2 instances (which are virtual machines), and you'll have them join the control plane. These Amazon EC2 instances will be your worker nodes. In this way, you'll have a multi-node Kubernetes cluster with a multi-worker setup. This result of this lab will be a full-featured multi-node Kubernetes cluster, which is an ideal solution with which to enter the coming chapters. Of course, we won't set up all of this ourselves; a lot of solutions exist in which to provision a fully configured Amazon EKS instance, and we will examine one of these solutions. Additionally, we will try not to spend a lot money on AWS ideally.

The cost of the infrastructure will depend on the AWS region where your cluster is going to be deployed:

- The control plane, which costs $0.74/hour
- 3 x t3.medium worker nodes (around 0.42 USD per instance depending on the AWS region where you'll deploy your cluster)

Because it's an hourly pricing model, remember to destroy everything from your AWS account when you are not using it in order to save money.

There is a prerequisite when it comes to working with Amazon EKS. You need to have an AWS account with a properly linked payment method. Indeed, similarly to GCP, AWS requires you to link a valid payment method prior to gaining access to their services. In this section, we assume that you already have an AWS account with a valid payment method linked to it.

Launching a multi-node Kubernetes cluster on Amazon EKS

There are multiple ways in which to get a working cluster with Amazon EKS. You can use CloudFormation, Terraform, AWS CLI commands, or even the web AWS console. Additionally, AWS offers a CLI tool to communicate with the Amazon EKS service, called `eksctl`, which can be used to set up Kubernetes clusters on Amazon EKS with just one or two commands. However, we're not going to use this solution because it would require you to install another tool on your machine dedicated to AWS, which is not the purpose of our discussion here.

Instead, in this tutorial, we are going to use AWS CloudFormation for the sake of simplicity. AWS has already put together a working CloudFormation template that allows you to deploy a Kubernetes cluster. You can find this template on the GitHub repository, called AWS Quickstart, at `https://github.com/aws-quickstart/quickstart-amazon-eks`.

AWS makes use of something called **virtual private clouds** (**VPCs**). In fact, everything you deploy on AWS is deployed inside of a VPC, and Amazon EKS is no exception. There are two possible solutions, as follows:

- You deploy your EKS cluster onto an existing VPC.
- You deploy your EKS cluster in a new VPC.

The first choice will work, but there is a risk that deploying the cluster into an existing VPC will overlap with existing resources. To not impact any applications that might already exist on your AWS account, we are going to deploy the Amazon EKS cluster on a new VPC. Fortunately for us, the CloudFormation template allows us to create a new VPC while provisioning the Amazon EKS cluster. I encourage you to create a new VPC since VPC creation is free. Bear in mind that you are still limited to five VPCs per AWS region when using them.

Let's sum up all of the different steps required to bootstrap a cluster on Amazon EKS:

1. Create a key pair on the Amazon EC2 service.
2. Launch the CloudFormation template in the Amazon CloudFormation service.
3. Create an IAM user in the Amazon IAM service.
4. Install and configure the AWS CLI with the IAM user created earlier.
5. Generate a `kubeconfig` file with the AWS CLI.

First, you need to create an EC2 key pair:

1. Please sign in to the AWS web console and open the Amazon EC2 service. You can find the service by searching for it in the search bar under the **Compute** menu on the home page:

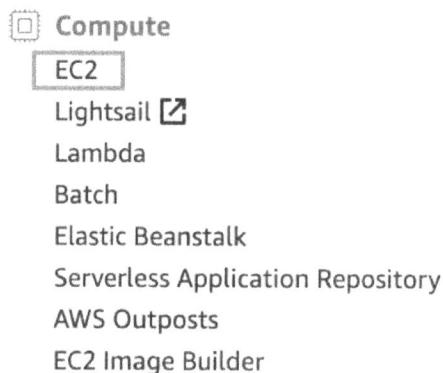

Figure 3.4 – The EC2 menu in the AWS web console

2. Once you have opened the EC2 service, select the **Key Pairs** tab on the left-hand side:

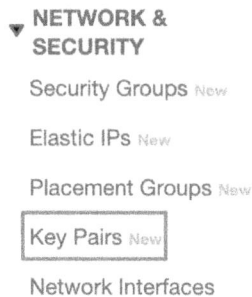

▼ **NETWORK & SECURITY**

Security Groups New

Elastic IPs New

Placement Groups New

Key Pairs New

Network Interfaces

Figure 3.5 – The Key Pairs menu in the EC2 web console

3. From this screen, click on the **Create Key** button. Give the new key pair an arbitrary name and then download the .pem file. This key pair is going to be useful for you to SSH to your worker nodes later.

4. The next step is to create the Kubernetes cluster itself. We are going to do that through the CloudFormation template that we talked about earlier. To do that, enter the CloudFormation web console that is accessible under the **Management & Governance** menu on the home page:

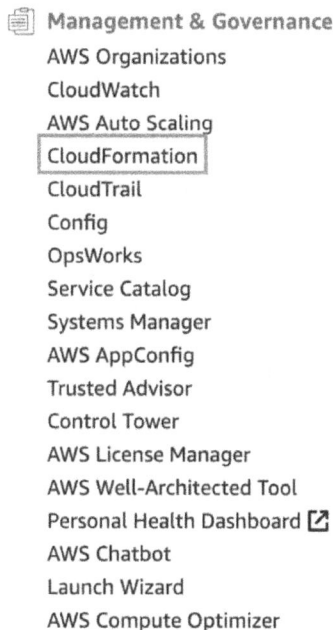

🗐 **Management & Governance**

AWS Organizations

CloudWatch

AWS Auto Scaling

CloudFormation

CloudTrail

Config

OpsWorks

Service Catalog

Systems Manager

AWS AppConfig

Trusted Advisor

Control Tower

AWS License Manager

AWS Well-Architected Tool

Personal Health Dashboard 🔗

AWS Chatbot

Launch Wizard

AWS Compute Optimizer

Figure 3.6 – The CloudFormation menu in the AWS web console

5. Select **Create a stack from an Amazon S3 URL** and enter the following URL into the input: `https://s3.amazonaws.com/aws-quickstart/quickstart-amazon-eks/templates/amazon-eks-entrypoint-new-vpc.template.yaml`. Please refer to the following screenshot:

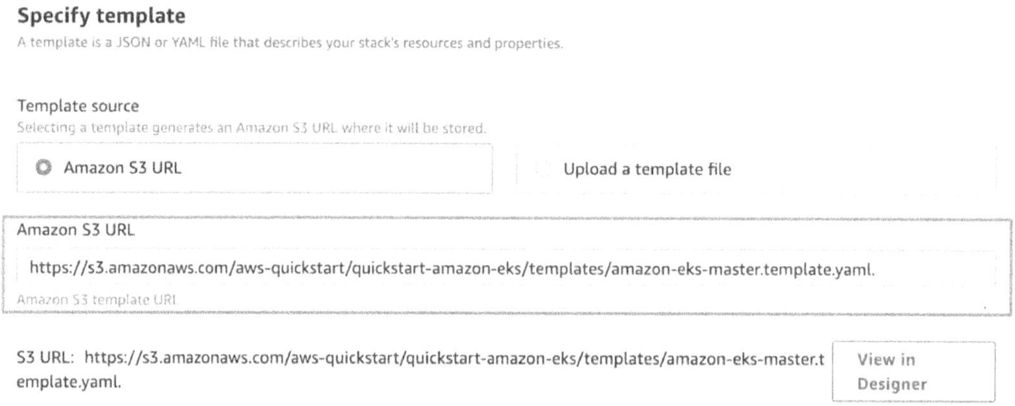

Specify template

A template is a JSON or YAML file that describes your stack's resources and properties.

Template source
Selecting a template generates an Amazon S3 URL where it will be stored.

● Amazon S3 URL	Upload a template file

Amazon S3 URL

https://s3.amazonaws.com/aws-quickstart/quickstart-amazon-eks/templates/amazon-eks-master.template.yaml.

Amazon S3 template URL

S3 URL: https://s3.amazonaws.com/aws-quickstart/quickstart-amazon-eks/templates/amazon-eks-master.template.yaml. | View in Designer |

Figure 3.7 – The S3 URL displayed in the CloudFormation stack creation web view

6. The next screen will enable you to define some variables and parameters that are needed by the CloudFormation template to launch your Amazon EKS cluster. You'll observe that a lot of variables are declared with a default value, but some of them must be set by you. Pay attention to the following input that should be displayed in the console:

   • **Stack name**: Set a logical name for CloudFormation to keep track of the resources created by the template. For example, you can set `eks_cluster`.

   • **Availability Zone**: Pick at least two availability zones available in your current AWS region.

   • **Allowed external access CIDR**: You can define your own IP address followed by a `/32` to allow your IP address to communicate with the EKS cluster. Otherwise, you can set `0.0.0.0/0` to allow remote communication from the whole internet if you do not know your IP address.

- **Number of Availability Zones**: Set 2 or 3 availability zones so that your EKS cluster will be highly available.

- **Provision bastion host**: You should set **enabled** if you do want to deploy a bastion host on your VPC. The bastion host will be an SSH jump machine to access the instances that will be launched deeper into your VPC.

- **SSH key name**: Select the name of the SSH key pair that you created in the previous step.

7. After you have filled in all of these options, select the **Next** button until you are asked to start the new stack creation. CloudFormation will begin provisioning your new Amazon EKS cluster; it can take approximately 15 minutes to complete, as the process is quite long. There is nothing left to do in CloudFormation. Please wait for the cluster to be fully provisioned before going any further.

8. After a few minutes, your new clusters will enter the ready state. We need to generate a new `kubeconfig` file in our local machine to be able to interact with the Amazon EKS cluster. Unfortunately for us, CloudFormation cannot do that. There are additional steps to take, and they require the usage of the AWS CLI command-line utility in order to generate a `kubeconfig` file. That's why we must install the AWS CLI:

- Use the following commands for Linux:

```
$ sudo apt-get update
$ sudo apt-get install awscli
```

- Use the following commands for macOS:

```
$ sudo apt-get update
$ sudo apt-get install awscli
```

- On Windows, download and launch the MSI installer accessible at `https://awscli.amazonaws.com/AWSCLIV2.msi`.

9. Once the AWS CLI has been installed, we need to retrieve AWS access keys from the AWS console in order to configure the AWS CLI. To get those keys, we must create a new IAM user in the Amazon IAM web console. Go to the AWS web console and open the **IAM** service that is underneath the **Security, Identity, & Compliance** category:

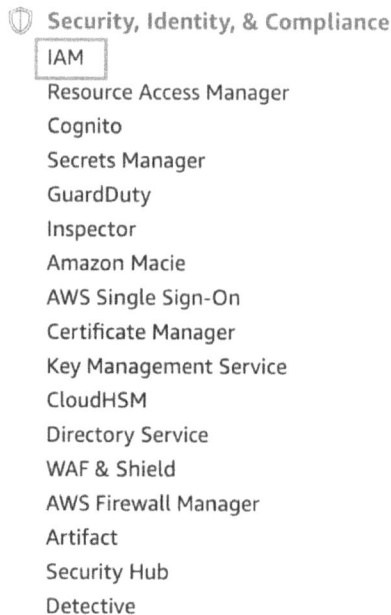

Security, Identity, & Compliance

IAM
Resource Access Manager
Cognito
Secrets Manager
GuardDuty
Inspector
Amazon Macie
AWS Single Sign-On
Certificate Manager
Key Management Service
CloudHSM
Directory Service
WAF & Shield
AWS Firewall Manager
Artifact
Security Hub
Detective

Figure 3.8 – The IAM menu in the AWS web console

10. Once you are in the AWS IAM console, select **Users** from the tab on the left-hand side:

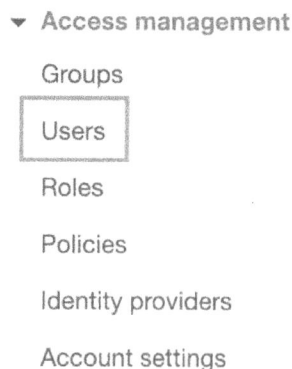

▼ Access management

Groups

Users

Roles

Policies

Identity providers

Account settings

Figure 3.9 – The Users menu in the IAM web console

11. In this new screen, click on the **Add user** button and fill in the different options as follows:

   - **User Name**: Select an arbitrary name. Let's name it eks_user.

   - **Access Type**: Select programmatic access.

12. Now you should be at the permission selection screen. AWS recommends that you respect the least privileged principle: that is, each user should strictly have access to the resource it needs to interact with. However, for the sake of simplicity, we will violate this principle and grant the AdministratorAccess policy to our eks_user.

13. Pick the **Attach existing policy directly** option and select **AdministratorAccess**, which should be displayed first in the drop-down list. From the next screen, you can set tags for the new user. Following this, review and create it.

14. When the user has been created, AWS gives you an access key ID and a secret access key. Grab these two values because we are going to need them. Run the following command to start the configuration of the AWS CLI:

```
$ aws configure -profile eks_user

AWS Access Key ID [None]: [Access Key ID]
AWS Secret Access Key [None]: [Secret access key]
Default region name [None]: [Your current AWS region]
Default output format [None]: json
```

15. Once you have configured the AWS CLI, export a new variable in your local shell to authenticate AWS CLI calls against your AWS account:

```
$ export AWS_PROFILE=eks_user
```

16. Now, you can run your first AWS CLI command. Let's list the Kubernetes clusters created in the Amazon EKS service. You should see the Kubernetes cluster created by the CloudFormation template:

```
$ aws eks list-clusters
{
        "clusters": [
            "EKS-XXXXXX"
    ]
}
```

17. Now that the AWS CLI has been fully configured, we can use it to generate the `kubeconfig` file. Run the following command. The name of the cluster required is the value that was outputted in the previous command:

```
$ aws eks update-kubeconfig --name EKS-XXXXX
Added new context arn:aws:eks:eu-west-1:XXX:cluster/
EKS-XXX to /root/.kube/config
```

18. As you can see, this command generates a `kubeconfig` file and directly places it in the correct directory, which is `$HOME/.kube/config`. You can now run your first `kubectl` command, which will list the nodes that are part of the Amazon EKS clusters:

```
$ kubectl get nodes
```

The preceding command should output the master and worker nodes running on the Amazon EKS cluster. This was challenging! But you have successfully set up a Kubernetes cluster using Amazon EKS.

This cluster is a production-ready, multi-node cluster. You can use it to practice on a real-world Kubernetes setup. However, bear in mind that it is not free, and you are spending money each second the service is running. That is why you should understand how to delete the Kubernetes cluster when it is not being used. We will explore this next.

Deleting the Kubernetes cluster on Amazon EKS

Unfortunately for us, there is no way to stop and resume a Kubernetes cluster running on the Amazon EKS service. The feature does not exist. The only thing we can do is to completely destroy it and then recreate it. This will work, but destroying it will result in a state loss. This means that you won't be able to restore the cluster to the state it had before its removal.

Deleting a cluster is super easy since we created it with a CloudFormation template. The easiest way to get rid of the cluster is to go to the stack list in the CloudFormation web console and remove the stack. As a result, CloudFormation will remove all of the resources that were created.

If you want to recreate the cluster, you can simply recreate it using the CloudFormation template once more. You won't have to reconfigure the AWS CLI since you have already done it once, but of course, you will need to regenerate a new `kubeconfig` file using the `aws eks update-kubeconfig` command, as demonstrated earlier in this chapter.

Bear in mind that you should destroy the cluster when you are not using it, for example, during nighttime. Removing the resources that you are not using is the best way to save money on your cloud bill.

Installing a Kubernetes cluster using Azure AKS

Lastly, I'd like to show you how you can provision a Kubernetes cluster on the Azure AKS service. Azure AKS is a service that is part of the Azure cloud provider, which is offered by Microsoft. AKS is the third major Kubernetes cloud offering around and is a competitor to Amazon EKS and Google GKE. Mostly, the service does the same job as the other two: it allows you to create Kubernetes clusters in just a few clicks directly on the Azure cloud. This solution might be good for you if you are willing to install a Kubernetes cluster without using your own machine or if Azure is simply your preferred cloud provider.

Launching a multi-node Kubernetes cluster on Azure AKS

Bootstrapping a cluster on Azure AKS is very easy. Similar to Google GKE, it has a command-line utility, called `az`, which can start a Kubernetes cluster in just one command. The `az` command line can also generate a `kubeconfig` file for us to allow Kubectl to communicate with our Kubernetes cluster on AKS:

1. Let's install the AKS command-line utility:

- Use the following commands for Linux:

```
$ sudo apt-get update
$ sudo apt-get install ca-certificates curl
apt-transport-https lsb-release gnupg
$ curl -sL https://packages.microsoft.com/keys/microsoft.
asc | gpg --dearmor |sudo tee /etc/apt/trusted.gpg.d/
microsoft.gpg > /dev/null
$ AZ_REPO=$(lsb_release -cs)
$ echo "deb [arch=amd64] https://packages.microsoft.com/
repos/azure-cli/ $AZ_REPO main" | sudo tee /etc/apt/
sources.list.d/azure-cli.list
$ sudo apt-get update
$ sudo apt-get install azure-cli
```

- Use the following commands for macOS:

  ```
  $ brew update && brew install azure-cli
  ```

- On Windows, you need to download and launch the MSI installer that is available in the official documentation from Microsoft. You can locate it at `https://docs.microsoft.com/fr-fr/cli/azure/install-azure-cli-windows?view=azure-cli-latest&tabs=azure-cli`.

2. Once the `az` command line has been installed, you need to configure it. The first thing to do is to configure authentication against your Azure account so that the `az` command line can issue a query against your Azure cloud:

   ```
   $ az login
   ```

3. This command will open your default web browser and ask you to authenticate your Azure web console again. Once authenticated, the `az` command line will detect it. The next step is to create a resource group.

   The Azure cloud works in the following way: everything you create in Azure must be created inside a logical unit called a resource group. Resource groups are a bit like projects in GCP. Additionally, the AWS cloud has something called resource groups, but they have a less important role in the AWS world.

4. Let's create a resource group in our Azure cloud in order to create our AKS cluster. Run the following command to create a resource group that is called `az_cluster`:

   ```
   $ az group create --name az_cluster --location
     francecentral
   ```

   This command will create a resource, called `az_cluster`, at the `francecentral` location. This location is fine to me, but you should pick one that is closer to your location. If you need to access all of the possible locations that Azure has, run the following command:

   ```
   $ az account list-locations
   ```

Once the resource group has been created, you can check for its presence in the Azure web console:

Name ↑	Subscription ↑↓	Location ↑↓	
az_cluster	Paiement à l'utilisation	France Central	•••

Showing 1 to 1 of 1 records.   No grouping ∨

Figure 3.10 – The az_cluster resource group displayed in the Azure web console

5. Now, we are ready to create the AKS cluster. Again, the `az` command line can achieve that with just one command:

```
$ az aks create --resource-group aks_cluster --name my_
aks_cluster --node-count 3 --generate-ssh-keys
```

6. The following command will create an AKS cluster, called my_aks_cluster, with three nodes in the resource group, called `az_cluster`, that we created earlier. We also ask Azure to generate SSH keys for us to SSH to our nodes. That command can take a few minutes to complete. However, after its completion, you will have a working Kubernetes cluster created on Azure AKS:

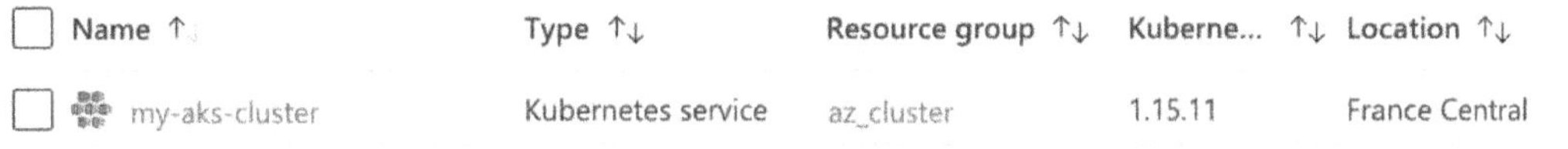

Name ↑	Type ↑↓	Resource group ↑↓	Kuberne... ↑↓	Location ↑↓
my-aks-cluster	Kubernetes service	az_cluster	1.15.11	France Central

Figure 3.11 – The AKS cluster displayed in the Azure web console

7. The last thing to do now is to configure a `kubeconfig` file in order to get Kubectl to communicate with the cluster on the AKS cluster. Fortunately for us, the `az` command line can do that easily. Run the following command to generate a `kubeconfig` file and place it in the correct directory:

```
$ az aks get-credentials --resource-group az_cluster
--name my_aks_cluster
```

8. Now, everything is ready for us to issue our first command to the AKS cluster. Let's run the following command to list the nodes that are part of the cluster:

```
$ kubectl get nodes
```

If everything is okay, you should be able to view the three nodes that we created earlier.

Stopping and deleting a Kubernetes cluster on Azure AKS

Like all cloud offerings, Azure AKS is not free. It's a commercial product. You should stop it and resume it when you are not using your cluster in order to avoid being billed too much. Unfortunately, it is impossible to stop and resume a cluster on AKS without losing its state. You must destroy it and then recreate it.

To remove an AKS cluster, we prefer to remove the resource group itself since a resource's life cycle is bound to its parent resource group. Deleting the resource group should delete the cluster. The following command will get rid of the AKS cluster:

```
$ az group delete --name az_cluster --yes --no-wait
```

Summary

This chapter was quite intense! You require a Kubernetes cluster to follow this book, and so, we examined five ways in which to set up Kubernetes clusters on different platforms. You learned about Minikube, which is the most common way to set up a cluster on a local machine. You also discovered Kind, which is a tool that can set up multi-node local clusters, which is a limitation of Minikube. Then, we looked at the three major Kubernetes cloud services, which are Google GKE, Amazon EKS, and Azure AKS. These three services allow you to create a Kubernetes cluster on the cloud for your practice and train with. This was just a quick introduction to these services, and we will have the opportunity to dive deeper into these services later. For the moment, simply pick the solution that is the best for you. Personally, I use both Kind and Amazon EKS, as they are my preferred tools.

In the next chapter, we are going to dive into Kubernetes by exploring the concept of Pods. The Pod resource is the most important resource that Kubernetes manages. We will learn how to create, update, and delete Pods. Additionally, we will look at how to provision them, how to get information from them, and how to update the containers they are running. We will deploy an NGINX Pod on a Kubernetes cluster and examine how we can access it from the outside. By the end of the next chapter, you will be capable of launching your first containers on your Kubernetes cluster through the usage of Pods. The cluster that you installed here will be very useful when you follow the real-world examples that are coming in the next chapter.

Section 2: Diving into Kubernetes Core Concepts

This section is the most important one since it is the first time that you will be introduced to Kubernetes core objects, also called primitives. You are going to be taught how Kubernetes manages its containerized applications. At the end of this section, you should be able to run Kubernetes-based workloads.

This part of the book comprises the following chapters:

4
Running Your Docker Containers

This chapter is probably the most important one in this book. Here, we are going to discuss the concept of **Pods**, which are the objects Kubernetes uses to launch your Docker containers. Pods are at the heart of Kubernetes and mastering them is essential.

In *Chapter 3*, *Installing Your First Kubernetes Cluster*, we said that the Kubernetes API defines a set of resources representing a computing unit. Pods are resources that are defined in the Kubernetes API that represent one or several Docker containers. We never create containers directly with Kubernetes, but we always create Pods, which will be *converted* into Docker containers on a worker node in our cluster.

At first, it can be a little difficult to understand the connection between Kubernetes Pods and Docker containers, which is why we are going to explain what Pods are and why we use them rather than Docker containers directly. A Kubernetes Pod can contain one or more Docker containers. In this chapter, however, we will focus on Kubernetes Pods, which contain only one Docker container. We will then have the opportunity to discover the Pods that contain several containers in the next chapter.

We will create, delete, and update Pods using the BusyBox image, which is a Linux-based image containing many utilities useful for running tests. We will also launch a Pod based on the NGINX Docker image to launch an HTTP server, before accessing the default NGINX home page via a feature `kubectl` exposes called port forwarding. It's going to be useful to access and test the Pods running on your Kubernetes cluster from your web browser.

Then, we will discover how to label and annotate our Pods to make them easily accessible. This will help us organize our Kubernetes cluster so that it's as clean as possible. Finally, we will discover two additional resources, which are jobs and Cronjobs. By the end of this chapter, you will be able to launch your first Docker containers managed by Kubernetes, which is the first step in becoming a Kubernetes master!

In this chapter, we're going to cover the following main topics:

- Let's explain the notion of Pods
- Launching your first BusyBox Pod
- Labeling and annotating your Pods
- Launching your first job
- Launching your first Cronjob

Technical requirements

Having a properly configured Kubernetes cluster is recommended to follow this chapter so that you can practice the commands shown as you read. Whether it's a minikube, Kind, GKE, EKS, or AKS cluster is not important. You also need a working `kubectl` installation on your local machine. Running the `kubectl get nodes` commands should output you at least one node:

```
$ kubectl get nodes
NAME    STATUS    ROLES    AGE    VERSION
 10.0.103.186.internal    Ready    <none>    6m    v1.17.12
```

You can have more than one node if you want, but at least one `Ready` node is required to have a working Kubernetes setup.

Let's explain the notion of Pods

In this section, we will explain the concept of Pods from a theoretical point of view. Pods have certain peculiarities that must be understood if you wish to master them well.

What are Pods?

When you want to create, update, or delete a container through Kubernetes, you do so through a Pod – you never interact with the containers directly. A Pod is a group of one or more containers that you want to launch on the same machine, in the same Linux namespace. That's the first rule to understand about Pods: they can be made up of one or more containers but all the containers that belong to the same Pod will be launched on the same worker node. A Pod cannot and *won't ever* span across multiple worker nodes: that's an absolute rule.

But why do we bother delegating the management of our Docker containers to this intermediary resource? After all, Kubernetes could have a container resource that would just launch a single container. The reason is that containerization invites you to think in terms of Linux processes rather than in terms of virtual machines. You may already know about the biggest and most recurrent Docker anti-pattern, which consists of using Docker containers as virtual machine replacements: in the past, you used to install and deploy all your processes on top of a virtual machine. But Docker containers are no virtual machine replacements, and they are not meant to run multiple processes.

Docker invites you to follow one golden rule: *there should be a one-to-one relationship between a Docker container and a Linux process.* That being said, modern applications are often made up of multiple processes, not just one, so in most cases, using only one Docker container won't suffice to run a full-featured microservice. This implies that the processes, and thus the containers, should be able to communicate with each other by sharing filesystems, networking, and so on. That's what Kubernetes Pods offer you: the ability to group your containers logically. All the containers/processes that make up a microservice should be grouped in the same Pod. That way, they'll be launched together and benefit from all the features when it comes to facilitating inter-process and inter-container communications.

To help you understand this, imagine you have a working WordPress blog on a virtual machine and you want to convert that virtual machine into a WordPress Pod to deploy your blog on your Kubernetes cluster. WordPress is one of the most common pieces of software and is a perfect example to illustrate the need for Pods. This is because WordPress requires multiple processes to work properly.

WordPress is a PHP application that requires both an HTTP server and a PHP interpreter to work. Let's list what Linux processes WordPress needs to work on:

- **An NGINX HTTP server**: It's a web application, so it needs an HTTP server running as a process to receive and serve server blog pages. NGINX is a good HTTP server that will do the job perfectly.

- **The FPM PHP interpreter**: It's a blog engine written in PHP, so it needs a PHP interpreter to work.

NGINX and PHP-FPM are two processes: they are two binaries that you need to launch separately, but they need to be able to work together. On a virtual machine, the job is simple: you just install NGINX and PHP-**Fast CGI Process Manager** (**FPM**) on the virtual machine and have both of them communicate through UNIX sockets. You can do this by telling NGINX that the Linux socket PHP-FPM is accessible thanks to the `/etc/nginx.config` configuration file.

In the Docker world, things become harder because running these two processes in the same container is an anti-pattern: you have to run two containers, one for each process, and you must have them communicate with each other and share a common directory so that they can both access the application code. To solve this problem, you have to use the Docker networking layer to have the NGINX container able to communicate with the PHP-FPM one. Then, you must use a volume mount to share the WordPress code between the two containers. You can do this with some Docker commands but imagine it now in production at scale, on multiple machines, on multiple environments, and so on. As you can imagine, that's the kind of problem the Kubernetes Pod resource solves.

Achieving inter-process communication is possible with bare Docker, but that's difficult to achieve at scale while keeping all the production-related requirements in mind. With tons of microservices to manage and spread on different machines, it would become a nightmare to manage all these Docker networks, volume mounts, and so on. That's why Kubernetes has the Pod resource. Pods are very useful because they wrap multiple containers and enable easy inter-process communication. The following are the core benefits Pods brings you:

- All the containers in the same Pod can reach each other through localhost.

- All the containers in the same Pod share the same network namespace.

- All the containers in the same Pod share the same port space.

- You can attach a volume to a Pod, and then mount the volume to underlying containers, allowing them to share directories and file locations.

With these benefits Kubernetes brings you, it would become super easy to provision your WordPress blog as you can create a Pod that will run two containers: NGINX and PHP-FPM. Since they both can access each other on localhost, having them communicating is super easy. You can then use a volume to expose WordPress's code to both containers.

The most complex applications will forcibly require several containers, so it's a good idea to group them in the same Pod to have Kubernetes launch them together. Keep in mind that the Pod is here for only one reason: to ease inter-container (or inter-process) communications at scale.

> **Important Note**
>
> That being said, it is not uncommon at all to have Pods that are only made up of one container. But in any case, you'll always have to use the Pod Kubernetes object to be able to interact with your containers in Kubernetes.

Lastly, please note that a Docker container that was launched manually on a machine managed by a Kubernetes cluster won't be seen by Kubernetes as a container it manages. It becomes a kind of *orphan* container outside of the scope of the orchestrator. Kubernetes only manages the container it has launched through its Pod object.

Each Pod gets an IP address

Containers inside a single Pod are capable of communicating with each other through localhost, but Pods are also capable of communicating with each other. At launch time, each Pod gets a single private IP address. Each Pod can communicate with any other Pod in the cluster by calling it through its IP address.

Kubernetes networking models allow Pods to communicate with each other directly without the need for a **Network Address Translation (NAT)** device. Keep in mind that they are not NAT gateways between Pods in your cluster.

Kubernetes uses a flat network model that is implemented by components called **Container Network Interface (CNIs)**.

How you should design your Pods

Here is the second golden rule about Pods: they are meant to be destroyed and recreated easily. Pods can be destroyed voluntarily or not. For example, if a given worker node running four Pods were to fail, each of the underlying containers would become inaccessible. Because of this, you should be able to destroy and recreate your Pods at will, without it affecting the stability of your application. The best way to achieve this is to respect two simple design rules when building your Pods:

- A Pod should contain everything required to launch a microservice.
- A Pod should be stateless (when possible).

When you start designing Pods on Kubernetes, it's hard to know exactly what a Pod should and shouldn't contain. It's pretty straightforward to explain: a Pod has to contain an application or a microservice. Take the example of our WordPress Pod, which we mentioned earlier: the Pod should contain the NGINX and PHP FPM containers, which are required to launch WordPress. If such a Pod were to fail, our WordPress would become inaccessible, but recreating the Pod would make WordPress accessible again because the Pod contains everything necessary to run WordPress.

That being said, every modern application makes use of database storage, such as Redis or MySQL. WordPress on its own does that too – it uses MySQL to store and retrieve your post. So, you'll also have to run a MySQL container somewhere. Two solutions are possible here:

- You run the MySQL container as part of the WordPress Pod.
- You run the MySQL container as part of a dedicated MySQL Pod.

Both solutions can be used, but the second is preferred. It's a good idea to decouple your application (here, this is WordPress, but tomorrow, it could be a microservice) from its database layer by running them in two separate Pods. Remember that Pods are capable of communicating with each other. You can benefit from this by dedicating a Pod to running MySQL and then giving its IP address to your WordPress blog.

By separating the database layer from the application, you improve the stability of the setup: the application Pod crashing will not affect the database.

To summarize, grouping the layers in the same Pods would cause three problems:

- Data durability
- Availability
- Stability

That's why you should keep your application Pods stateless, by storing their states in an independent Pod. The data layer can be considered an application on its own and has its own treatment decoupled from the application code itself. To achieve that, you should run them in separate Pods.

Now, let's launch our first Pod. Creating a WordPress Pod would be too complex for now, so let's start easy by launching some NGINX Pods and see how Kubernetes manages the Docker container.

Launching your first Pods

In this section, we will explain how to create our first Pods in our Kubernetes cluster. Pods have certain peculiarities that must be understood to master them well.

We are not going to create a resource on your Kubernetes cluster at the moment; instead, we are simply going to explain what Pods are. In the next section, we'll start building our first Pods.

Creating a Pod with imperative syntax

In this section, we are going to create a Pod based on the NGINX image. We need two parameters to create a Pod:

- The Pod's name, which is arbitrarily defined by you
- The Docker image(s) to build its underlying container(s)

As with almost everything on Kubernetes, you can create Pods using either of the two syntaxes available: the imperative syntax and the declarative syntax. As a reminder, the imperative syntax is to run `kubectl` commands directly from a terminal, while with declarative syntax, you must write a YAML file containing the configuration information for your Pod, and then apply it with the `kubectl create -f` command.

To create a Pod on your Kubernetes cluster, you have to use the `kubectl run` command. That's the simplest and fastest way to get a Pod running on your Kubernetes cluster. Here is how the command can be called:

```
$ kubectl run nginx-Pod --image nginx:latest
```

In this command, the Pod's name is set to `nginx-Pod`. This name is important because it is a pointer to the Pod: when you need to run the `update` or `delete` command on this Pod, you'll have to specify that name to tell Kubernetes which Pod the action should run on. The `--image` flag will be used to build the Docker container that this Pod will run.

Here, you are telling Kubernetes to build a Pod based on the `nginx:latest` Docker image hosted on Docker Hub. This `nginx-Pod` Pod contains only one container based on this `nginx:latest` image: you cannot specify multiple images here; this is a limitation of the imperative syntax.

If you want to build a Pod containing multiple containers built from several different Docker images, then you will have to go through the declarative syntax and write a YAML file.

Creating a Pod with declarative syntax

Creating a Pod with declarative syntax is simple too. You have to create a YAML file containing your Pod definition and apply it against your Kubernetes cluster using the `kubectl create -f` command.

Here is the content of the `nginx-Pod.yaml` file, which you can create on your local workstation:

```
apiVersion: v1
kind: Pod
metadata:
  name: nginx-Pod
spec:
  containers:
    - name: nginx-container
      image: nginx:latest
```

Try to read this file and understand its content. YAML files are only key/value pairs. The Pod's name is `nginx-Pod`, and then we have an array of containers in the `spec:` part of the file containing only one container created from the `nginx:latest` image. The container itself is named `nginx-container`.

Once the `nginx-Pod.yaml` file has been saved, run the following command to create the Pod:

```
$ kubectl create -f nginx-Pod.yaml
Pod/nginx-Pod created
$ kubectl apply -f nginx-Pod.yaml # this command works too!
```

If a Pod called `nginx-Pod` already exists in your cluster, this command will fail. Remember that Kubernetes cannot run two Pods with the same name: the Pod's name is the unique identifier and is used to identify the Pods. Try to edit the YAML file to update the Pod's name and then apply it again.

Reading the Pod's information and metadata

At this point, you should have a running Pod on your Kubernetes cluster. Here, we are going to try to read its information. At any time, we need to be able to retrieve and read information regarding the resources that were created on your Kubernetes cluster; this is especially true for Pods. Reading the Kubernetes cluster can be achieved using two `kubectl` commands: `kubectl get` and `kubectl describe`. Let's take a look at them:

- `kubectl get`: The `kubectl get` command is a list operation; you use this command to list a set of objects. Do you remember when we listed the nodes of your cluster after all the installation procedures described in the previous chapter? We did this using `kubectl get nodes`. The command works by requiring you to pass the object type you want to list. In our case, it's going to be the `kubectl get Pods` operation. In the upcoming chapters, we will discover other objects, such as `configmaps`. To list them, you'll have to type `kubectl get configmaps`; the same goes for the other object types. `kubectl get` does not require you to know the name of a precise resource, because it's intended to list.

- `kubectl describe`: The `kubectl describe` command is quite different. It's intended to retrieve a complete set of information for one specific object that's been identified from both its kind and object name. You can retrieve the information of our previously created Pod by using `kubectl describe Pods nginx-Pod`.

  Calling this command will return a full set of information available about that specific Pod, such as its IP address and so on.

Now, let's look at some more advanced options about listing and describing objects in Kubernetes.

Listing the objects in JSON or YAML

The -o option is one of the most useful options offered by the kubectl command line. This one has some benefits you must be aware of. That option allows you to customize the output of the kubectl command line. By default, using the kubectl get Pods command will return a list of the Pods in your Kubernetes cluster in a formatted way so that the end user can see it easily. You can also retrieve this information in JSON format or YAML format by using the -o option:

```
$ kubectl get Pods -o yaml # In YAML format
$ kubectl get Pods -o json # In JSON format
# If you know the Pod name, you can also get a specific Pod

$ kubectl get Pods <POD_NAME> -o yaml
# OR
$ kubectl get Pods <POD_NAME> -o json
```

This way, you can retrieve and export data from your Kubernetes cluster in a scripting-friendly format.

Backing up your resource using the list operation

You can also use these flags to back up your Kubernetes cluster. Imagine a situation where you created a Pod using the imperative way, so you don't have the YAML declaration file stored in your computer. If the Pod fails, it's going to be hard to recreate it. The -o option helps us retrieve the YAML declaration file of a resource that's been created in Kubernetes, even if we created it using the imperative way. To do this, run the following command:

```
$ kubectl get Pods/nginx-Pod -o yaml > nginx-Pod.yaml
```

This way, you have a YAML backup of the nginx-Pod resource as it is running on your cluster. If someone goes wrong, you'll be able to recreate your Pod easily. Pay attention to the nginx-Pod section of this command. To retrieve the YAML declaration, you need to specify which resource you are targeting. By redirecting the output of this command to a file, you get a nice way to retrieve and back up the configuration of the object inside your Kubernetes cluster.

Getting more information from the list operation

It's also worth mentioning the -o wide format, which is going to be very useful for you: using this option allows you to expand the default output to add more data. By using it on the Pods object, for example, you'll get the name of the worker node where the Pod is running:

```
$ kubectl get Pods -o wide
# The worker node name is displayed
```

Keep in mind that the -o option can take a lot of different parameters and that some of them are much more advanced, such as jsonpath, which allows you to directly execute kind of sort operations on top of a JSON body document to retrieve only specific information, just like the jq library you used previously if you have already written some Bash scripts that deal with JSON parsing.

Accessing a Pod from the outside world

At this point, you should have a Pod containing an NGINX HTTP server on your Kubernetes cluster. You should now be able to access it from your web browser. However, this is a bit complicated.

By default, your Kubernetes cluster does not expose the Pod it runs to the internet. For that, you will need to use another resource called a service, which we will cover in more detail in *Chapter 7, Exposing Your Pods with Services*. However, kubectl does offer a command for quickly accessing a running container on your cluster called kubectl port-forward. This is how you can use it:

```
$ kubectl port-forward Pod/nginx-Pods 8080:80
Forwarding from 127.0.0.1:8080 -> 80
Forwarding from [::1]:8080 -> 80
```

This command is quite easy to understand: I'm telling kubectl to forward port 8080 on my local machine (the one running kubectl) to port 80 on the Pod identified by Pod/nginx-Pod.

Kubectl then outputs a message, telling me that it started to forward my local 8080 port to the 80 one of the Pod. If you get an error message, it's probably because your local port 8080 is currently being used. Try to set a different port or simply remove the local port from the command to let kubectl choose a local port randomly:

```
$ kubectl port-forward Pod/nginx-Pods :80
```

Now, you can launch your browser and try to reach the `http://`
`localhost:<localport>` address, which in my case is `http://localhost:80`:

Figure 4.1 – The NGINX default page running in a Pod and accessible on localhost, which indicates the
port-forward command worked!

Entering a container inside a Pod

When a Pod is launched, you can access the Pods it contains. Under **Docker**, the
command to execute a command in a running container is called `docker exec`.
Kubernetes copies this behavior via a command called `kubectl exec`. Use the
following command to access our NGINX container inside `nginx-Pod`, which we
launched earlier:

```
$ kubectl exec -ti Pods nginx-Pod bash
```

After running this command, you will be inside the NGINX container. You can do
whatever you want here, just like with any other container. The preceding command
assumes that the `bash` binary is installed in the container you are trying to access.
Otherwise, the `sh` binary is generally installed on a lot of containers and might be used to
access the container. Don't be afraid to path a full binary path, like so:

```
$ kubectl exec -ti Pods nginx-Pod /bin/bash
```

Now, let's discover how to delete a Pod from a Kubernetes cluster.

Deleting a Pod

Deleting a Pod is super easy. You can do so using the `kubectl delete` command. You need to know the name of the Pod you want to delete. In our case, the Pod's name is `nginx-Pod`. Run the following command:

```
$ kubectl delete Pods nginx-Pod
$ # or...
$ kubectl delete Pods/nginx-Pod
```

If you do not know the name of the Pod, remember to run the `kubectl get Pods` command to retrieve the list of the Pod and find the one you want to delete.

There is also something you must know: if you have built your Pod with declarative syntax and you still have its YAML configuration file, you can delete your Pod without having to know the name of the container because it is contained in the YAML file.

Run the following command to delete the Pod using the declarative syntax:

```
$ kubectl delete -f nginx-Pod.yaml
```

After you run this command, the Pod will be deleted in the same way.

> **Important Note**
> Remember that all containers belonging to the Pod will be deleted. The container's life cycle is bound to the life cycle of the Pod that launched it. If the Pod is deleted, the containers it manages will be deleted. Remember to always interact with the Pods and not with the containers directly.

With that, we have reviewed the most important aspects of Pod management, such as launching a Pod with the imperative or declarative syntax, deleting a Pod, and also listing and describing them. Now, I will introduce one of the most important aspects of Pod management in Kubernetes: labeling and annotating.

Labeling and annotating the Pods

We will now discuss another key concept of Kubernetes: labels and annotations. Labels are key/value pairs that you can attach to your Kubernetes objects. Labels are meant to tag your Kubernetes objects with key/value pairs defined by you. Once your Kubernetes objects have been labeled, you can build a custom query to retrieve specific Kubernetes objects based on the labels they hold. In this section, we are going to discover how to interact with labels through `kubectl` by assigning some labels to our Pods.

What are labels and why do we need them?

Labels are key/value pairs that you can attach to your created objects, such as Pods. What label you define for your objects is up to you – there is no specific rule regarding this. These labels are attributes that will allow you to organize your objects in your Kubernetes cluster: once your objects have been labeled, you can list and query them using the labels they hold. To give you a very concrete example, you could attach a label called `environment = prod` to some of your Pods, and then use the `kubectl get Pods` command to list all the Pods within that environment. So, you could list all the Pods that belong to your production environment in one command:

```
$ kubectl get Pods --label "environment=production"
```

As you can see, it can be achieved using the `--label` parameter, which can be shortened using its `-l` equivalent:

```
$ kubectl get Pods --l "environment=production"
```

This command will list all the Pods holding a label called `environment` with a value of `production`. Of course, in our case, no Pods will be found since none of the ones we created earlier are holding this label. You'll have to be very disciplined about labels and not forget to set them every time you create a Pod or another object, and that's why we are introducing them quite early in this book: not only Pods but almost every object in Kubernetes can be labeled, and you should take advantage of this feature to keep your cluster organized and clean.

You use labels not only to organize your cluster but also to build relationships between your different Kubernetes objects: you will notice that some Kubernetes objects will read the labels that are carried by certain Pods and perform certain operations on them based on the labels they carry. If your Pods don't have labels or they are misnamed or contain the wrong values, some of these mechanisms might not work as you expect.

On the other hand, using labels is completely arbitrary: there is no particular naming rule, nor any convention Kubernetes expects you to follow. Thus, it is your responsibility to use the labels as you wish and build your own convention. If you are in charge of the governance of a Kubernetes cluster, you should enforce the usage of mandatory labels and build some monitoring rules to quickly identify non-labeled resources.

Keep in mind that labels are limited to 63 characters: they are intended to be short. Here are some label ideas you could use:

- `environment` (`prod`, `dev`, `uat`, and so on)
- `stack` (`blue`, `green`, and so on)

- `tier` (`frontend` and `backend`)
- `app_name` (`wordpress`, `magento`, `mysql`, and so on)
- `team` (`business` and `developers`)

Labels are not intended to be unique between objects. For example, perhaps you would like to list all the Pods that are part of the production environment. Here, several Pods with the same label key and value pair can exist in the cluster at the same time without posing any problem – it's even recommended if you want your list query to work. For example, if you want to list all the resources that are part of the prod environment, a label environment such as = `prod` should be created on multiple resources. Now, let's look at annotations, which are another way we can assign metadata to our Pods.

What are annotations and how do they differ from labels?

Kubernetes also uses another type of metadata called **annotations**. Annotations are very similar to labels as they are also key/value pairs. However, annotations do not have the same use as labels. Labels are intended to identify resources and build relationships between them, while annotations are used to provide contextual information about the resource that they are defined on.

For example, when you create a Pod, you could add an annotation containing the email of the support team to contact if this app does not work. This information has its place in an annotation but has nothing to do with a label.

While it is highly recommended that you define labels wherever you can, you can omit annotations: they are less important to the operation of your cluster than labels. Be aware, however, that some Kubernetes objects or third-party applications often read annotations and use them as configuration. In this case, their usage of annotations will be explained explicitly in their documentation.

Adding a label

In this section, we will learn how to add and remove labels and annotations from Pods. We will also learn how to modify the labels of a Pod that already exists on a cluster.

You can add a label when creating a Pod. Let's take the Pod based on the NGINX image that we used earlier. We will recreate it here with a label called `tier`, which will contain the `frontend` value. Here is the `kubectl` command to run for that:

```
$ kubectl run nginx-Pod --image nginx --label "tier=frontend"
```

As you can see, a label can be assigned using the `--label` parameter. You can add multiple labels by repeatedly using the `--label` parameter, like this:

```
$ kubectl run nginx-Pod --image nginx --label "tier=frontend"
--label "environment=prod"
```

Here, the `nginx` Pod will be created with two labels.

The `--label` flag has a short version called `-l`. You can use this to make your command shorter and easier to read. Here is the same command with the `-l` parameter used instead of `--label`:

```
$ kubectl run nginx-Pod --image nginx -l "tier=frontend" -l
"environment=prod"
```

Another important thing to notice is that labels can be defined with declarative syntax. Labels can be appended to a YAML Pod definition. Here is the same Pod, holding the two labels we created earlier, but this time, it's been created with the declarative syntax:

```
# ~/labelled_Pod.yaml
apiVersion: v1
kind: Pod
metadata:
  name: nginx-Pod
  labels:
    environment: prod
    tier: frontend
spec:
  containers:
    - name: nginx-container
      image: nginx:latest
```

Consider the file that was created at `~/labelled_Pod.yaml`. The following `kubectl` command would create the Pod the same way as it was created previously:

```
$ kubectl create -f ~/labelled_Pod.yaml
```

Consider the file created at ~/labelled_Pod.yaml. The following kubectl command would create the Pod the same way as it was created previously. This time, running the command we used earlier should return at least one Pod – the one we just created:

```
$ kubectl get Pods --label "environment=production"
```

Now, let's learn how we can list the labels attached to our Pod.

Listing labels attached to a Pod

```
$ kubectl get Pods --show-labels
```

There is no dedicated command to list the labels attached to a Pod, but you can make the output of kubectl get Pods a little bit more verbose. By using the --show-labels parameter, the output of the command will include the labels attached to the Pods.

This command does not run any kind of query based on the labels; instead, it displays the labels themselves as part of the output. It can be useful for debugging. Of course, this option can be chained with other options, such as -o wide:

```
$ kubectl get Pods --show-labels -o wide
```

This command would then show you the list of the Pods and their labels, as well as the node that the Pod has been scheduled to run on.

Adding or updating a label to/of a running Pod

Now that we've learned how to create Pods with labels, we'll learn how to add labels to a running Pod. You can add, create, or modify the labels of a resource at any time using the kubectl label command. Here, we are going to add another label to our nginx Pod. This label will be called stack and will have a value of blue:

```
$ kubectl label Pods nginx-Pod stack=blue
```

This command only works if the Pod has no label called stack. When the command is executed, it can only add a new tag and not update it. This command will update the Pod by adding a label called stack with a value of blue. Run the following command to see that the change was applied:

```
$ kubectl get Pods nginx-Pod --show-labels
```

To update an existing label, you must append the `--overwrite` parameter to the preceding command. Let's update the `stack=blue` label to make it `stack=green`; pay attention to the `overwrite` parameter:

```
$ kubectl label Pods nginx-Pod stack=green --overwrite
```

Here, the label should be updated. The `stack` label should now be equal to `green`. Run the following command to show the Pod and its labels again:

```
$ kubectl get Pods nginx-Pod --show-labels
```

That command will only list `nginx-Pod` and display its label as part of the output in the terminal.

> **Important Note**
>
> Adding or updating labels using the `kubectl label` command might be dangerous. As we mentioned earlier, you'll build relationships between different Kubernetes objects based on labels. By updating them, you might break some of these relationships and your resources might start to behave not as expected. That's why it's better to add labels when a Pod is created and keep your Kubernetes configuration immutable. It's always better to destroy and recreate rather than update an already running configuration.

The last thing we must do is learn how to delete a label attached to a running Pod.

Deleting a label attached to a running Pod

Just like we added and updated labels of a running Pod, we can also delete them. The command is a little bit trickier. Here, we are going to remove the label called `stack`, which we can do by adding a minus symbol (`-`) right after the label name:

```
$ kubectl label nginx-Pod stack-
```

Adding that minus symbol at the end of the command might be quite strange, but running `kubectl get Pods --show-labels` again should show that the `stack` label is now gone:

```
$ kubectl get Pods nginx-Pod --show-labels
```

Adding an annotation

Let's learn how to add annotations to a Pod. It won't take long to cover this because it works just like it does with labels:

```
# ~/annotated_Pod.yaml
apiVersion: v1
kind: Pod
metadata:
  annotations:
    tier: webserver
  name: nginx-Pod
  labels:
    environment: prod
    tier: frontend
spec:
  containers:
    - name: nginx-container
      image: nginx:latest
```

Here, I simply added the `tier=webserver` annotation, which can help me identify that this Pod is running an HTTP server. Just keep in mind that it's a way to add additional metadata.

The name of an annotation can be prefixed by a DNS name. This is the case for Kubernetes components such as `kube-scheduler`, which must indicate to cluster users that this component is part of the Kubernetes core. The prefix can be omitted completely, as shown in the preceding example.

Launching your first job

Now, let's discover another Kubernetes resource that is derived from Pods: the Job resource. In Kubernetes, a computing resource is a Pod, and everything else is just an intermediate resource that manipulates Pods.

This is the case for the `Job` object, which is an object that will create one or multiple Pods to complete a specific computing task, such as running a Linux command.

What are jobs?

A job is another kind of resource that's exposed by the Kubernetes API. In the end, a job will create one or multiple Pods to execute a command defined by you. That's how jobs work: they launch Pods. You have to understand the relationship between the two: jobs are not independent of Pods, and they would be useless without Pods. In the end, the two things they are capable of are launching Pods and managing them. Jobs are meant to handle a certain task and then exit. Here are some examples of typical use cases for a Kubernetes job:

- Taking a backup of a database

- Sending an email

- Consuming some messages in a queue

These are tasks you do not want to run forever. You expect the Pods to be terminated once they have completed their task. This is where the Jobs resource will help you.

But why bother using another resource to execute a command? After all, we can create one or multiple Pods directly that will run our command and then exit.

This is true. You can use a Pod based on a Docker image to run the command you want and that would work fine. However, jobs have mechanisms implemented at their level that allow them to manage Pods in a more advanced way. Here are some things that jobs are capable of:

- Running Pods multiple times

- Running Pods multiple times in parallel

- Retrying to launch the Pods if they encountered any errors

- Killing a Pod after a specified number of seconds

Another good point is that a job manages the labels of the Pods it will create so that you won't have to manage the labels on those Pods directly.

All of this can be done without using jobs, but this would be very difficult to manage. Fortunately for us, the Jobs resource exists, and we are going to learn how to use it now.

Creating a job with restartPolicy

Since creating a job might require some advanced configurations, we are going to focus on declarative syntax here. This is how you can create a Kubernetes job through YAML. We are going to make things simple here; the job will just echo `Hello world`:

```
# ~/hello-world-job.yaml
apiVersion: batch/v1
kind: Job
metadata:
  name: hello-world-job
spec:
  template:
    metadata:
      name: hello-world-job
    spec:
      restartPolicy: OnFailure
      containers:
      - name: hello-world-container
        image: busybox
        command: ["/bin/sh", "-c"]
        args: ["echo 'Hello world'"]
```

Pay attention to the `kind` resource, which tells Kubernetes that we need to create a job and not a Pod, as we did previously. Also, notice `apiVersion:`, which also differs from the one that's used to create the Pod. You can create the job with the following command:

```
$ kubectl create -f hello-world-job.yaml
```

As you can see, this job will create a Pod based on the Docker `busybox` image. This will run the `echo 'Hello World'` command. Lastly, the `restartPolicy` option is set to `OnFailure`, which tells Kubernetes to restart the Pod or the container in case it fails. If the entire Pod fails, a new Pod will be relaunched. If the container fails (the memory limit has been reached or a non-zero exit code occurs), the individual container will be relaunched on the same node because the Pod will remain untouched, which means it's still scheduled on the same machine.

The `restartPolicy` parameter can take two options:

- `Never`
- `OnFailure`

Setting it to `Never` will prevent the job from relaunching the Pods, even if it fails. When debugging a failing job, it's a good idea to set `restartPolicy` to `Never` to help with debugging. Otherwise, new Pods might be recreated over and over, making your life harder when it comes to debugging.

In our case, there is little chance that our job was not successful since we only want to run a simple `Hello world`. To make sure that our job worked well, we can read its log. To do that, we need to retrieve the name of the Pod it created using the `kubectl get Pods` command. Then., we can use the `kubectl logs` command, as we would do with any Pods:

```
$ kubectl logs
```

Here, we can see that our job has worked well since we can see the `Hello world` message displayed in the log of our Pod. However, what if it had failed? Well, this depends on `restartPolicy` – if it's set to `Never`, then nothing would happen and Kubernetes wouldn't try to relaunch the Pods.

However, if `restartPolicy` was set to `OnFailure`, Kubernetes would try to restart the job after 10 seconds and then double that time on each new failure. 10 seconds, then 20 seconds, then 40 seconds, then 80 seconds, and so on. After 6 minutes, Kubernetes would give up.

Understanding the job's backoffLimit

By default, the Kubernetes job will try to relaunch the failing Pod 6 times during the next 6 minutes after its failure. You can change this limitation by changing the `backoffLimit` option. Here is the updated YAML file:

```
# ~/hello-world-job.yaml
apiVersion: batch/v1
kind: Job
metadata:
  name: hello-world-job
spec:
  backoffLimit: 3
  template:
    metadata:
      name: hello-world-job
    spec:
      restartPolicy: OnFailure
```

```
   containers:
   - name: hello-world-container
     image: busybox
     command: ["/bin/sh", "-c"]
     args: ["echo 'Hello world'"]
```

This way, the job will only try to relaunch the Pods twice after its failure.

Running a task multiple times using completions

You can also instruct Kubernetes to launch a job multiple times using the Job object.

You can do this by using the completions option to specify the number of times you want a command to be executed. The number of completions will create five different Pods that will be launched one after the other. Once one Pod has finished, the next one will be started. Here is the updated YAML file:

```
apiVersion: batch/v1
kind: Job
metadata:
  name: hello-world-job
spec:
  completions: 10
  template:
    metadata:
      name: hello-world-job
    spec:
      restartPolicy: OnFailure
      containers:
      - name: hello-world-container
        image: busybox
        command: ["/bin/sh", "-c"]
        args: ["echo 'Hello world'; sleep 3"]
```

The `completions` option was added here. Also, please notice that the `args` section was updated by us adding the `sleep 3` option. Using this option will make the task sleep for 3 seconds before completing, giving us enough time to notice the next Pod being created. Once you've applied this configuration file to your Kubernetes cluster, you can run the following command:

```
$ kubectl get Pods --watch
```

The `watch` mechanism will update your `kubectl` output when something new arrives, such as the creation of the new Pods being managed by your Kubernetes. If you want to wait for the job to finish, you'll see 10 Pods being created with a 3-second delay between each.

Running a task multiple times in parallel

The `completions` option ensures that the Pods are created one after the other. You can also enforce parallel execution using the `parallelism` option. If you do that, you can get rid of the `completions` option. Here is the updated YAML file:

```
apiVersion: batch/v1
kind: Job
metadata:
  name: hello-world-job
spec:
  parallelism: 5
  template:
    metadata:
      name: hello-world-job
    spec:
      restartPolicy: OnFailure
      containers:
      - name: hello-world-container
        image: busybox
        command: ["/bin/sh", "-c"]
        args: ["echo 'Hello world'; sleep 3"]
```

Please notice that the `completions` option is now gone and that we replaced it with `parallelism`. The job will now launch five Pods at the same time and will have them run in parallel.

Terminating a job after a specific amount of time

You can also decide to terminate a Pod after a specific amount of time. This can be very useful when you are running a job that is meant to consume a queue, for example. You could poll the messages for 1 minute and then automatically terminate the processes. You can do that using the activeDeadlineSeconds parameter. Here is the updated YAML file:

```
apiVersion: batch/v1
kind: Job
metadata:
  name: hello-world-job
spec:
  backoffLimit: 3
  activeDeadlineSeconds: 60
  template:
    metadata:
      name: hello-world-job
    spec:
      restartPolicy: OnFailure
      containers:
      - name: hello-world-container
        image: busybox
        command: ["/bin/sh", "-c"]
        args: ["echo 'Hello world'"]
```

Here, the job will terminate after 60 seconds, no matter what happens. It's a good idea to use this feature if you want to keep a process running for an exact amount of time and then terminate it.

What happens if a job succeeds?

If your job is completed, it will remain created in your Kubernetes cluster and will not be deleted automatically: that's the default behavior. The reason for this is that you can read its logs a long time after its completion. However, keeping your jobs created on your Kubernetes cluster that way might not suit you. You can delete the jobs automatically and the Pods they created by using the `ttlSecondsAfterFinished` option, but keep in mind that this feature is still in alpha as of Kubernetes version 1.12. Here is the updated YAML file for implementing this solution:

```yaml
apiVersion: batch/v1
kind: Job
metadata:
  name: hello-world-job
spec:
  ttlSecondsAfterFinished: 30
  template:
    metadata:
      name: hello-world-job
    spec:
      restartPolicy: OnFailure
      containers:
      - name: hello-world-container
        image: busybox
        command: ["/bin/sh", "-c"]
        args: ["echo 'Hello world'"]
```

Here, the jobs are going to be deleted 30 seconds after their completion. If you do not want to use this option or it's not available in your Kubernetes version – for example, you are running a Kubernetes version before 1.12 – then you'll need to delete them manually. This is what we are going to discover now.

Deleting a job

Let's learn how to delete a Kubernetes job manually. Keep in mind that the Pods that are created are bound to the life cycle of their parent. Deleting a job will result in deleting the Pods they manage.

Start by getting the name of the job you want to destroy. In our case, it's `hello-world-job`. Otherwise, use the `kubectl get jobs` command to retrieve the correct name. Then, run the following command:

```
$ kubectl delete jobs hello-world-job
```

If you want to delete the jobs but not the Pods it created, you need to add the `--cascade=false` parameter to the `delete` command:

```
$ kubectl delete jobs hello-world-job --cascade=false
```

Thanks to this command, you can get rid of all the jobs that will be kept on your Kubernetes cluster once they've been completed.

Launching your first Cronjob

To close this first chapter on Pods, I suggest that we discover another new Kubernetes resource called **Cronjob**.

What are Cronjobs?

The name Cronjob can mean two different things and it is important to not get confused:

- The UNIX cron feature
- The Kubernetes Cronjob resource

Historically, Cronjobs are command scheduled using the cron UNIX feature, which is the most robust way to schedule the execution of a command in UNIX systems. This idea was later taken up in Kubernetes.

In Kubernetes, you are not going to schedule the execution of a command but the execution of a Pod. You can do that using the Cronjob resource.

Be careful because even though the two ideas are similar, they don't work the same at all. On UNIX and other derived systems such as UNIX, you schedule commands by editing a file called `Crontab`, which is usually found in `/etc/crontab`. In the world of Kubernetes, things are different: you are not going to schedule the execution of commands but the execution of Job resources, which themselves will create Pod resources. You can achieve this by manipulating a new kind of resource called Cronjob. Keep in mind that the Cronjob object you'll create will create Jobs objects.

Think of it as a kind of wrapper around the Job resource: in Kubernetes, we call that a controller. Cronjob can do everything the Job resource is capable of because it is nothing more than a wrapper around the Job resource, according to the `cron` expression specified.

The good news is that the Kubernetes Cronjob resource is using the `cron` format inherited from UNIX. So, if you have already written some Cronjobs on a Linux system, mastering Kubernetes Cronjobs will be super straightforward.

But first, why would you want to execute a Pod? The answer is simple; here are some concrete use cases:

- Taking database backups regularly every Sunday at 1 A.M.
- Clearing cached data every Monday at 4 P.M.
- Sending a queued email every 5 minutes
- Various maintenance operations to be executed regularly

The use cases of Kubernetes Cronjobs do not differ much from their UNIX counterparts – they are used to answer the same need, but they do provide the massive benefit of allowing you to use your already configured Kubernetes cluster to schedule regular jobs using your Docker images and your already existing Kubernetes cluster. In the end, the whole idea is to schedule the execution of your commands thanks to Jobs and Pods in a Kubernetes way.

Creating your first Cronjob

It's time to create your first Cronjob. Let's do this using declarative syntax. First, let's create a `cronjob.yaml` file and place the following YAML content into it:

```
apiVersion: batch/v1beta1
kind: CronJob
metadata:
  name: hello-world-cronjob
spec:
  schedule:
  ttlSecondsAfterFinished: 30
  template:
    metadata:
      name: hello-world-job
    spec:
      restartPolicy: OnFailure
```

```
    containers:
    - name: hello-world-container
      image: busybox
      command: ["/bin/sh", "-c"]
      args: ["echo 'Hello world'"]
```

Before applying this file to the Kubernetes cluster, let's start to explain it. There are two important things to notice here:

- The schedule key, which lets you input the `cron` expression
- The `jobTemplate` section, which is exactly what you would input in a job YAML manifest

Let's explain these two keys quickly before applying the file.

Understanding the schedule

The schedule key allows you to insert an expression in a `cron` format such as Linux. Let's explain how these expressions work: if you already know these expressions, you can skip these explanations.

A `cron` expression is made up of five entries separated by white space. From left to right, these entries correspond to the following:

- Minutes
- Hour
- Day of the month
- Month
- Day of the week

Each entry can be filled with an asterisk, which means *every*. You can also set several values for one entry by separating them with a , . You can also use a – to input a range of values. Let me show you some examples:

- "10 11 * * *" means "At 11:10 every day of every month."
- "10 11 * 12 *" means "At 11:10 every day of December."
- "10 11 * 12 1" means "At 11:10 of every Monday of December."
- "10 11 * * 1,2" means "At 11:10 of every Monday and Tuesday of every month."
- "10 11 * 2-5 *" means "At 11:10 every day from February to May."

Here are some examples that should help you understand how cron works. Of course, you don't have to memorize the syntax: most people help themselves with documentation or cron expression generators online. If this is too complicated, feel free to use this kind of tool; it can help you confirm that your syntax is valid before you deploy the object to Kubernetes.

Understanding the role of the jobTemplate section

We will cover this shortly to make you understand an important concept regarding Kubernetes. If you've been paying attention to the structure of the YAML file, you may have noticed that the `jobTemplate` key contains the definition of a Job object. The reason is simple: when we use the Cronjob object, we are simply delegating the creation of a Job object to the Cronjob object. This is why the YAML file of the Cronjob object requests that we provide it with a `jobTemplate`.

Therefore, the Cronjob object is a resource that only manipulates another resource.

In Kubernetes, a lot of things work like this, and this is important to remember. Later, we will discover many objects that will allow us to create Pods so that we don't have to do it ourselves. These special objects are called **controllers**: they manipulate other Kubernetes resources by obeying their own logic. Moreover, when you think about it, the Job object is itself a controller since, in the end, it only manipulates Pods by providing them with its own features, such as the possibility of running Pods in parallel.

In a real context, you should always try to create Pods using these intermediate objects as they provide additional and more advanced management features.

Try to remember this rule: the basic unit in Kubernetes is a Pod, but you can delegate the creation of Pods to many other objects. In the rest of this section, we will continue to discover *naked* Pods. Later, we will learn how to manage their creation and management via controllers.

Controlling the Cronjob execution deadline

For some reason, a Cronjob may fail to execute. In this case, Kubernetes cannot execute the Job at the moment it is supposed to start.

Managing the history limits of jobs

When a Cronjob completes, whether it's successful or not, history is kept on your Kubernetes cluster. The history setting can be set at the Cronjob level, so you can decide whether to keep the history of each Cronjob or not, and in case you want to keep it, how many entries should be kept for succeeded and failed jobs.

Let's learn how to do this.

Creating a Cronjob

If you already have the YAML manifest file, creating a Cronjob object is easy. You can do so using the `kubectl create` command:

```
$ kubectl create -f ~/cronjob.yaml
cronjob/hello-world-cronjob created
```

With that, the Cronjob has been created on your Kubernetes cluster. It will launch a scheduled Pod, as configured in the YAML file.

Deleting a Cronjob

Like any other Kubernetes resource, deleting a Cronjob can be achieved through the `kubectl delete` command. Like before, if you have the YAML manifest, it's easy:

```
$ kubectl delete -f ~/cronjob.yaml
cronjob/hello-world-cronjob deleted
```

With that, the Cronjob has been destroyed by your Kubernetes cluster. No scheduled jobs will be launched anymore.

Summary

We have come to the end of this chapter on Pods and how to create them; I hope you enjoyed it. You've learned how to use the most important objects in Kubernetes: Pods.

The knowledge you've developed in this chapter is part of the essential basis for mastering Kubernetes: all you will do in Kubernetes is manipulate Pods, label them, and access them. In addition, you saw that Kubernetes behaves like a traditional API, in that it executes CRUD operations to interact with the resources on the cluster. In this chapter, you learned how to launch Docker containers on Kubernetes, how to access these containers using `kubectl` port forwarding, how to add labels and annotations to Pods, how to delete Pods, and how to launch and schedule jobs using the Cronjob resource.

Just remember this rule about Docker container management: any container that will be launched in Kubernetes will be launched through the object. Mastering this object is like mastering most of Kubernetes: everything else will consist of automating things around the management of Pods, just like we did with the Cronjob object; you have seen that the Cronjob object only launches Job objects that launch Pods. If you've understood that some objects can manage others, but in the end, all containers are managed by Pods, then you've understood the philosophy behind Kubernetes, and it will be very easy for you to move forward with this orchestrator.

Also, I invite you to add labels and annotations to your Pods, even if you don't see the need for them right away. Know that it is essential to label your objects well to keep a clean, structured, and well-organized cluster.

However, you still have a lot to discover when it comes to managing Pods because so far, we have only seen Pods that are made up of only one Docker container. The greatest strength of Pods is that they allow you to manage multiple containers at the same time, and of course, to do things properly, there are several design patterns that we can follow to manage our Pods when they are made of several containers.

In the next chapter, we will learn how to manage Pods that are composed of several containers. While this will be very similar to the Pods we've seen so far, you'll find that some little things are different and that some are worth knowing. First, you will learn how to launch multi-container Pods using `kubectl` (hint: `kubectl` will not work), then how to get the containers to communicate with each other. After that, you will learn how to access a specific container in a multi-container Pod, as well as how to access logs from a specific container. Finally, you will learn how to share volumes between containers in the same Pod.

As you read the next chapter, you will learn about the rest of the fundamentals of Pods in Kubernetes. So, you'll get an overview of Pods while we keep moving forward by discovering additional objects in Kubernetes that will be useful for deploying applications in our clusters.

5
Using Multi-Container Pods and Design Patterns

Running complex applications on Kubernetes will require that you run not one but several containers in the same Pods. The strength of Kubernetes lies in its ability to create Pods made up of several containers: these Pods are capable of managing multiple containers at once. We will focus on those Pods in this chapter by studying the different aspects of hosting several containers in the same Pod, as well as having these different containers communicate with each other.

So far, we've only created Pods running a single container: those were the simplest forms of Pods, and you'll use them Pods to manage the simplest of applications. We also discovered how to update and delete them by running simple **Create, Read, Update, Delete (CRUD)** operations against those Pods using the `kubectl` command-line tool.

Besides mastering the basics of CRUD operations, you also learned how to access a running Pod inside a Kubernetes cluster.

In this chapter, we will push all of this one step forward and discover how to manage Pods when they are meant to launch not one but several containers: the good news is that everything you learned previously will also be valid for multi-container Pods. Things won't differ much in terms of raw Pod management because updating and deleting Pods is not different, no matter how many containers the Pod contains.

Besides those basic operations, we are also going to cover how to access a specific container inside a multi-container Pod and how to access its logs. When a given Pod contains more than one container, you'll have to run some specific commands with specific arguments to access it, and that's something we are going to cover in this chapter.

We will also discover some important design patterns such as ambassador, sidecar, and adapter containers. You'll need to learn these architectures to effectively manage multi-container Pods. You'll also learn how to deal with volumes from Kubernetes. Docker also provides volumes, but in Kubernetes, they are used to share data between containers launched by the same Pod, and this is going to be an important part of this chapter. After this chapter, you're going to be able to launch complex applications inside Kubernetes Pods.

In this chapter, we're going to cover the following main topics:

- Understanding what multi-container Pods are
- Sharing volumes between containers in the same pod
- The ambassador design pattern
- The sidecar design pattern
- The adapter design pattern

Technical requirements

You will require the following prerequisites for this chapter:

- A working `kubectl` command-line utility
- A local or cloud-based Kubernetes cluster to practice with

Understanding what multi-container Pods are

In this section, we'll learn about the core concepts of Pods for managing several containers at once by discussing some concrete examples of multi-container Pods.

Then, we will create and delete a Pod made up of at least two and discover how to access its logs. After that, we'll learn how to access a specific container within a Pod containing multiple containers. Finally, we will learn how to access the logs of a specific container within a running Pod.

Concrete scenarios where you need multi-container Pods

You should group your containers into a Pod when they need to be tightly linked. More broadly, a Pod must correspond to an application or a process running in your Kubernetes cluster. If your application requires multiple containers to function properly, then those containers should be launched and managed through a single Pod.

When the containers are supposed to work together, you should group them into a single Pod. Keep in mind that a Pod cannot span across multiple worker nodes. So, if you create a Pod containing several containers, then all these containers will be created on the same worker node and the same Docker daemon installation.

To understand where and when to use multi-container Pods, take the example of two simple applications:

- **A log forwarder**: In this example, imagine that you have deployed a web server such as NGINX that stores its logs in a dedicated directory. You might want to collect and forward these logs. For that, you could deploy something like a Splunk forwarder as a container within the same Pod as your NGINX server. These log forwarding tools are used to forward logs from a source to a destination location, and it is very common to deploy agents such as Splunk, Fluentd, or Filebeat to grab logs from a container and forward them to a central location such as an ElasticSearch cluster. In the Kubernetes world, this is generally achieved by running a multi-container Pod with one container dedicated to running the application, and another one dedicated to grabbing the logs and sending them elsewhere. Having these two containers managed by the same Pod would ensure that they are launched on the same node as the log forwarder and at the same time.

- **A proxy server**: Another typical use case of a multi-container Pod would be an application where you have an NGINX web server acting as a reverse proxy in front of an application. It is very common to use middleware such as NGINX to route web traffic to your actual web application by following some custom rules. By bundling the two containers in the same Pod, you'll get two Pods running in the same node. You could also run a third container in the same Pod to forward the logs that are emitted by the two others to a central logging location! This is because Kubernetes has no limit on the number of containers you can have in the same Pod, so long as you have enough computing resources to run them all.

In general, every time several of your containers work together and are tightly coupled, you should have them in a multi-container Pod. Just with these two examples, it's easy to understand why such Pods as so powerful. Most of the Pods you'll launch in while working with Kubernetes will probably handle more than one container.

Now, let's discover when to not create a multi-container Pod.

When not to create a multi-container Pod

Pods are especially useful when they are managing several containers, but they should not be seen as the go-to solution every time you need to set up a container.

The really basic golden rule when designing a multi-container is to keep in mind that all the containers that are declared in the Pod will be scheduled on the same worker node.

Now, let's discover how to create multi-container Pods. As you can imagine, we will have to use the `kubectl` command-line tool!

Creating a Pod made up of two containers

In the previous chapter, we discovered two syntaxes for manipulating Kubernetes in the past:

- The imperative syntax
- The declarative syntax

Most of the Kubernetes objects we are going to discover in this book can be created or updated using these two methods, but unfortunately, this is not the case for multi-container Pods.

When you need to create a Pod containing multiple containers, you will need to go through the declarative syntax. This means that you will have to create a YAML file containing the declaration of your Pods and all the containers it will manage, and then apply it through `kubectl create -f file.yaml` or `kubectl apply -f file.yaml`.

You cannot create Pods with multiple containers with `kubectl create` or `kubectl run`.

Consider the following YAML manifest file stored in `~/multi-container-Pod.yaml`:

```
# ~/multi-container-Pod.yaml
apiVersion: v1
kind: Pod
metadata:
  name: nginx-Pod
spec:
  containers:
    - name: nginx-container
      image: nginx:latest
    - name: busybox-container
      image: busybox:latest
```

This YAML manifest will create a Kubernetes Pod made up of two containers: one based on the `nginx:latest` image and the other one based on the `busybox:latest` image.

As you can see, there is no dedicated `kind: resource` for multi-container Pods – just like when we created a single-container Pod, we are only using `kind: Pod`.

To create it, use the following command:

```
$ kubectl create -f multi-container-Pod.yaml
Pods/multi-container-Pod created
```

This will result in the Pod being created. The kubelet on the elected worker node will have the Docker daemon to pull both images and instantiate two Docker containers.

To check if the Pod was correctly created, we can run `kubectl get Pods`:

```
$ kubectl get Pods
```

Do you remember the role of `kubelet` from *Chapter 2, Kubernetes Architecture – From Docker Images to Running Pods*? This component runs on each node that is part of your Kubernetes cluster and is responsible for converting pod manifests received from `kube-apiserver` into actual containers.

This `kubelet` launches your actual Docker containers on the worker nodes, and it is the only component of Kubernetes that is directly interacting with the Docker daemon.

> **Important Note**
>
> Keep this important note in mind because this is one is extremely important: all the containers that are declared in the same Pod will be scheduled, or launched, on the same worker node or Docker daemon. Pods cannot span multiple machines. All containers that are part of a Pod will be launched on the same worker node!
>
> This is something extremely important: containers in the same Pod are meant to live together. If you terminate a Pod, all its containers will be killed together, and when you create a Pod, Kubelet will, at the very least, attempt to create all its containers together.
>
> High availability is generally achieved by replicating multiple Pods over multiple nodes.

From a Kubernetes perspective, applying this file results in a fully working multi-container Pod made up of two containers, and we can make sure that the Pod is running with the two containers by running a standard `kubectl get Pods` command to fetch the Pod list from `kube-apiserver`.

Do you see the column that states 2/2? This is the number of containers inside the Pod. Here, this is saying that the two containers that are part of this Pod were successfully launched!

What happens when Kubernetes fails to launch one container in a Pod?

Kubernetes keeps track of all the containers that are launched in the same Pod. But it often happens that a specific container cannot be launched. One of the most popular causes for this issue is when a typo is introduced on one of the Docker images or tags specified in the Pod definition. Let's introduce such a typo in the YAML manifest to demonstrate how Kubernetes reacts when some containers of a specific Pod cannot be launched.

In the following example, I have defined a Docker image that does not exist at all for the NGINX container; note the `nginx:i-do-not-exist` Docker tag:

```
# ~/failed-multi-container-Pod.yaml
apiVersion: v1
kind: Pod
metadata:
  name: failed-multi-container-Pod
spec:
  containers:
    - name: nginx-container
      image: nginx:i-do-not-exist
    - name: busybox-container
      image: busybox:latest
```

Now, we can apply the following container using the `kubectl create -f failed-multi-container-Pod.yaml` command:

```
$ kubectl create -f failed-multi-container-Pod.yaml
Pod/failed-multi-container-Pod created
```

Here, you can see that the Pod was effectively created. This is because even if there's a failing Docker tag, the YAML remains valid from a Kubernetes perspective. So, Kubernetes simply creates the Pod and persists the entry into `etcd`, but we can easily imagine that `kubelet` will encounter an error when it launches the `docker pull` command to retrieve the image from Docker Hub.

Let's check the status of the Pod using `kubectl get Pods`:

```
$ kubectl create -f failed-multi-container-Pod.yaml
NAME                         READY     STATUS
RESTARTS     AGE
failed-multi-container-Pod   0/2       CrashLoopBackOff   4
2m23s
```

As you can see, the status of the pod is `CrashLoopBackOff`. This means that Kubernetes is constantly crashing when trying to launch the pod and retries again and again. To find out why it's crashing, you have to describe the Pod using the `kubectl describe Pods/failed-multi-container-Pod` command:

```
$ kubectl describe Pods/failed-multi-container-Pod


Warning  Failed      3m59s (x2 over 4m19s)   kubelet
Failed to pull image "nginx:i-do-not-exist": rpc error: code
= Unknown desc = Error response from daemon: manifest for
nginx:i-do-not-exist not found: manifest unknown: manifest
unknown
   Warning  Failed      3m59s (x2 over 4m19s)   kubelet
Error: ErrImagePull
   Normal   Pulling     3m59s (x3 over 4m19s)   kubelet
Pulling image "busybox:latest"
   Normal   Created     3m58s (x3 over 4m17s)   kubelet
Created container busybox-container
```

It's a little bit hard to read, but by following this log, you can see that `busybox-container` is okay since `kubelet` has succeeded in creating it, as shown by the last line of the preceding output. But there's a problem with the other container; that is, `nginx-container`.

Here, you can see that the output error is `ErrImagePull` and as you can guess, it's saying that the container cannot be launched because the `docker pull` command fails to retrieve the `nginx:i-do-not-exist` Docker tag.

So, Kubernetes does the following:

1. First, it creates the entry in `etcd` if the Pod of the YAML file is valid.
2. Then, it simply tries to launch the container.
3. If an error is encountered, it will try to launch the failing container again and again.

If any other container works properly, it's fine. But your Pod will never enter the `Running` status because of the crashing container. After all, your app certainly needs the failing container to work properly; otherwise, that container should not be there at all!

Now, let's learn how to delete a multi-container Pod.

Deleting a multi-container Pod

When you want to delete a Pod containing multiple containers, you have to go through the kubectl delete command, just like you would for a single-container Pod.

Them, you have two choices:

- You specify the path to the YAML manifest file that's used by using the -f option.
- You delete the Pod without using its YAML path if you know its name.

The first way consists of specifying the path to the YAML manifest file. You can do so using the following command:

```
$ kubectl delete -f multi-container-Pod.yaml
```

Otherwise, if you already know the Pod's name, you can do this as follows:

```
$ kubectl delete Pods/multi-Pod
$ # or equivalent
$ kubectl delete Pods multi-Pod
```

To figure out the name of the Pods, you can use the kubectl get commands:

```
$ kubectl get Pods
NAME                          READY    STATUS
RESTARTS     AGE
failed-multi-container-Pod    0/2      CrashLoopBackOff    9
22m
```

When I ran them, only failed-multi-container-Pod was created in my cluster, so that's why you can just see one line in my output. Keep in mind that these commands are working at the Pods level, not the containers level. Do not pass the name of a container as this wouldn't work at all.

Here is how you can delete failed-multi-container-Pod imperatively without specifying the YAML file that created it:

```
$ kubectl delete Pods/failed-multi-container-Pod
Pod "failed-multi-container-Pod" deleted
```

After a few seconds, the Pod is removed from the Kubernetes cluster and all its containers are removed from the Docker daemon and the worker node.

The amount of time that's spent before the command is issued and the Pod's name being deleted and released is called the grace period. Let's discover how to deal with it!

Understanding the Pod deletion grace period

One important concept related to deleting Pods is what is called the grace period. It is a concept that has something to do with how Kubernetes releases the name of the Pod during Pod deletion. Both single-container Pods and multi-container Pods have this grace period, which can be observed when you delete them. This grace period can be ignored by passing the `--grace-period=0 --force` option to the `delete` command.

The whole idea is that you cannot have two Pods with the same name running at the same time on your cluster, because the Pods' names are unique identifiers: that's why we use this as a parameter to identify a specific Pod when running the `kubectl delete` command, for example.

When the deletion is forced by setting a grace period to `0` with the `--force` flag, the Pod's name is immediately released and becomes available for another Pod to take it. While during an unforced deletion, the grace period is respected, and the Pod's name is released after it is effectively deleted:

```
$ kubectl delete Pods/multi-container-Pods --grace-period=0 -
force
warning: Immediate deletion does not wait for confirmation that
the running resource has been terminated. The resource may
continue to run on the cluster indefinitely.
Pod "multi-container-Pod" force deleted
```

Keep in mind that this command should be used carefully if you don't know what you are doing. Forcefully deleting a Pod shouldn't be seen as the norm because as the output states, `kubectl` you cannot be sure that the Pod was effectively deleted. If, for some reason, the Pod could not be deleted, it might run indefinitely, so do not run this command if you are not sure of what to do.

Now, let's discover how to access a specific container that is running inside a multi-container Pod.

Accessing a specific container inside a multi-container Pod

When you have several containers in the same Pod, you can access each of them individually. Here, we will access the NGINX container of our multi-container Pods. Let's start by recreating it because we deleted it in our previous example:

```
$ kubectl create -f multi-container-Pod.yaml
Pod/multi-container-Pod created
```

To access a running container, you need to use the kubectl exec command, just like you need to use docker exec to launch a command in an already created container when using Docker without Kubernetes.

This command will ask for two important parameters:

- The Pod that wraps the container you want to target
- The name of the container itself, as entered in the YAML manifest file

We already know the name of the Pod because we can easily retrieve it with the kubectl get command. In our case, the Pod is named multi-container-Pod.

However, we don't have the container's name because there is no kubectl get containers command that would allow us to list the running containers. This is why we will have to use the kubectl describe Pods/multi-container-Pod command to find out what is contained in this Pod:

```
$ kubectl describe Pods/multi-container-Pod
```

This command will show the names of all the containers contained in the targeted Pod. Here, we can see that our Pod is running two containers: one called busybox-container and another called nginx-container. The one we need is nginx-container.

Additionally, the following is a little command for listing all the container names contained in a dedicated Pod:

```
$ kubectl get Pods/multi-container-Pod -o jsonpath="{.items[*].
spec.containers[*].name}"
```

This command will spare you from using the describe command. However, it makes use of jsonpath, which is an advanced feature of kubectl: this command might look strange but it mostly consists of a sort filter that's applied against the command.

`jsonpath` filters are not easy to get right, so feel free to add this command as a bash alias or note it somewhere because it's a useful one.

In any case, we can now see that we have those two containers inside the `multi-container-Pod` Pod:

- `nginx-container`
- `busybox-container`

Now, let's access `nginx-container`. You have the name of the targeted container in the targeted Pod, so use the following command to access the Pod:

```
$ kubectl exec -ti multi-container-Pod --container nginx-
container -- /bin/bash
#
```

After running this command, you'll be inside `nginx-container`. Let's explain this command a little bit. `kubectl exec` does the same as `docker exec`.

When you run this command, you get the shell of the container, called `nginx-container`, inside the multi-container Pod, at which point you will be ready to run commands inside this very specific container on your Kubernetes cluster.

The main difference from the single container Pod situation is the `--container` option (the `-c` short option works too). You need to pass this option to tell `kubectl` what container you want to reach.

Now, let's discover how to run commands in the containers running in your Pods!

Running commands in containers

One powerful aspect of Kubernetes is that you can, at any time, access the containers running on your Pods to execute some commands. We did this previously, but did you know you can also execute any command you want directly from the `kubectl` command-line tool?

First, we are going to recreate the Pod containing the NGINX and Busybox containers:

```
$ kubectl create -f multi-container-Pod.yaml
Pod/multi-container-Pod created
```

To run a command in a container, you need to use `kubectl exec`, just like we did previously. But this time, you have to remove the `-ti` parameter to prevent `kubectl` from attaching to your running terminal session.

Here, I'm running the `ls` command to list files in `nginx-container` from the `multi-container-Pod` Pod:

```
$ kubectl exec Pods/multi-container-Pod --container nginx-
container-1 -- ls
```

This command will ask for two important parameters:

- The name of the container, as specified in the YAML file
- The name of the Pod that contains it

You can omit the container name but if you do so, `kubectl` will use the default first one. In our case, the default one will be `nginx-container` because it was the first one to be declared in the YAML manifest file.

Once you have entered all these parameters, you have to input the command you want to run from the rest with a double dash (`--`).

The name of the container, as well as the name of the Pod that contains it, will need to be provided. We already know the name of the Pod because we can easily retrieve it with the `kubectl get` command. In our case, the Pod is called `multi-container-Pod`.

We will now discover how to override the commands that are run by your containers.

Overriding the default commands run by your containers

When using Docker, you have the opportunity to write files called `Dockerfile` to build Docker images. `Dockerfile` makes use of two keywords to tell us what commands and arguments the containers that were built with this image will launch when they're created using the `docker run` command.

These two keywords are `ENTRYPOINT` and `CMD`:

- `ENTRYPOINT` is the main command the Docker container will launch.
- `CMD` is used to replace the parameters that are passed to the `ENTRYPOINT` command.

For example, a classic `Dockerfile` that should be launched to run the `sleep` command for `30` seconds would be written like this:

```
# ~/Dockerfile
FROM busybox:latest
ENTRYPOINT ["sleep"]
CMD ["30"]
```

This is just plain old Docker and you should be familiar with these concepts. As you may already know, the CMD argument is what you can pass to the `docker run` command. If you build this image using this `Dockerfile` using the `docker build` command, you'll end up with a Busybox image that just runs the `sleep` command (ENTRYPOINT) when `docker run` for `30` seconds (the CMD argument).

Thanks to the CMD instruction, you can override the default `30` seconds like so:

```
$ docker run my-custom-ubuntu:latest 60
$ docker run my-custom-ubuntu:latest # Just sleep for 30
seconds
```

Kubernetes, on the other hand, allows us to override both ENTRYPOINT and CMD thanks to YAML pod definition files. To do so, you must append two optional keys to your YAML configuration file: `command` and `args`.

This is a very big benefit Kubernetes brings you because you can decide to append arguments to the command that's run by your container's `Dockerfile`, just like the CMD arguments does with bare Docker, or completely override ENTRYPOINT!

Here, I'm going to write a new manifest file that will override the default ENTRYPOINT and CMD parameters of the `busybox` image to make the `busybox` container sleep for 60 seconds. Here is how to proceed:

```
# ~/nginx-busybox-with-custom-command-and-args.yaml
apiVersion: v1
kind: Pod
metadata:
  name: nginx-busybox-with-custom-command-and-args
spec:
  initContainers:
  - name: my-init-container
    image: busybox:latest
    command: ["sleep", "15"]
```

```
containers:
  - name: nginx-container
    image: nginx:latest
  - name: busybox-container
    image: busybox:latest
    command: ["sleep"] # Corresponds to the ENTRYPOINT
    args: ["60"] # Corresponds to CMD
```

This is a bit tricky to understand because what `Dockerfile` calls `ENTRYPOINT` corresponds to the `command` argument in the YAML manifest file, and what `Dockerfile` calls `CMD` corresponds to the `args` configuration key in the YAML manifest file.

What if you omit one of them? Kubernetes will default to what is inside the Docker image. If you omit the `args` key in the YAML, then Kubernetes will go for the `CMD` provided in the `Dockerfile`, while if you omit the `command` key, Kubernetes will go for the `ENTRYPOINT` declated in the `Dockerfile`. Most of the time, or at least if you're comfortable with your container's `ENTRYPOINT`, you're just going to override the `args` file (the `CMD  Dockerfile` instruction).

Now, let's discover another feature: `initContainers`! In the next section, you'll see another way to execute some additional side containers in your Pod to configure the main ones.

Introducing initContainers

`initContainers` is a feature provided by Kubernetes Pods to run setup scripts before the actual containers start. You can think of them as additional sides containers you can define in your Pod YAML manifest file: they will run first when the Pod is created. Then, once they complete, the Pod starts creating its main containers.

You can execute not one but several `initContainers` in the same Pod, but when you define lots of them, keep in mind that they will run one after another, not in parallel. Once an `initContainer` completes, the next one starts, and so on. In general, `initContainers` are used to pull application code from a Git repository and expose it to the main containers using volume mounts or to run start-up scripts.

Since `initContainers` can have their own Docker images. You can offload some configuration to them by keeping your main containers images as small as possible, thus increasing the whole security of your setup by removing unnecessary tools from your main container images. Here is a YAML manifest that introduces an `initContainer`:

```
# ~/nginx-with-init-container.yaml
apiVersion: v1
kind: Pod
metadata:
  name: nginx-with-init-container
spec:
  initContainers:
  - name: my-init-container
    image: busybox:latest
    command: ["sleep", "15"]
  containers:
    - name: nginx-container
      image: nginx:latest
```

As you can see from this YAML file, `initContainer` runs the `busybox:latest` image, which will sleep for 15 seconds before completing. Once the execution of `initContainer` is complete, Kubernetes will create the NGINX container.

> **Important Note**
> Note that Kubernetes cannot launch the main containers if `initContainer` fails. That's why it is really important to not see `initContainer` as something optional or that could fail. They will be forcibly executed if they are specified in the YAML manifest file, and if they fail, the main containers will never be launched!

Let's create the Pod. After, we will run the `kubectl get Pods -w` command for `kubectl` to watch for a change in the Pod list. The output of the command will be updated regularly, showing the change in the Pod's status. Please note the `status` command, which is saying that an `initContainer` is running!

```
$ kubectl create -f nginx-with-init-container.yaml
Pod/nginx-with-init-container created

$ kubectl get Pods -w
```

NAME	READY	STATUS	RESTARTS	AGE
nginx-with-init-container	0/1	Init:0/1	0	3s
nginx-with-init-container 17s	0/1	PodInitializing	0	
nginx-with-init-container 19s	1/1	Running	0	

As you can see, `Init:0/1` indicates that `initContainer` is being launched. After its completion, the Init: prefix disappears for the next statuses, indicating that we are done with `initContainer` and that Kubernetes is now creating the main container – in our case, the NGINX one!

Use `initContainer` wisely when you're building your Pods! And remember: if you can avoid using them, do so. You are not forced to use them, but they can be really helpful for running configuration scripts or to pull something from external servers before you launch your actual containers! Now, let's learn how to access the logs of a specific container inside a running Pod!

Accessing the logs of a specific container

When using multiple containers in a single Pod, you can retrieve the logs of a dedicated container inside the Pod. The proper way to proceed is by using the `kubectl logs` command.

The most common way a containerized application exposes its logs is by sending them to `stdout`, which is basically what Docker displays when you run the `kubectl logs` command.

The `kubectl logs` command is capable of streaming the `stdout` property of a dedicated container in a dedicated Pod and retrieving the application logs from the container. For it to work, you will need to know the name of both the precise container and its parent Pod, just like when we used `kubectl exec` to access a specific container.

Please read the previous section, *Accessing a specific container inside a multi-container Pod*, to discover how:

```
$ kubectl logs -f Pods/multi-container-Pods --container nginx-
container
```

Please note the `--container` option (the `-c` short option works too), which specifies the container you want to retrieve the logs for. Note that it also works the same for `initContainers`: you have to pass its name to this option to retrieve its logs.

> **Important Note**
>
> Remember that if you do not pass the `--container` option, you will retrieve all the logs from all the containers that have been launched inside the Pod. Not passing this option is useful in the case of a single-container Pod, but you should consider this option every time you use a multi-container Pod.

There are other multiple useful options you need to be aware of when it comes to accessing the logs of a container in a Pod. You can decide to retrieve the logs written in the last 2 hours by using the following command:

```
$ kubectl logs --since=2h Pods/multi-container-Pods --container
nginx-container
```

Also, you can use the `--tail` option to retrieve the most recent lines of a log's output. Here's how to do this:

```
$ kubectl logs --tail=30 Pods/multi-container-Pods --container
nginx-container
```

Here, we are retrieving the 30 most recent lines in the log output of `nginx-container`.

Now, you are ready to read and retrieve the logs from your Kubernetes Pods, regardless of whether they are made up of one or several containers!

In this section, we discovered how to create, update, and delete multi-container Pods. We also discovered how to force the deletion of a Pod. We then discovered how to access a specific container in a Pod, as well as how to retrieve the logs of a specific container in a Pod. Though we created an NGINX and a Busybox container in our Pod, they are relatively poorly linked since they don't do anything together. To remediate that, we will now learn how to deal with volumes so that we can share files between our two containers.

Sharing volumes between containers in the same Pod

In this section, we'll learn what volumes are from a Kubernetes point of view and how to use them. Docker also has a notion of volumes but it differs from Kubernetes volumes: they answer the same need but they are not the same.

In this section, we will discover what Kubernetes volumes are, why they are useful, and how they can help us when it comes to Kubernetes volumes.

What are Kubernetes volumes?

We are going to answer a simple problem. Our multi-container Pods are currently made up of two containers: an NGINX one and a Busybox one. We are going to try to share the log directory in the NGINX container with the Busybox container by mounting the log directory of NGINX in the directory of the Busybox container. This way, we will create a relationship between the two containers to have them share a directory.

Kubernetes has two kinds of volumes:

- Volumes, which we will discuss here.

- `PersistentVolume`, which is a more advanced feature we will discuss later in *Chapter 9, Persistent Storage in Kubernetes*.

Keep in mind that these two are not the same. `PersistentVolume` is a resource of its own, whereas "volumes" is a Pod configuration. As the name suggests, `PersistentVolume` is persistent, whereas volumes are not supposed to be. But keep in mind that this is not always the case!

Simply put, volumes are storage-bound to the Pod's life cycle: when you create a Pod, you'll have the opportunity to create volumes and attach them to the container(s) inside the Pods. Volumes are nothing more than storage attached to the life cycle of the Pod. As soon as the Pod is deleted, the volumes that were created with it are deleted too.

Even though they are not limited to this use case and this is not always true, you can consider volumes as a particularly great way to share a directory and files between containers running in the same Pod.

> **Important Note**
>
> Remember that volumes are bound to the Pod's life cycle, not the container's life cycle. If a container crashes, the volume would survive because if a container crashes, it won't cause its parent Pod to crash, and thus, no volume will be deleted. So long as a Pod is alive, its volumes are too.

When Docker introduced the concept of volumes, it was just shared directories you could mount to a container. Kubernetes also built its volume feature around this idea and used volumes as shared directories.

But Kubernetes also brought support for a lot of drivers, which helps integrate Pods' volumes with external solutions. For example, an AWS EBS volume can be used as a Kubernetes volume. Here are some solutions among the most common ones that can be used as Kubernetes volumes:

- `awsElasticBlockStore`
- `azureDisk`
- `gcePersistentDisk`
- `glusterfs`
- `hostPath`
- `emptyDir`
- `nfs`
- `persistentVolumeClaim` (when you need to use a `PersistentVolume`, which is outside the scope of this chapter)

That is why I said it is not true to say that a volume is fully bound to the life cycle of a Pod. For example, a pod volume backed by an AWS EBS could survive a Pod being deleted because the backend provider (here, this is AWS) might have its own way of managing the storage life cycle. That is why we are going to purely focus on the simplest form of volumes for now.

> **Important Note**
> Please note that using external solutions to manage your Kubernetes volumes will require you to follow those external solutions' requirements. For example, using an AWS EBS volume as a Kubernetes volume will require your Pods to be executed on a Kubernetes worker node, which would be an EC2 instance. The reason for this is that AWS EBS volumes can only be attached to EC2 instances. Thus, a Pod exploiting such a volume would need to be launched on an EC2 instance.

We are going to discover the two most common volume drivers here: `emptyDir` and `hostPath`. We will also talk about `persistentVolumeClaim` because this one is going to a little be special compared to the other volumes and will be fully discovered in *Chapter 9, Persistent Storage in Kubernetes*

Now, let's start discovering how to share files between containers in the same Pod using volumes with the `emptyDir` volume type!

Creating and mounting an emptyDir volume

The emptyDir volume type is certainly the most common volume type that's used. As the name suggests, it is simply an empty directory that is initialized at Pod creation that you can mount to the location of each container running in the Pod.

It is certainly the easiest and simplest way to have your container share data between them. Let's create a Pod that will manage two containers.

In the following example, I am creating a Pod that will launch two containers, and just like we had previously, it's going to be an NGINX container and a Busybox container. I'm going to override the command that's run by the Busybox container when it starts to prevent it from completing. That way, we will get it running indefinitely as a long process and we will be able to launch additional commands to check if our emptyDir has been initialized correctly.

Both containers will have a common volume mounted at /var/i-am-empty-dir-volume/, which will be our emptyDir volume, initialized in the same Pod. Here is the YAML file for creating the Pod:

```yaml
# ~/two-nginx-with-emptydir-Pod.yaml
apiVersion: v1
kind: Pod
metadata:
  name: two-containers-with-empty-dir
spec:
  containers:
    - name: nginx-container
      image: nginx:latest
      volumeMounts:
      - mountPath: /var/i-am-empty-dir-volume
        name: empty-dir-volume
    - name: busybox-container
      image: busybox:latest
      command: ["/bin/sh"]
      args: ["-c", "while true; do sleep 30; done;"] # Prevents busybox from exiting after completion
      volumeMounts:
      - mountPath: /var/i-am-empty-dir-volume
        name: empty-dir-volume
  volumes:
```

```
    - name: empty-dir-volume # name of the volume
      emptyDir: {} # Initialize an empty directory # The path on
the worker node.
```

Note that the object we will create in our Kubernetes cluster will become more and more complex as we go through this example, and as you can imagine, most complex things cannot be achieved with just imperative commands. That's why you are going to see more and more examples relying on the YAML manifest file: you should start to take up the habit of trying to read them to figure out what they do.

That being said, we can now apply the manifest file using the following `kubectl create -f` command:

```
$ kubectl create -f two-containers-with-emptydir-Pod.yaml
Pod/two-containers-with-empty-dir created
```

Now, we can check that the Pod is successfully running by issuing the `kubectl get Pods` command:

```
$ kubectl get Pods
NAME                                READY   STATUS    RESTARTS
AGE
two-containers-with-empty-dir       2/2     Running   0
47s
```

Now that we are sure the Pod is running and that both the NGINX and Busybox containers have been launched, we can check that the directory can be accessed in both containers by issuing the `ls` command.

If the command is not failing, as we saw previously, we can run the `ls` command in the containers by simply running the `kubectl exec` command. As you may recall, the command takes the Pod's name and the container's name as arguments. We are going to run it twice to make sure the volume is mounted in both containers:

```
$ kubectl exec two-containers-with-empty-dir --container nginx-
container -- ls /var
i-am-empty-dir-volume

$ kubectl exec two-containers-with-empty-dir --container
busybox-container -- ls /var
i-am-empty-dir-volume
```

As you can see, the `ls /var` command is showing the name in both containers! This means that `emptyDir` was initialized and mounted in both containers correctly.

Now, let's create a file in one of the two containers. The file should be immediately visible in the other container, proving that the volume mount is working properly!

In the following command, I am simply creating a `.txt` file called `hello-world.txt` in the mounted directory:

```
$ kubectl exec two-containers-with-empty-dir --container nginx-
container -- /bin/sh -c "echo 'hello world' >> /var/i-am-empty-
dir-volume/hello-world.txt"
```

```
$ kubectl exec two-containers-with-empty-dir --container nginx-
container -- cat /var/i-am-empty-dir-volume/hello-world.txt
```

```
hello world
$ kubectl exec two-containers-with-empty-dir --container
busybox-container -- cat /var/i-am-empty-dir-volume/hello-
world.txt
```

```
hello world
```

As you can see, I used `nginx-container` to create the `/var/i-am-empty-dir-volume/hello-world.txt` file, which contains the `hello-world` string. Then, I simply used the `cat` command to access the file from both containers; you can see that the file is accessible in both cases. Again, remember that `emptyDir` volumes are completely tied to the life cycle of the Pod. If the Pod declaring it is destroyed, then the volume is destroyed too, along with all its content, and it will become impossible to recover!

Now, we will discover another volume type: the `hostPath` volume. As you can imagine, it's going to be a directory that you can mount on your containers that is backed by a path on the host machine – the worker node running the Pod!

Creating and mounting a hostPath volume

The `hostPath` volume is also a common volume type. As its name suggests, it will allow you to mount a directory in the host machine to containers in your Pod! The host machine is the Kubernetes worker node executing the Pod. Here are some examples:

- If your cluster is based on Minikube (a single-node cluster), the host is your local machine.

- On Amazon EKS, the host machine will be an EC2 instance.

- In a Kubeadm cluster, the host machine is generally a standard Linux machine.

The host machine is the machine running the Pod, and you can mount a directory from the filesystem of the host machine to the Kubernetes Pod!

In the following example, I'll be working on a Kubernetes cluster based on Minikube, so `hostPath` will be a directory that's been created on my computer that will then be mounted in a Kubernetes Pod.

> **Important Note**
>
> Using the `hostPath` volume type can be useful, but you have to keep in mind that it creates a strong relationship between the worker node and the Pods running on top of it. In the Kubernetes world, you can consider it as an anti-pattern.
>
> The whole idea behind Pods is that they are supposed to be easy to delete and rescheduled on another worker node without problems. Using `hostPath` will create a tight relationship between the Pod and the worker node, and that could lead to major issues if your Pod were to fail and be rescheduled on a node where the required path on the host machine is not present.

Now, let's discover how to create `hostPath`.

Let's imagine that I have a file on my worker node on `worker-node/nginx.conf` and I want to mount it on `/var/config/nginx.conf` on the `nginx` container.

Here is the YAML file to create the setup. As you can see, I declared a `hostPath` volume at the bottom of the file that defines a path that should be present on my host machine. Now, I can mount it on any container that needs to deal with the volume in the `containers` block:

```
# ~/multi-container-Pod-with-host-path.yaml
apiVersion: v1
kind: Pod
```

```
metadata:
  name: multi-container-Pod-with-host-path
spec:
  containers:
    - name: nginx-container
      image: nginx:latest
      volumeMounts:
      - mountPath: /var/config
        name: my-host-path-volume
    - name: busybox-container
      image: busybox:latest
      command: ["/bin/sh"]
      args: ["-c", "while true; do sleep 30; done;"] # Prevents
busybox from exiting after completion
  volumes:
  - name: my-host-path-volume
    hostPath:
      path: /tmp # The path on the worker node.
```

As you can see, mounting the value is just like what we did with the `emptyDir` volume in the previous section regarding the `emptyDir` volume type. By using a combination of volumes at the Pod level and `volumeMounts` at the container level, you can mount a volume on your containers.

You can also mount the directory on the `busybox` container so that it gets access to the directory on the host.

Before running the YAML manifest file, though, you need to create the path on your host and create the necessary file:

```
$ echo "Hello World" >> /tmp/hello-world.txt
```

Now that the path exists on the host machine, we can apply the YAML file to our Kubernetes cluster and, immediately after, launch a `kubectl get Pods` command to check that the Pod was created correctly:

```
$ kubectl create -f multi-container-Pod-with-host-path.yaml
Pod/multi-container-Pod-with-host-path

$ kubectl get Pods
```

NAME	READY	STATUS	RESTARTS
multi-container-Pod-with-host-path 92s	2/2	Running	0

Everything seems good! Now, let's echo the file that should be mounted at `/var/config/hello-world.txt`.

At beginning of this chapter, we discovered the different aspects of multi-container Pods! We discovered how to create, update, and delete multi-container Pods, as well as `initContainers`, access logs, override Docker commands directly from the pod's resources, and how to share directories between containers using the most two basic volumes. Now, we are going to put a few architecting principles together and discover some notions related to multi-container Pods called "patterns."

The ambassador design pattern

When designing a multi-container Pod, you can decide to follow some architectural principles to build your Pod. Some typical needs are answers by these design principles, and the ambassador pattern is one of them.

Here, we are going to discover what the ambassador design pattern is, how to build an ambassador container in Kubernetes Pods, and look at a concrete example of them.

What is the ambassador design pattern?

In essence, the ambassador design pattern applies to multi-container Pods. We can define two containers in the same Pod:

- The first container will be called the main container.
- The other container will be called the ambassador container.

In this design pattern, we assume that the main container might have to access external services to communicate with them. For example, you can have an application that must interact with a SQL database that is living outside of your Pod, and you need to reach this database to retrieve data from it.

This is the typical use case where you can deploy an adapter container alongside the main container, next to it, in the same Pod. The whole idea is to get the ambassador to proxy the requests ran by the main container to the database server. The ambassador container will be essentially a SQL proxy. Every time the main container wants to access the database, it won't access it directly but rather create a connection to the ambassador container that will play the role of a SQL proxy.

> **Important Note**
> Running an ambassador container is fine, but only if the external API is not living in the same Kubernetes cluster. To run requests on another pod, Kubernetes provides strong mechanics called services. We will have the opportunity to discover them in *Chapter 7, Exposing Your Pods with Services.*

But why would you need a proxy to access external databases? Here are some concrete benefits this design pattern can bring you:

- Offloading SQL configuration
- Management of SSL/TLS certificates

Please note that having an ambassador proxy is not limited to a SQL proxy but this example is demonstrative of what this design pattern can bring you. Note that the ambassador proxy is only supposed to be called for outbound connections from your main container to something else, such as data storage or an external API. It should not be seen as an entry point to your cluster! Now, let's quickly discover how to create an ambassador SQL with a YAML file.

A simple example of an ambassador multi-container Pod

Note that we know about ambassador containers, let's learn how to create one with Kubernetes. The following YAML manifest file creates a Pod that makes creates two containers:

- `nginx-app`, derived from the `nginx:latest` image
- `Sql-ambassador-proxy`, created from the `mysql-proxy:latest` Docker image:

```
# ~/ nginx-with-ambassador.yaml
apiVersion: v1
kind: Pod
```

```
metadata:
  name: nginx-with-ambassador
spec:
  containers:
    - name: mysql-proxy-ambassador-container
      image: mysql-proxy:latest
      ports:
        - containerPort: 3306
      env:
      - name: DB_HOST
        value: mysql.xxx.us-east-1.rds.amazonaws.com
    - name: nginx-container
      image: nginx:latest
```

As you can imagine, it's going to be the developer's job to get the application code running in the NGINX container to query the ambassador instead of the Amazon RDS endpoint. As the ambassador container can be configured from environment variables, it's going to be easy for you to just input the configuration variables in `ambassador-container`.

> **Important Note**
>
> Do not get tricked by the order of the containers in the YAML file. The fact that the ambassador container appears first does not make it the *main* container of the Pod. This notion of the *main* container does not exist at all from a Kubernetes perspective – both are plain Docker containers that run in parallel with no concept of a hierarchy between them. Here, we just access the Pod from the NGINX container, which makes it the most important one.

Remember that the ambassador running in the same Pod as the NGINX container makes it accessible from NGINX on `localhost:3306`!

The sidecar design pattern

The sidecar design pattern is an extremely useful one. It is good for when you want to extend the features of your main containers with features it would normally not be able to achieve on its own.

Just like we did for the ambassador container, we're going to explain exactly what it is by covering some examples. Then, we're going to discover some concrete examples.

What is the sidecar design pattern?

Think of the sidecar container as an extension or a helper for your main container. Its main purpose is to extend the main container to bring it a new feature, but without changing anything about it. Unlike the ambassador design pattern, the main container may even not be aware of the presence of a sidecar.

Just like the ambassador design pattern, the sidecar design pattern makes use of at least two containers:

- The main container, the one that is running the application
- The sidecar container, the one that is bringing something additional to the first one

You may have already guessed, but this pattern is especially useful when you want to run monitoring or log forwarder agents.

There are three things to understand when you want to build a sidecar that is going to forward your logs to another location:

- You must locate the directory where your main containers write their logs.
- You must create a volume to make this directory accessible to the log forwarder sidecar.
- You must launch the sidecar container with the proper configuration.

Based on these concepts, the main container remains unchanged, and even if the sidecar fails, it wouldn't have an impact on the main container, which could continue to work.

Now, we're going to use an example YAML file for the sidecar design pattern.

A simple example of a sidecar multi-container Pod

Just like the ambassador design pattern, the sidecar makes use of multi-container Pods. We will define two containers in the same Pod.

The adapter design pattern

The last design pattern that we are going to discover here is the adapter design pattern. As its name suggests, it's going to *adapt* an entry from a source format to a target format.

What is the adapter design pattern?

The adapter design pattern is the last paradigm we are going to discover in this chapter. As with the ambassador and sidecar design patterns, this one expects that you run at least two containers:

- The first one is the main container.

- The second one is the adapter container.

This design pattern is helpful and should be used whenever the main containers emit data in a format, A, that should be sent to another application that is expecting the data in another format, B. As the name suggests, the adapter container is here to *adapt*.

Again, this design pattern is especially well-suited for log or monitoring management. Imagine a Kubernetes cluster where you have dozens of applications running; they are writing logs in Apache format that you need to convert into JSON so that they can be indexed by a search engine. This is exactly where the adapter design pattern comes into play. Running an adapter container next to the application containers will help you get these logs adapted to the source format before they are sent somewhere else.

Just like for the sidecar design pattern, this one can only work if both the containers in your Pod are accessing the same directory using volumes.

A simple example of an adapter multi-container Pod

In this example, I'm going to use a Pod that is using an adapter container with a shared directory mounted as a Kubernetes volume.

This Pod is going to run two containers:

- `nginx-app`, derived from the `nginx:latest` image

- `adapter-container`, created from the `ubuntu:latest` Docker image:

```
# ~/ nginx-with-ambassador.yaml
apiVersion: v1
kind: Pod
metadata:
  name: nginx-with-ambassador
spec:
  containers:
    - name: mysql-proxy-ambassador-container
      image: mysql-proxy:latest
```

```
    ports:
        - containerPort: 3306
    env:
    - name: DB_HOST
      value: mysql.xxx.us-east-1.rds.amazonaws.com
  - name: nginx-container
    image: nginx:latest
```

Please note that the container is mounting the same directory to provide access to both containers.

Summary

This chapter was quite a big one, but you should now have a good understanding of what pods are and how to use them, especially when it comes to managing multiple Docker containers in the same Pod. Since microservice applications are often made up of several containers and not just one, it's going to be difficult to manage only a single-container Pod in your cluster.

I recommend that you focus on mastering the declarative way of creating Kubernetes resources. As you have noticed in this chapter, the key to achieving the most complex things with Kubernetes resides in writing YAML files. One example is that you simply cannot create a multi-container Pod without writing YAML files.

This chapter completes the previous one: *Chapter 4, Running Your Docker Containers*. You need to understand that everything we will do with Kubernetes will be Pod management because everything in Kubernetes revolves around them. Keep in mind that containers are never created directly, but always through a pod object, and that all the containers within the same Pod are created on the same worker node. If you understand that, then you can continue to the next chapter!

The next chapter is going to introduce two of the most important objects, ConfigMaps and Secrets, as we continue to dig into the core concepts of Kubernetes. In Kubernetes, we consider that applications and their configurations should be treated as two completely distinct things to improve application portability: that is why we have the pod resource, which lets us create the application container, and the ConfigMaps and Secrets objects, which are there to help us inject configuration data into our pods.

6

Configuring Your Pods Using ConfigMaps and Secrets

The last two chapters, entitled *Chapter 4, Running Your Docker Containers*, and *Chapter 5, Using Multi-Container Pods and Design Patterns*, introduced you to launching Docker containers using Kubernetes. At this point, you know that whenever you need to launch a container on Kubernetes, you will need to do so using Pods. This was the key concept for you to understand and assimilate. We also learned that Kubernetes is nothing more than a REST API that describes resources types we call **Kind**. When created against an API, each instance of a Kind will result in a computing resource being provisioned on a worker node. A Pod is one of these resources, and when they're created against the worker node, this results in Docker containers.

In this chapter, we'll learn about two new Kubernetes objects: **ConfigMaps** and **Secrets**.

These are two very important objects or resources that allow you to configure the apps that run in your Pods. ConfigMaps and Secrets enable you to decouple the applications running in your Pods from their configuration values. Kubernetes has its own way of managing configuration values and that's what we will learn in this chapter. We will also explain why it is so important to treat application code and configuration values as two different and independent things, and the benefit you'll gain from doing so.

ConfigMaps and Secrets have a ton of mechanisms, such as being able to be injected into containers running in Pods in the form of environment variables or volumes to mount on your Pods. That's what we're going to look at in this chapter. Modern applications require you to treat application configuration as a first-class citizen to make applications blind to the environment they are running on. This way, you'll be able to build super-portable applications in a Kubernetes-friendly way by discovering how to expose configuration values to your running Pods.

In this chapter, we're going to cover the following main topics:

- Understanding what ConfigMaps and Secrets are
- Configuring your Pods using the ConfigMap object
- Managing sensitive configuration with the Secret object

Technical requirements

For this chapter, you will need the following:

- A working Kubernetes cluster (local or cloud-based, though this is not important)
- A working kubectl CLI configured to communicate with the Kubernetes cluster

You can get these two prerequisites by following *Chapter 2, Kubernetes Architecture – From Docker Images to Running Pods*, and *Chapter 3, Installing Your First Kubernetes Cluster*, to get a working Kubernetes cluster and a properly configured kubectl client, respectively.

Understanding what ConfigMaps and Secrets are

Configuration management is a consideration that must always be taken seriously in any IT project. Kubernetes has made configuration a first-class passenger by creating two resources specifically designed to manage configuration: ConfigMaps and Secrets.

In the next section, we will explain what ConfigMaps and Secrets are and how they are used to manage configuration in Kubernetes.

Decoupling your application and your configuration

When we use Kubernetes, we want our applications to be as portable as possible. A good way to achieve this is to decouple the application from its configuration. Back in the old days, configuration and application were the same things: since the application code was designed to work only on one environment, configuration values were often bundled within the application code itself, so the configuration and application code were tightly coupled.

Having both application code and configuration values treated as the same thing reduces the portability of an application. Nowadays, things have changed a little bit and we must be able to update the application configuration because we want to make our application as portable as possible, enabling us to deploy applications in multiple environments flawlessly.

Consider the following problem:

1. You deploy a Java application to the development environment for testing.
2. After the tests, the app is ready for production, and you need to deploy it. However, the MySQL endpoint for production is different from the one in development.

There are two possibilities here:

*   The configuration and application code are not decoupled, and the MySQL is hardcoded and bundled within the application code: you are stuck and need to rebuild the whole app after editing the application code.
*   The configuration and application code are decoupled. That's good news for you as you can simply override the MySQL endpoint as you deploy to production.

That's the key to the concept of portability: the application code should be independent of the infrastructure it is running on. The best way to achieve this is to decouple the application code from its configuration.

Let's look at some typical examples of the types of configuration values you should decouple from the app:

*   API keys to access an Amazon S3 bucket
*   The password of the MySQL server used by your application

- The endpoint of a Redis cluster used by your application
- Pre-computed values such as JWT token private keys

All of these values are likely to change from one environment to another, and the applications that are launched in your Pods should be able to load a different configuration, depending on the environment they are launched on. This is why we will seek to systematically maintain a barrier between our applications and the configurations they consume. By doing this, we can treat them as two completely different entities in our Kubernetes cluster. The best way to achieve this is by considering our application and the configurations it uses as two different entities.

This is why Kubernetes suggests using the ConfigMap and Secrets objects, which are designed to carry your configuration data. Then, you will need to attach these ConfigMaps and Secrets when you create your Pods.

> **Important Note**
> Please avoid including your configuration values as part of your Docker images. Your Dockerfile should build your application but not configure it. By including the container configuration at build time, you create a strong relationship between your application and how it's configured, which reduces the portability of your container.

ConfigMaps are meant to hold non-critical configuration values, whereas Secrets are globally the same but meant to hold sensitive configuration values, such as database passwords and so on.

So, you can imagine a ConfigMap and Secret for each environment and for each application, which would contain the parameters that the application needs to function in a specific context and environment. The whole idea to understand is that ConfigMaps and Secrets behave like repositories for key/value pairs. These key/value pairs can contain plain values called **literal** or full configuration files such as YAML, TOML, and so on. Then, on Pod creation, you can pick the name of a ConfigMap or a Secret and link it to your Pod so that the configuration values are exposed to the containers running into it.

You always proceed in this order:

1. Create a ConfigMap or a Secret.
2. Fill it with your configuration values.
3. Create a Pod referencing the ConfigMap or Secret.

By following these steps, you can make your Pods portable between environments, which means you are fully compliant with a common DevOps good practice.

Now that we've explained why it is important to decouple application code and configuration values, it is time to explain why and how to achieve this in a Kubernetes-friendly way.

Understanding how Pods consume ConfigMaps and Secrets

Outside of Kubernetes, modern containerized applications consume their configuration in two ways:

- As OS environment variables
- As configuration files

The reason for this is that overriding an environment variable is super easy with Docker and all programming languages offer functions to easily read environment variables. Configuration files can easily be shared and mounted as Docker volumes between containers.

Back in the Kubernetes world, ConfigMaps and Secrets are following these two methods. Once created in your Kubernetes cluster, these items can be consumed in one of two ways:

- Included as environment variables in the container running in your Pods
- Mounted as Kubernetes volumes, just like any other volume

You can decide to inject one value from a ConfigMap or a Secret as an environment variable or inject all the values into a ConfigMap as environment variables; the same goes for Secrets.

ConfigMap and Secrets can also behave as volume mounts. When you mount a ConfigMap as a volume, you can inject all the values it contains in a directory into your container. If you store the full configuration files in your ConfigMap, using this feature to override a configuration directory becomes incredibly easy.

After this introduction to ConfigMap and Secrets, you should now understand why they are so important when it comes to configuring an application. Mastering them is crucial if you intend to work with Kubernetes cleanly and solidly. As we mentioned earlier in this chapter, ConfigMaps are used to store *unsecured* configuration values, whereas Secrets are used for more sensitive configuration data such as hashes or database passwords.

Since the two objects don't behave the same, let's look at them separately. First, we are going to discover how ConfigMaps work. We will discover Secrets after.

Configuring your Pods using ConfigMaps

In this section, we will learn how to list, create, delete, and read ConfigMaps. Then, we will learn how to attach them to our Pods so that their values are injected into our Pods in the form of environment variables or volumes.

Listing ConfigMaps

Listing the ConfigMaps that were created in your cluster is fairly straightforward and can be accomplished using kubectl , just like any other object in Kubernetes. You can do this by using the full resource name, which is configmaps:

```
$ kubectl get configmaps
```

Alternatively, you can use the shorter alias, which is cm:

```
$ kubectl get cm
```

These two commands are equivalent. At this point, kubectl may return a few default ConfigMaps or an error message saying that no configmaps were found. This is because some cloud services create default ConfigMaps for internal operations while others don't – it depends on where your Kubernetes cluster is running.

Creating a ConfigMap

Like other Kubernetes objects, ConfigMaps can be created imperatively or declaratively. Just like we did to create Pods, imperative methods consist of issuing a command, whereas declarative methods consist of creating a YAML file and then applying it to the cluster to create the resource. Nothing changes here.

You can decide to create an empty ConfigMap and then add values to it, or create a ConfigMap directly initialized with values. The following command will create an empty ConfigMap called my-first-configmap via the imperative method:

```
$ kubectl create configmap my-first-configmap
configmap/my-first-configmap created
$ # kubectl create cm my-first-configmap also works...
```

Once this command has been executed, you can type the `kubectl get cm` command once again to see your new `configmap`:

```
$ kubectl get cm
NAME                  DATA    AGE
my-first-configmap    0       42s
```

Now, we are going to create a new empty ConfigMap, but this time, we are going to create it with the declarative method. This way, we're going to have to create a YAML file and apply it through `kubectl`.

The following content should be placed in a file called `~/my-second-configmap.yaml`:

```
# ~/my-second-configmap.yaml
apiVersion: v1
kind: ConfigMap
metadata:
  name: my-second-configmap
```

Once this file has been created, you can apply it to your Kubernetes cluster using the `kubectl create -f` command:

```
$ kubectl create -f ~/my-second-configmap.yaml
configmap/my-second-configmap created
```

You can type the `kubectl get cm` command once more to see your new `configmap` added next to the one you created earlier:

```
$ kubectl get cm
NAME                   DATA    AGE
my-fist-configmap      0       106s
my-second-configmap    0       18s
```

Please note that the output of the `kubectl get cm` command also returns the number of keys each `configmap` contains in the DATA column. For now, it's zero, but in the following examples, you'll see that we can fill a `configmap` when it's created, so DATA will reflect the number of keys we put in `configmap`.

Creating a ConfigMap from literal values

Having an empty ConfigMap is quite useless, so let's learn how to create a ConfigMap with values inside it. Let's do this imperatively: adding the `-from-literal` flag to the `kubectl create` cm command.

Here, I'm going to create a ConfigMap called `my-third-configmap` with a key named `color` and its value set to `blue`:

```
$ kubectl create cm my-third-configmap --from-
literal=color=blue
configmap/my-third-configmap created
```

Also, be aware that you can create a ConfigMap with multiple parameters at once: you just need to add as much configuration data as you want to `configmap` by chaining several from-literals as you need in your command:

```
$ kubectl create cm my-fourth-configmap --from-
literal=color=blue --from-literal=version=1 --from-
literal=environment=prod
configmap/my-fourth-configmap created
```

Here, we are creating a ConfigMap with three configuration values inside it. Now, you can list your ConfigMaps once more using this command. You should see the few additional ones you just created.

Please note that the DATA column in the return of `kubectl get cm` is now reflecting the number of configuration values inside each `configmap`:

```
$ kubectl get cm
NAME                    DATA    AGE
my-first-configmap      0       9m30s
my-fourth-configmap     3       6m23s
my-second-configmap     0       8m2s
my-third-configmap      1       7m9s
```

It is also possible to create the same `configmap` declaratively. Here is the equivalent of this as a YAML configuration file that is ready to be applied against the cluster. Please note the new data YAML key, which contains all the configuration values:

```
# ~/my-fifth-configmap.yaml
apiVersion: v1
kind: ConfigMap
```

```
metadata:
  name: my-fifth-configmap
data:
  color: "blue"
  version: "1"
  environment: "prod"
```

Now, let's learn how to store entire configuration files inside a ConfigMap.

Storing entire configuration files in a ConfigMap

As we mentioned earlier, it's also possible to store complete files inside a ConfigMap – you are not restricted to literal values. The trick is to give the path of a file stored in your filesystem to the `kubectl` command line. The content of the file will then be taken by kubectl and used to populate a parameter in `configmap`.

Having the content of a configuration file stored in a ConfigMap is super useful because you'll be able to mount your ConfigMaps in your Pods, just like you can do with volumes.

The good news is that you can mix literal values and files inside a ConfigMap. Literal values are meant to be short strings, whereas files are just treated as longer strings: they are not two different data types. Here, a sixth ConfigMap is created with a literal value, just like it was previously, but now, we are also going to store the content of a file in it.

Let's create a file called `configfile.txt` in the `$HOME/configfile.txt` location with arbitrary content:

```
$ echo "I'm just a dummy config file" >> $HOME/configfile.txt
```

Here, that configuration file has the `.txt` extension but it could be a `.yaml`, `.toml`, `.rb`, or any other configuration format your application can use.

Now, we need to import that file into a ConfigMap, so let's create a brand new ConfigMap to demonstrate this. You can do this using the `--from-file` flag, which can be used together with the `--from-literal` flag in the same command:

```
$ kubectl create cm my-sixth-configmap --from-
literal=color=yellow --from-file=$HOME/configfile.txt
configmap/my-sixth-configmap created
```

Let's run the `kubectl get cm` command once more to make sure our sixth `configmap` is created. The command will show that it contains two configurations values; in our case, the one created from a literal and the other one created from the content of a file:

```
$ kubectl get cm
NAME                    DATA    AGE
my-fifth-configmap      3       11m
my-first-configmap      0       25m
my-fourth-configmap     3       16m
my-second-configmap     0       24m
my-sixth-configmap      2       38s
my-third-configmap      1       23m
```

As you can see, `my-sixth-configmap` contains two pieces of data: the literal and the file.

Now, let's create a seventh ConfigMap. Just like the sixth one, it's going to contain a literal and a file, but this time, we're going to create it declaratively.

The YAML format allows you to use multiple lines with the | symbol. We're using this syntax as part of our declaration file:

```
# ~/my-seventh-configmap.yaml
apiVersion: v1
kind: ConfigMap
metadata:
  name: my-seventh-configmap
data:
  color: "green"
  configfile.txt: |
    I'm another configuration file.
```

Let's apply this YAML file to create our `configmap` with the `kubectl create` command:

```
$ kubectl create -f my-seventh-configmap.yaml
configmap/my-seventh-configmap created
```

Just to make sure, let's list the ConfigMaps in our cluster using `kubectl get cm` to make sure our seventh `configmap` has been created and contains two values. So, let's run the `kubectl get cm` command once more:

```
$ kubectl get cm
NAME                    DATA    AGE
my-fifth-configmap      3       91m
my-first-configmap      0       105m
my-fourth-configmap     3       96m
my-second-configmap     0       104m
my-seventh-configmap    2       49s
my-sixth-configmap      2       80m
my-third-configmap      1       103m
```

Now, let's discover the last possible way to create a ConfigMap; that is, from an `env` file.

Creating a ConfigMap from an env file

As you can guess, you can create a ConfigMap from an `env` file imperatively using the `--from-env-file` flag.

An `env` file is a `key=value` format file where each key is separated by a line break. This is a configuration format that's used by some applications, so Kubernetes brought a way to generate a ConfigMap from an existing `env` file. This is especially useful if you have an already existing application you want to migrate into Kubernetes.

Here is a typical env file:

```
# ~/my-env-file.txt
hello=world
color=blue
release=1.0production=true
```

By convention, env files are named `.env`, but it's not mandatory. So long as the file is formatted correctly, Kubernetes will be able to generate a ConfigMap based on the parameters.

You can use the following command to import the configuration in the env file as a ConfigMap into your Kubernetes cluster:

```
$ kubectl create cm my-eigth-configmap --from-env-file=my-env-
file.txt
configmap/my-eight-configmap created
```

Lastly, let's list the ConfigMaps in our cluster to check that our new ConfigMap was created with three configuration values:

```
$ kubectl get cm my-eight-configmap
NAME                    DATA    AGE
My-eight-configmap      3       1m
my-fifth-configmap      3       91m
my-first-configmap      0       105m
my-fourth-configmap     3       96m
my-second-configmap     0       104m
my-seventh-configmap    2       49s
my-sixth-configmap      2       80m
my-third-configmap      1       103m
```

As you can see, the new configmap is now available in the cluster, and it was created with the three parameters that were present in the env file. That's a solid way to import your env files into Kubernetes ConfigMaps.

Important Note

Remember that ConfigMaps are not meant to contain sensitive values. Data in ConfigMaps are not encrypted and that's why you can view them with just a kubectl describe cm command. For anything that requires privacy, you'll have to use the Secret object and not the ConfigMap one.

Now, let's discover how to read the values inside a ConfigMap.

Reading values inside a ConfigMap

So far, we've only listed the ConfigMaps to retrieve the number of keys in them. Let's take this a little bit further: you can read actual data inside a ConfigMap, not just get the number of them. This is useful if you want to debug a ConfigMap or if you're not confident about what kind of data is stored in them.

The data in a ConfigMap is not meant to be sensitive so that you can read and retrieve it easily from kubectl: it will be displayed in the terminal's output.

You can read the value in a ConfigMap with the `kubectl describe` command. I will run this command against the `my-fourth-configmap` ConfigMap since it's the one that contains the most data. The output is quite big, but as you can see, the two pieces of configuration data are displayed clearly:

```
$ kubectl describe cm my-fourth-configmap
Name:          my-fourth-configmap
Namespace:     default
Labels:        <none>
Annotations:   <none>
Data
====
color:
----
blue
environment:
----
prod
version:
----
1
Events:   <none>
```

The `kubectl describe cm` command returns these kinds of results. Expect to receive results similar to this one and not results in a computer-friendly format such as JSON or YAML.

As the data is displayed clearly in the terminal output, keep in mind that any user of the Kubernetes cluster will be able to retrieve this data directly by typing the `kubectl describe cm` command, so be careful to not set any sensitive value into a ConfigMap.

Now, let's discover how we can inject ConfigMap data into running Pods as environments variables.

Linking ConfigMaps as environment variables

In this section, we're going to bring our ConfigMaps to life by linking them to Pods. First, we will focus on injecting ConfigMaps as environment variables. Here, we want the environment variables of a container within a Pod to come from the values of a ConfigMap.

You can do this in two different ways:

- **One given value in a given ConfigMap**: You can set the value of an environment variable based on the parameters contained in one or multiple ConfigMaps.

- **All the values contained in a given ConfigMap**: You take one ConfigMap and inject all the values it contains into an environment at once.

The first method brings you flexibility, but it can become more difficult to manage and harder to keep things organized.

The second way is good if you are creating one ConfigMap per Pod specification or application so that each app has a ConfigMap ready to be deployed with.

> **Important Note**
> Please note that it's impossible to link a ConfigMap to a Pod with the kubectl `imperative` method. The reason is that it's impossible to create a Pod referencing a ConfigMap directly from the `kubectl run` command. You will have to write some YAML files to use your ConfigMaps in your Pods.

Earlier in this chapter, we created a ConfigMap called `my-third-configmap` that contains a parameter called `color` with a value of `blue`. In this example, we will create a Pod with the `nginx:latest` image, and we will link `my-third-configmap` to the Pod so that the NGINX container is created with an environment variable called `COLOR` with a value set to `blue`, according to what we have in the ConfigMap. Here is the YAML manifest to achieve that. Pay attention to the `env:` key in the `container` spec:

```
# ~/nginx-Pod-with-configmap.yaml
apiVersion: v1
kind: Pod
metadata:
  name: nginx-Pod-with-configmap
spec:
  containers:
    - name: nginx-container-with-configmap
```

```
    image: nginx:latest
    env:
      - name: COLOR #Any other name works here.
        valueFrom:
          configMapKeyRef:
            name: my-third-configmap
            key: color
```

Now, we can create this Pod using the `kubectl create` command:

```
$ kubectl create -f ~/nginx-Pod-with-configmap.yaml
Pod/nginx-Pod-with-configmap created
```

Now that our NGINX Pod has been created, let's launch the `env` command in it to list all the environments variables that are available in the container. As you may have guessed, we will issue the `env` Linux command in this specific container by calling the `kubectl exec` command. Here is the command and the output to expect:

```
$ kubectl exec Pods/nginx-Pod-with-configmap -- env
PATH=/usr/local/sbin:/usr/local/bin:/usr/sbin:/usr/bin:/sbin:/
bin
HOSTNAME=nginx-Pod-with-configmap
COLOR=blue
KUBERNETES_PORT_443_TCP_PROTO=tcp
KUBERNETES_PORT_443_TCP_PORT=443
KUBERNETES_PORT_443_TCP_ADDR=172.20.0.1
KUBERNETES_SERVICE_HOST=172.20.0.1
KUBERNETES_SERVICE_PORT=443
KUBERNETES_SERVICE_PORT_HTTPS=443
KUBERNETES_PORT=tcp://172.20.0.1:443
KUBERNETES_PORT_443_TCP=tcp://172.20.0.1:443
NGINX_VERSION=1.19.2
NJS_VERSION=0.4.3
PKG_RELEASE=1~buster
HOME=/root
```

You should see the `COLOR` environment variable in the output if your ConfigMap has been linked to your Pod correctly.

Now, we are going to remove the Pod and discover the second way of injecting configmaps as environment variables. We're going to inherit all the parameters from a ConfigMap, not a single one. Start by removing the `nginx-Pod-with-configmap` Pod:

```
$ kubectl delete Pods/nginx-Pod-with-configmap
Pod/nginx-Pod-with-configmap deleted
```

Now, we are going to link another ConfigMap, the one called `my-fourth-configmap`. This time, we don't want to retrieve a single value in this ConfigMap, but all the values inside of it. Here is the updated YAML Pod manifest. This time, we don't use individual env keys, but an `envFrom` key in our `container` spec:

```
# ~/nginx-Pod-with-configmap.yaml
apiVersion: v1
kind: Pod
metadata:
  name: nginx-Pod-with-configmap
spec:
  containers:
    - name: nginx-container-with-configmap
      image: nginx:latest
      envFrom:
        - configMapRef:
            name: my-fourth-configmap
```

Once the manifest file is ready, you can recreate the NGINX Pod:

```
$ kubectl create -f nginx-Pod-with-configfile.yaml
Pod/nginx-Pod-with-configmap created
```

Now, let's run the `env` command once more in the `nginx` container using the `kubectl` `exec` command to list the environment variables:

```
$ kubectl exec Pods/nginx-Pod-with-configmap -- env

PATH=/usr/local/sbin:/usr/local/bin:/usr/sbin:/usr/bin:/sbin:/
bin
HOSTNAME=nginx-Pod-with-configmap
color=blue
environment=prod
```

```
version=1
KUBERNETES_PORT=tcp://172.20.0.1:443
KUBERNETES_PORT_443_TCP=tcp://172.20.0.1:443
KUBERNETES_PORT_443_TCP_PROTO=tcp
KUBERNETES_PORT_443_TCP_PORT=443
KUBERNETES_PORT_443_TCP_ADDR=172.20.0.1
KUBERNETES_SERVICE_HOST=172.20.0.1
KUBERNETES_SERVICE_PORT=443
KUBERNETES_SERVICE_PORT_HTTPS=443
NGINX_VERSION=1.19.2
NJS_VERSION=0.4.3
PKG_RELEASE=1~buster
HOME=/root
```

Note that the three parameters that were declared in the `my-fourth-configmap` ConfigMap have been set as environment variables in the container, but this time, you don't have control over how the environment variables are named in the container: their names are directly inherited from the parameter key names in the ConfigMap.

Now, it's time to learn how to mount a ConfigMap as a volume in a container.

Mounting a ConfigMap as a volume mount

Earlier in this chapter, we created two ConfigMaps that store dummy configuration files. Kubectl allows you to mount a ConfigMap inside a Pod as a volume. This is especially useful when the ConfigMap contains the content of a file that you want to inject into a container's filesystem.

Just like when we injected environment variables, we can't do this imperatively: we need to write a YAML manifest file. Here, we are going to mount a ConfigMap called `my-sixth-configmap` as a volume mount to the `nginx-Pod-with-configmap` Pod we used earlier.

Start by deleting the Pod from the cluster, if it still exists:

```
$ kubectl delete Pods/nginx-Pod-with-configmap
Pod "nginx-Pod-with-configmap" deleted
```

Now, let's update our Pod manifest file and add the ConfigMap as a volume mount to our nginx Pod:

```
# ~/nginx-Pod-with-configmap-volume.yaml
apiVersion: v1
kind: Pod
metadata:
  name: nginx-Pod-with-configmap-volume
spec:
  volumes:
    - name: configuration-volume
      configMap:
        name: my-sixth-configmap # Configmap name goes here
  containers:
    - name: nginx-container-with-configmap
      image: nginx:latest
      volumeMounts:
        - name: configuration-volume # match the volume name
          mountPath: /etc/conf
```

Here, we declared a volume named `configuration volume` at the same level as the containers, and we told Kubernetes that this volume was built from a ConfigMap. The referenced ConfigMap (here, `my-sixth-configmap`) must be present in the cluster when we apply this file. Then, at the container level, we mounted the volume we declared earlier on `path /etc/conf:`. The parameter in the ConfigMap should be present at the specified location.

Let's apply this file to create a new ConfigMap with the volume attached to our cluster:

```
$ kubectl create -f nginx-Pod-with-configmap-volume.yaml
Pod "nginx-Pod-with-configmap-volume" created
```

Run the `ls` command in the container to make sure that the directory has been mounted:

```
$ kubectl exec Pods/nginx-Pod-with-configmap-volume -- ls /etc/
conf
color
configfile.txt
```

Here, the directory has been successfully mounted and both parameters that were created in the ConfigMap are available in the directory as plain files.

Let's run the cat command to make sure both files hold the correct values:

```
$ kubectl exec Pods/nginx-Pod-with-configmap-volume -- cat /
etc/conf/color
yellow
```

```
$ kubectl exec Pods/nginx-Pod-with-configmap-volume -- cat /
etc/conf/configfile.txt
I'm just a dummy config file
```

Good! Both files contain the values that were declared earlier when we created the ConfigMap! For example, you could store a virtual host NGINX configuration file and have them mounted to the proper directory so that NGINX could serve your website based on the configuration values hosted in a ConfigMap. That's how you can override the default configuration and cleanly manage your app in Kubernetes. Now, you have a really strong and consistent interface for managing and configuring the containers running in Kubernetes.

Next, we will learn how to delete and update a ConfigMap.

Deleting a ConfigMap

Deleting a ConfigMap is very easy. However, be aware that you can delete a ConfigMap even if its values are used by a container. Once the Pod has been launched, it's independent of the ConfigMap object:

```
$ kubectl delete cm my-first-configmap
configmap "my-first-configmap" deleted
```

Regardless of whether the ConfigMap's values are used by the container, it will be deleted as soon as this command is entered. *Note that the ConfigMap cannot be recovered*, so please think twice before removing a ConfigMap you have created imperatively since you won't be able to recreate it. Unlike declaratively created ConfigMaps, its content is not stored in a YAML file.

Also, I recommend that you are careful when removing your ConfigMaps, especially if you delete ConfigMaps that are used by running Pods. If your Pod were to crash, you won't be able to relaunch it without updating the manifest file: the Pods would look for the missing ConfigMap you deleted.

Updating a ConfigMap

There is no real way to update a ConfigMap once it's been created. To update a ConfigMap, you'll have to delete it, update its declarative YAML file by adding or removing the parameters you want, and then recreate it.

We are now done with ConfigMaps. You should be able to manage them like a pro now!

Managing sensitive configuration with the Secret object

The Secret object is a resource that allows you to configure applications running on Kubernetes. Secrets are extremely similar to ConfigMaps and can be used together. The difference is that Secrets are encoded and intended to host sensitive data such as passwords, tokens, or private API keys, while ConfigMaps are intended to host non-sensitive configuration data. Other than that, Secrets and ConfigMaps mostly behave the same.

Let's start by discovering how to list the Secrets that are available in your Kubernetes cluster.

Listing Secrets

Like any other Kubernetes resource, you can list secrets using the `kubectl get` command. The resource identifier is a secret here:

```
$ kubectl get secret
```

Just like with ConfigMaps, the DATA column tells you the number of sensitive parameters that have been hashed and saved in your secret.

The `--from-literal` flag can be used to fill a Secret object at creation time, the same as for ConfigMaps.

Creating a Secret imperatively with --from-literal

You can create a Secret imperatively or declaratively – both methods are supported. Let's start by discovering how to create a Secret imperatively. Here, we want to store a database password, `my-db-password`, in a Secret object in our Kubernetes cluster. You can achieve that imperatively with kubectl by adding the `--from-literal` flag to the `kubectl create secret` command:

```
$ kubectl create secret generic my-first-secret --from-
literal='db_password=my-db-password'
```

Now, run on the `kubectl get secrets` command to retrieve the list of Secrets in your Kubernetes cluster. The new Secret should be displayed:

```
$ kubectl create secret generic my-first-secret --from-
literal='db_password=my-db-password'
```

Now, let's figure out how to create a Secret declaratively.

Creating a Secret declaratively with a YAML file

It is also possible to create a secret declaratively from a YAML file. However, you will have to include an additional step: you will have to encode your secret parameter in a `base64`. The reason is we would need to include your secret value in a YAML file, and since these values are supposed to be sensitive, we do not want the hardcoded value to appear in the YAML file.

When you use `--from-literal`, Kubernetes will encode your strings in `base64` itself, but when you create a Secret from a YAML manifest file, you will have to handle this step yourself.

So, let's start by converting the `my-db-password` string into `base64`:

```
$ echo 'my-db-password' | base64
bXktZGItcGFzc3dvcmQK
```

`bXktZGItcGFzc3dvcmQK` is the `base64` representation of the `my-db-password` string. And that's what we will need to write in our YAML file. Here is the content of the YAML file for creating the Secret object properly:

```
# ~/secret-from-file.yaml
apiVersion: v1
kind: Secret
metadata:
```

```
   name: my-second-secret
type: Opaque
data:
   db_password: bXktZGItcGFzc3dvcmQK
```

Once this file has been stored on your system, you can create the secret using the `kubectl create` command:

```
$ kubectl create -f ~/secret-from-file.yaml
```

Now, we can make sure that the secret has been created properly by listing the secrets in our Kubernetes cluster:

```
$ kubectl get secret
```

Now, let's discover another Kubernetes feature: the ability to create a secret with values from a file.

Creating a Secret with content from a file

We can create a Secret with values from a file, the same as we did with ConfigMaps. We start by creating a file that will contain our secret value. Let's say that we have to store a password in a file and import it as a Secret object in Kubernetes:

```
echo -n 'mypassword' > ./password.txt
```

After running this command, we have a file called `password.txt` that contains a string, `mypassword`, that is supposed to be our secret value. The `-n` flag is being used here to ensure that `password.txt` does not contain any extra blank lines at the end of the text.

Now, let's run the `kubectl create secret` command by passing the location of `password.txt` to the `--from-file` flag. This will result in a new secret containing a `base64` representation of the `mypassword` string being created:

```
$ kubectl create secret generic mypassword --from-file=./
password.txt
secret/mypassword created
```

This new secret is now available in your Kubernetes cluster! Now, let's learn how to read a Kubernetes Secret.

Reading a Secret

Because Secrets are supposed to host sensitive data, the kubectl output won't show you the secret decoded data. You'll simply have access to the key. Why? Let's take a look:

- To prevent the secret from being accidentally opened by someone who shouldn't be able to open it

- To prevent the secret from being displayed as part of a terminal output, which could result in it being logged somewhere

Because of these securities, you simply won't be able to retrieve the actual content of a secret, but you can still grab information about its size and so on.

You can do this using the `kubectl describe` command, just like we did earlier for ConfigMaps. As we mentioned previously, ConfigMaps and Secrets are very similar; they almost behave the same:

```
$ kubectl describe secret/mypassword


Name: db-user-pass
Namespace: default
Type: Opaque
Data

===
password: 10 bytes
```

Do not get confused if your output is a little different than this one. If you receive something similar, it means that the new secret is available in your Kubernetes cluster and that you successfully retrieved its data!

Now that we know how to create and read a secret, let's learn how to inject one into our Pods.

Consuming a Secret as an environment variable

In the same way that we could inject the values of a ConfigMap into a Pod in the form of environment variables, we can do the same with secrets. Returning to the example with our NGINX container, we are going to retrieve the db_password value of the my-first-secret secret and inject it as an environment variable into the Pod. Here is the updated YAML manifest. Again, everything occurs under env: key:

```
# ~/nginx-Pod-with-secret-env-variable.yaml
apiVersion: v1
metadata:
  name: nginx-Pod-with-secret-env-variable
  namespace: default
spec:
  containers:
    - name: nginx-container
      image: nginx:latest
      env:
    - name: PASSWORD_ENV_VAR # Name of env variable
      valueFrom:
        secretKeyRef:
          name: mypassword # Name of secret object
          key: password # Name of key in secret object
```

Now, you can apply this file using the kubectl create command:

```
kubectl create -f nginx-Pod-with-secret-env-variable.yaml
Pod/nginx-Pod-with-secret-env-variable created
```

Now, run the env command to list the environment variables in your container:

```
$ kubectl exec Pods/nginx-Pod-with-secret-env-variable - env /
...
PASSWORD_ENV_VAR=mypassword
...
```

As you can see, the mypassword string is available as the environment variable. Now, let's how to consume a secret as a volume mount.

Another example to know about when reading a secret from a Pod is the `envFrom` YAML key. When using this key, you'll read all the values from a Secret and get them as environment variables in the Pod all at once. It works the same as for the ConfigMap object. Here is the preceding example but updated with an `envFrom` key:

```yaml
# ~/nginx-Pod-with-secret-envfrom.yaml
apiVersion: v1
metadata:
  name: nginx-Pod-with-secret-env-variable
  namespace: default
spec:
  containers:
    - name: nginx-container
      image: nginx:latest
      envFrom:
      - secretRef:
        name: mysecret # Name of the secret object
```

Using this, all the keys in the Secret object will be used as environment variables within the Pod. Note that if a key name cannot be used as an environment variable name, then it will be simply ignored!

Consuming a Secret as a volume mount

You can mount secrets as volume to your Pods, but you can only do so declaratively. So, you'll have to write some YAML files to do this successfully.

You have to start from a YAML manifest file that will create a Pod. Here is a YAML file that mounts a secret called `mypassword` in the `/etc/passwords-mounted-path/password` location:

```yaml
# ~/nginx-Pod-with-secret-volume.yaml
ApiVersion: v1
Metadata:
  Name: nginx-Pod-with-secret-volume
Spec:
  containers:
    - name: nginx-container
      Image: nginx:latest
      VolumeMounts:
```

```
    - name: mysecretVolume # Name of the volume
        MountPath: /etc/password-mounted-path
  Volumes:
  - name: mysecretVolume # Name of the volume
    Secret:
      SecretName: mypassword # name of secret
```

Once you have created this file on your filesystem, you can apply the YAML file using `kubectl`:

```
kubectl create -f nginx-Pod-with-secret-volume.yaml
Pod/nginx-Pod-with-secret-volume created
```

Please make sure that the `mypassword` secret exists before you attempt to create the secret.

Finally, you can run a command inside `nginx-container` using the `kubectl exec` command to check if the volume containing the secret was set up correctly:

```
$ kubectl exec Pods/nginx-Pod-with-secret-volume -- cat /etc/
password-mounted-path/password
mypassword
```

As you can see, the `mypassword` string is displayed correctly: the secret was correctly mounted as a volume!

Deleting a Secret

Deleting a secret is very simple and can be done via the `kubectl delete` command:

```
$ kubectl delete secret my-first-secret
secret "my-first-secret" deleted
```

Now, let's learn how to update an existing secret in a Kubernetes cluster.

Updating a Secret

Lastly, note that there is no clean and consistent way to modify a secret once it has been created through kubectl.

To update a secret, you will need to delete it and then recreate it with the new values you need by following the examples given previously. Make sure that the volume is not referenced by the YAML manifest file as it may result in an error when you create Pods.

Summary

This chapter was one of the most important ones because we discovered how to properly configure our contained applications in Kubernetes.

In particular, we explained why it is so important to decouple your applications from their configuration. Kubernetes in particular emphasizes this principle by differentiating the applications, which are Pods, and their configurations, which are ConfigMaps and Secrets. Use these three objects wisely to build portable applications between environments.

In the next chapter, we will continue discovering Kubernetes by tackling another central concept of Kubernetes, which are Services. Services are Kubernetes objects that allow you to expose your Pods to each other, but also the internet: this is a very important network concept for Kubernetes, and mastering it is essential to use the orchestrator correctly. Fortunately, mastering Services is not very complicated and the next chapter will explain how to achieve this. You will learn how to associate ports of the container with the ports of the worker node it is running on, and also how to associate a static IP with your Pods so that they can always be reached at the same address for other Pods in the cluster.

7
Exposing Your Pods with Services

After reading the previous chapters, you now know how to deploy Docker applications on Kubernetes by building Pods that can contain one container or multiple containers for more complex applications. You also know that it is possible to decouple applications from their configuration by using Pods and configmaps together, and Kubernetes is also capable of storing your sensitive configurations thanks to Secret objects.

The good news is that with these three resources, you can start deploying applications on Kubernetes properly and get your first app running on Kubernetes. However, you are still missing something important: you need to be able to expose Pods to end users or even to other Pods within the Kubernetes cluster. This is where Kubernetes services come in, and that's the concept we're going to discover now!

In this chapter, we'll learn about a new Kubernetes resource kind called Service. Since Kubernetes services are a quite big topic with a lot of things to cover, this chapter is going to be quite big with a lot of information. But after you master these services, you're going to be able to expose your Pods and get your end users to your apps!

Services are also a key concept to master **high availability** (**HA**) and redundancy in your Kubernetes setup. In a word: it is crucial to master them to be effective with Kubernetes!

In this chapter, we're going to cover the following main topics:

- Why would you want to expose your Pods?
- The `NodePort` service
- The `ClusterIP` service
- The `LoadBalancer` service
- Implementing `ReadinessProbe`
- Securing your Pods using the `NetworkPolicy` object

Technical requirements

To follow along with the examples in this chapter, make sure you have the following:

- A working Kubernetes cluster (whether this is local or cloud-based has no importance)
- A working `kubectl` **command-line interface** (**CLI**) configured to communicate with the Kubernetes cluster

Why would you want to expose your Pods?

In the previous chapters, we discussed the microservice architecture, which offers to expose your functionality through **REpresentational State Transfer** (**REST**) **application programming interfaces** (**APIs**). These APIs rely completely on the **HyperText Transfer Protocol** (**HTTP**) protocol, which means that your microservices must be accessible via the web, and thus via an **Internet Protocol** (**IP**) address on the network.

In this section, we will explain what Services are in Kubernetes. Next, we'll explain what they're used for and how they can help you expose your Pod-launched microservices.

Understanding Pod IP assignment

To understand what Services are, we need to talk about Pods for a moment once again. On Kubernetes, everything is Pod management: Pods host your applications, and they have a special property. Kubernetes assigns them a private IP address as soon as they are created on your cluster. Keep that in mind because it is super important: each Pod created in your cluster has its unique IP address assigned by Kubernetes.

To illustrate this, we'll start by creating an nginx Pod. We're using an nginx Pod here, but in fact, it would be the same for any type of Pod, no matter which Docker image is used to create it.

Let's do that using the declarative way with the following **YAML Ain't Markup Language (YAML)** definition:

```
# ~/new-nginx-pod.yaml
apiVersion: v1
kind: Pod
metadata:
  name: new-nginx-pod
spec:
  containers:
  - image: nginx:latest
    name: new-nginx-container
    ports:
    - containerPort: 80
```

As you can see from the previous YAML, this Pod called `new-nginx-pod` has nothing special and will just launch a container named `new-nginx-container` based on the `nginx:latest` Docker image.

Once we have this YAML file, we can apply it using the following command to get the Pod running on our cluster:

```
$ kubectl create -f ~/new-nginx-pod.yaml
pod/new-nginx-pod.yaml created
```

As soon as this command is called, the Pod gets created on the cluster, and as soon as the Pod is created on the cluster, Kubernetes will assign it an IP address that will be unique.

Let's now retrieve the IP address Kubernetes assigned to our Pod. To do that, we can use the `kubectl get pods` command to list the Pods. In the following code snippet, please note the usage of the `-o wide` option that will display the IP address as part of the output:

```
$ kubectl get pods -o wide
NAME            READY   STATUS    AGE   IP             NODE
new-nginx-pod   1/1     Running   5s    10.0.102.212   wkn1
```

Please note that the IP address is added to the output of the command only because we're adding the -o wide option. Otherwise, the IP address is not present in the shorter output version. In our case, the IP address is 10.0.102.212. This IP will be unique on my cluster and is assigned to this unique Pod.

Of course, if you're following along on your cluster, you will have a different IP. This IP is a private IP **version 4** (**v4**). It only exists in the Kubernetes cluster. If you try to type this IP into your web browser, you won't get anything because this idea does not exist on the public internet—it only exists within your Kubernetes cluster.

Depending on the cloud platform and **container network interface** (**CNI**) you use, this network might be an **Amazon Web Services** (**AWS**), **Google Cloud Platform** (**GCP**), or Azure **virtual private cloud** (**VPC**), or the basic flat network used by Kubernetes. In all cases, this principle is the same: each Pod is assigned its own IP address, and that is an absolute rule.

At this point, you must understand that all Pods get a unique IP address as soon as they are created on the cluster and that this IP address is a private one. If that sounds clear to you, we can move on with this important topic: we will now discover that this IP address assignment is dynamic, and the issues it can cause at scale.

Understanding Pod IP assignment is dynamic

Now that you're aware that Kubernetes assigns an IP address to each Pod when it is created, you must know that its IP address assignment is a dynamic one.

Indeed, these IP addresses are not static, and if you delete and recreate a Pod, you're going to see that the Pod will get a new IP address that's totally different from the one used before. And if that is the case, even if the Pod is recreated with the exact same YAML configuration.

To demonstrate that, let's delete the Pod and recreate it using the same YAML file, as follows:

```
$ kubectl delete -f ~/new-nginx-pod.yaml
pod/new-nginx-pod.yaml deleted
$ kubectl create -f ~/new-nginx-pod.yaml
pod/new-nginx-pod.yaml created
```

We can now run once more the `kubectl get pods -o wide` command to figure out that the new IP address is not the same as before, as follows:

```
$ kubectl get pods -o wide
NAME             READY    STATUS     AGE    IP               NODE
new-nginx-pod    1/1      Running    5s     10.0.102.104     wkn1
```

Now, the IP address is `10.0.102.104`. This IP is different from the one we had before, `10.0.102.212`.

As you can see, when a Pod is destroyed and then recreated, even if you recreate it with the same name and the same configuration, it's going to have a different IP address.

The reason is that technically, it is not really the same Pod but two different Pods: that is why Kubernetes assigns two completely different IP addresses. Now, imagine you wrote a script to do some actions on that nginx Pod using its IP address to communicate with it. If that Pod gets deleted and recreated for some reason, then your script is broken.

Never hardcode a pod's IP addresses in your application code

In a world where you were in a production environment, this would be a real problem: microservice applications are designed to interact with each other through the HTTP protocol, which relies on **Transmission Control Protocol (TCP)**/IP.

So, in any form, you need to be able to find a way to establish a reliable way to retrieve a pod's IP addresses at any time.

The golden rule here is to never rely on the IP address of a Pod to access it directly. As Pods can be easily deleted, recreated, or rescheduled to another worker node, hardcoding IP addresses is a bad idea: if a Pod is destroyed and then recreated, any application that needs to communicate with it would no longer be able to call it because the IP assigned would resolve nothing.

There are very concrete cases that we can give where this problem can arise, as follows:

- A Pod running an *A* microservice has a dependency and calls a *B* microservice that is running as another Pod on the same Kubernetes cluster.
- An application running as a Pod needs to retrieve some data from a MySQL server also running as a Pod on the same Kubernetes cluster.
- An application uses a Redis cluster as a caching engine deployed in multiple Pods on the same cluster.

- Your end users access an application by calling an IP address, and that IP changes because of a Pod failure.

- You set up a **Domain Name System** (**DNS**) server that uses a pod's IP address as DNS entries.

Any time you have an interconnection between services or—more widely—any network communication, this problem will arise.

The solution to this issue is the usage of the Service resource kind.

This new object will act as an intermediate object that will remain on your cluster. Services are not meant to be destroyed, and they recreate often. In fact, they can remain on your cluster long-term without causing any issue. They are meant to act as a proxy in between. In fact, they are the core object for networking and load balancing in Kubernetes.

Understanding how services route traffic to Pods

Kubernetes services are not Docker containers or Pods. Kubernetes services are resources running within your Kubernetes cluster, and they are used to create appropriate IPTABLES to ensure the traffic is properly redirected to backend Pods.

The idea is simple: just as Pods get a dynamic IP address at creation time, each service gets a static DNS that can be resolved from anywhere on the cluster. That static DNS entry will never change if the service remains on the cluster. You'll simply tell each service which Pod they should serve traffic to, and that's pretty much it.

In one word: you can consider a service as a proxy with a static DNS name you place in front of your Pods to serve traffic to. This way, you get a static and reliable way to access your Pods, as depicted in the following screenshot:

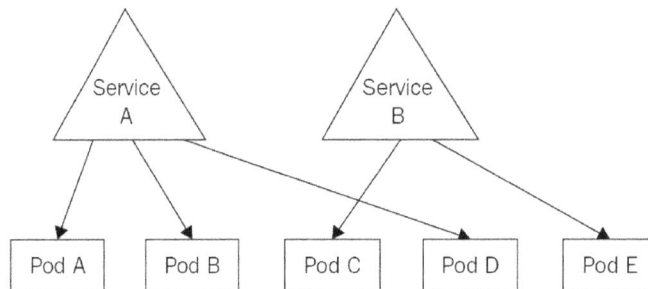

Figure 7.1 – Service A is exposing Pods A, B, and D, whereas Service B is exposing Pods C and E

In fact, Services are deployed to Kubernetes as a resource kind, and just as with most Kubernetes objects, you can deploy them to your cluster using interactive commands or declarative YAML files.

To keep things simple and to sum up: you can consider Services to be load balancers internal to your Kubernetes cluster, built to provide a consistent and static network gateway to your Pod.

When you create a Service, you'll have to give it a name. This name will be used by Kubernetes to build a DNS name that all Pods on your cluster will be able to call. This DNS entry will resolve to your Service, which is supposed to remain on your cluster. The only part that is quite tricky at your end will be to give a list of Pods to expose to your services: we will discover how to do that in this chapter. Don't worry—it's just configuration based on labels and selectors.

Once everything is set up, you can just reach the Pods by calling the service. This service will receive the requests and forward them to the Pods. And that's pretty much it!

Understanding round-robin load balancing in Kubernetes

Kubernetes services, once configured properly, can expose one or several Pods. When multiple Pods are exposed by the same Pod, the requests are evenly load-balanced to the Pods behind the service using the round-robin algorithm, as illustrated in the following screenshot:

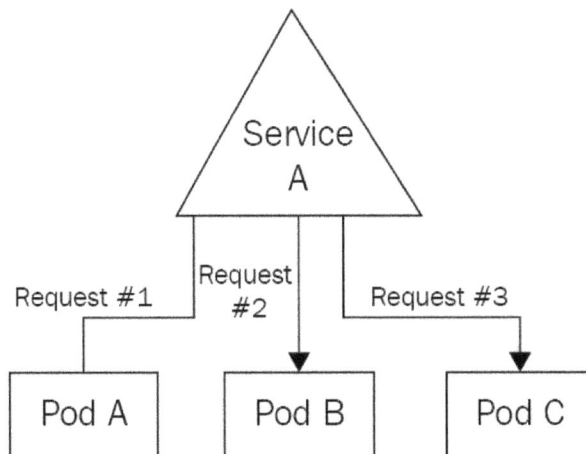

Figure 7.2 – Service A proxies three requests to the three Pods behind it. At scale, each service will receive 33% of the requests received by the service

Scaling applications becomes easy. Adding more Pods behind the service will be enough. As the Kubernetes service has round-robin logic implemented, it can proxy requests evenly to the Pods behind it.

If the preceding Pod had four Pods, then, each of them would receive roughly 25% of all the requests the service received. If 50 Pods were behind the service, each of them would receive roughly 2% of all the requests received by the service. That's pretty much it. All you need to understand is that services are behaving like load balancers you can find on the cloud by following the round-robin algorithm.

Let's now discover how you can call a service in Kubernetes from another Pod.

Understanding how to call a service in Kubernetes

When you create a Service in Kubernetes, it will be attached two very important things, as follows:

- An IP address that will be unique and specific to it (just as Pods get their own IP)
- An automatically generated DNS name that won't change and is static

You'll be able to use any of the two in order to reach the service, which will then forward your request to the Pod it is attached to. Most of the time, though, you'll call the service by its generated domain, which is easy to determine and predictable. Let's discover how Kubernetes assigns DNS names to services.

Understanding how DNS names are generated for services

The DNS name generated for a service is derived from its name. For example, if you create a Service named `my-app-service`, its DNS name will be `my-app-service.default.svc.cluster.local`.

This one is quite complicated, so let's break it into smaller parts, as follows:

`my-app-service`	This is basically the name you gave to your service. It is the first part of the DNS name.
`default`	`default` is the name of the Kubernetes namespace where the service is declared. We did not introduce Kubernetes namespaces so for the moment, this value will be `default`... as we're working on the default namespace by default.
`svc`	`svc` means service.
`cluster.local`	`cluster.local` is the top-level DNS. It means that the name will only resolve when you're within your Kubernetes cluster. These DNS names should only be used for calls between pods within the same cluster.

This table should help you distinguish all the parts of the domain name. The two moving parts are the two first ones, which are basically the service name and the namespace where it lives. The DNS name will always end with the `.svc.local` string.

So, at any moment, from anywhere on your cluster, if you try to use `curl` or `wget` to call the `my-app-service.default.svc.cluster.local` address, you know that you'll reach your service.

That name will resolve to the service as soon as it's executed from a Pod within your cluster. But by default, services won't proxy to anything if they are not configured to retrieve a list of Pods they will proxy. We will now discover how to do that!

How services get a list of the Pods they service traffic to

When working with services in Kubernetes, you will often come across the idea of *exposing* your Pods.

Indeed, this is the terminology Kubernetes uses to tell that a service is proxying network traffic to Pods. We say a service is exposing Pods. That terminology is everywhere: your colleagues may ask you one day: "*Which service is exposing that Pod?*" The following screenshot shows Pods being exposed:

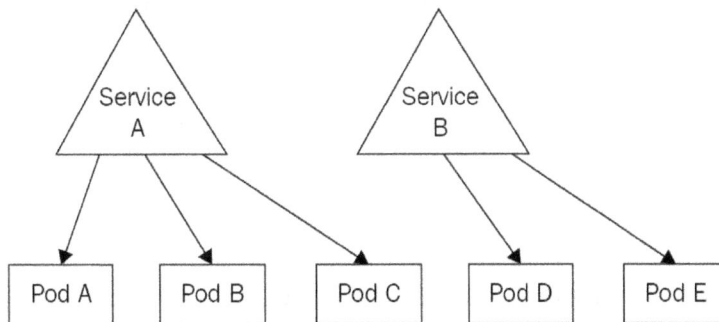

Figure 7.3 – Here, Pods A, B, and C are exposed by service A, whereas Pods D and E are exposed by service B

That terminology is also the one used by `kubectl` and the one used in Kubernetes official documentation in general. You can successfully create a Pod and a service to expose it using `kubectl` in literally one command, using the `--expose` parameter. The following example will create an nginx Pod with a service exposing it.

For the sake of this example, we will also need to provide a port to the command to tell on which port the service will be accessible:

```
$ kubectl run nginx --image nginx:latest --expose  --port 80
service/nginx created
pod/nginx created
```

Let's now list the Pods and the services using kubectl to demonstrate that the following command created both objects:

```
$ kubectl get pods
NAME     READY    STATUS     RESTARTS    AGE
nginx    1/1      Running    0           2m44s


$ kubectl get services
NAME     TYPE         CLUSTER-IP        EXTERNAL-IP    PORT(S)    AGE
nginx    ClusterIP    10.98.191.187     <none>         80/TCP
2m28s
nginx                                   ClusterIP      10.98.191.187
<none>                   80/TCP        2m9s
```

As you can see based on the output of the command, both objects were created.

We've said earlier that Services can find Pods they have to expose based on Pods' labels. That being said, the nginx Pod we just created surely has some labels. To show them, let's run the kubectl get pods nginx --show-labels command. In the following snippet, pay attention to the --show-labels parameter, which will display the labels as part of the output:

```
$ kubectl get pods nginx --show-labels
NAME     READY    STATUS     RESTARTS    AGE       LABELS
nginx    1/1      Running    0           3m27s     run=nginx
```

As you can see, an nginx Pod was created with a label called run, with a value of nginx. Let's now describe the nginx service. It should have a selector that matches this label. The code is illustrated here:

```
$ kubectl describe svc nginx
Name:            nginx
Namespace:       default
Labels:          <none>
```

```
Annotations:          <none>
Selector:             run=nginx
Type:                 ClusterIP
IP:                   10.98.191.187
Port:                 <unset>    80/TCP
TargetPort:           80/TCP
Endpoints:            172.17.0.8:80
```

I did remove some lines of the output to make it shorter. But you can clearly see that the service has a line called `Selector` that matches the label assigned to the `nginx` Pod. This way, the link between the two objects is made. We're now 100% sure that the service can reach the `nginx` Pod and that everything should work normally.

Though it works, I strongly advise you to never do that in production. Indeed, services are very customizable objects, and the `--expose` parameter is hiding a lot of their features. Instead, you should really use declarative syntax and tweak the YAML to fit precise needs.

Let's demonstrate that by using the `dnsutils` Docker image.

Using the dnsutils Docker image to debug your services

One very nice image to discover is the `dnsutils` one. It is extremely useful to debug your service.

Indeed, as your services are created within your cluster, it is often hard to access them, especially if our Pod is meant to remain accessible only within your cluster, or if your cluster has no internet connectivity, and so on.

In this case, it is good to deploy a Pod in your cluster with just some binaries installed into it to run basic networking commands such as `wget`, `nslookup`, and so on. One image that meets this need well is the `dnsutils` one. You can easily test your services using this image.

Here, we're going to curl the `nginx` Pod home page by calling the service that is exposing the Pod. That service's name is just `nginx`. Hence, we can forge the DNS name Kubernetes assigned to it: `nginx.default.svc.cluster.local`.

If you try to reach this **Uniform Resource Locator** (**URL**) from a Pod with your cluster, you should successfully reach the nginx home page. Let's run the following command to launch the dnsutils Pod on our cluster:

```
$ kubectl apply -f https://k8s.io/examples/admin/dns/dnsutils.
yaml
pod/dnsutils created
```

Let's now run a kubectl get pods command in order to verify the Pod was launched successfully, as follows:

```
$ kubectl get pods/dnsutils
NAME        READY    STATUS     RESTARTS    AGE
dnsutils    1/1      Running    0           56s
```

That's perfect! Let's now run a nslookup command against the service DNS name, as follows:

```
$ kubectl exec -ti dnsutils -- nslookup  nginx.default.svc.
cluster.local
Server:         10.96.0.10
Address:    10.96.0.10 # This address is only resolvable from
within the Kubernetes cluster, via local kube-dns or CoreDNS.

Name: nginx.default.svc.cluster.local
Address: 10.98.191.187
```

Everything looks good. Let's now run a wget command to check if we can retrieve the nginx home page, as follows:

```
$ kubectl exec -ti dnsutils -- wget nginx.default.svc.cluster.
local
Connecting to nginx.default.svc.cluster.local
(10.98.191.187:80)
index.html              100% |*******************************|
615    0:00:00 ETA
```

The wget command seems to have downloaded the index.html file properly. I can now display its content using cat, like this:

```
$ kubectl exec -ti dnsutils -- cat index.html
<html>
```

```
<head>
<title>Welcome to nginx!</title>
<style>
html { color-scheme: light dark; }
body { width: 35em; margin: 0 auto;
font-family: Tahoma, Verdana, Arial, sans-serif; }
</style>
</head>
<body>
<h1>Welcome to nginx!</h1>
```

Everything is perfect here! We successfully called the nginx service by using the dnsutils service, as illustrated in the following screenshot:

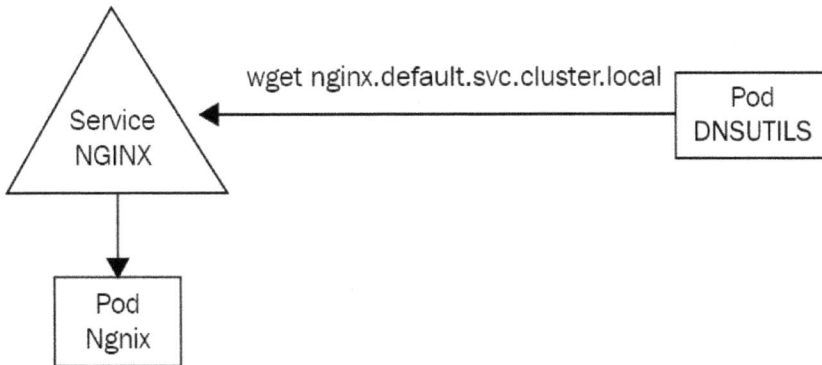

Figure 7.4 – The dnsutils Pod is used to run wget against the nginx service to communicate with the nginx Pod behind the service

Keep in mind that you need to deploy a dnsutils Pod inside the cluster to be able to debug the service. Indeed, the nginx.default.svc.cluster.local DNS name is not a public one and can be only accessible from within the cluster.

Let's explain why you should not use the --expose flag to expose your Pods now.

Why you shouldn't use the --expose flag

The --expose flag should not be used to create services because you won't get much control over how the service gets created. By default, --expose will create a ClusterIP service, but you may want to create a NodePort service.

Defining the service type is also possible using the imperative syntax, but in the end, the command you'll have to issue is going to be very long and complex to understand. That's why I encourage you to not use the `--expose` flag and stick with declarative syntax for complex objects such as services.

Let's now discuss how DNS names are generated in Kubernetes when using services.

Understanding how DNS names are generated for services

If you understood everything we said before, you now know that Kubernetes services-to-pod communication relies entirely on labels on the Pod side and selectors on the service side.

If you don't use both correctly, communication cannot be established between the two.

The workflow goes like this:

1. You create some Pods and you set some labels arbitrarily.
2. You create a service and configure its selector to match the Pods' labels.
3. The service starts and looks for Pods that match its selector.
4. You call the service through its DNS or its IP (DNS is way easier).
5. The service forwards the traffic to one of the Pods that matches its labels.

If you look at the previous example achieved using the imperative style with the `--expose` parameter, you'll notice that the Pod and the services were respectively configured with proper labels (on the Pod side) and selector (on the service side), which is why the Pod is successfully exposed.

We're almost done with Services theory, but I strongly advise you to not use the `--expose` parameter. The reason is that services are complex objects with a lot of configurations accessible behind the scenes, and to access all the possible options, it is better to write YAML files.

Besides that, you must understand now that there is not one but several types of services in Kubernetes. Indeed, the `Service` object can be configured to achieve different types of Pod exposure, and the `--expose` parameter only allows for the simplest one. Using the `--expose` parameter is good for tests, but do remember that you should always use YAML files to really have access to all features of Kubernetes.

That being said, let's now explain the different types of services in Kubernetes.

Understanding the different types of services

There are several types of services in Kubernetes.

Although there is only one kind called Service in Kubernetes, that kind can be configured differently to achieve different results.

Fortunately for us, no matter which type of service you choose, the goal remains the same: to expose your Pods using a single static interface.

Each type of service has its own function and its own use, so basically, there's one service for each use case. A service cannot be of multiple types at once, but you can still expose the same Pods by two services' objects with different types… as long as the services' objects are named differently so that Kubernetes can assign different DNS names.

In this chapter, we will discover the three main types of services, as follows:

- `NodePort`: This one binds a port from an ephemeral port range of the host machine (the worker node) to a port on the Pod, making it available publicly. By calling the port of the host machine, you'll reach the associated Kubernetes Pod. That's the way to reach your Pods for traffic coming from outside your cluster.

- `ClusterIP`: The `ClusterIP` service is the one that should be used for private communication between Pods within the Kubernetes cluster. This is the one we experimented with in this chapter and is the one created by the `--expose` flag by default. This is certainly the most used of them all because it allows inter-communication between Pods: as its name suggests, it has a static IP that is set cluster-wide. By reaching its IP address, you'll be redirected to the Pod behind it. If more than one Pod is behind it, the `ClusterIP` service will provide a load-balancing mechanism following the round-robin algorithm. Even if this service type is called `ClusterIP`, we generally call it by its generated DNS name.

- `LoadBalancer`: The service of the `LoadBalancer` type is quite complex. It will detect the cloud computing platform the Kubernetes cluster is running on and will create a load balancer on the cloud. If you're running on AWS, it will create an **Elastic Load Balancer** (**ELB**). If you run vanilla Kubernetes, it will just create nothing… This service is useless outside of a cloud platform. In general, people tend to prefer to use other services such as Terraform to provision their cloud infrastructure, making this service certainly the least used of them all.

As all of this might seem a little bit unclear to you, let's immediately dive into the first type of service—the `NodePort` one.

As mentioned earlier, this one is going to be super useful to access our Pods from outside the cluster, by attaching Pods to the worker node's port.

The NodePort service

Now that we are aware of the theory behind the concept of services in Kubernetes, we will start by discovering the first type, which is NodePort.

NodePort is a Kubernetes service type designed to make Pods reachable from a port available on the host machine, the worker node. In this section, we're going to discover this type of port and be fully focused on NodePort services!

Why do you need NodePort services?

We're now going to discover exactly where the NodePort services sit and what they do in your Kubernetes cluster.

The first thing to understand is that NodePort services allow us to access a Pod running on a worker node, on a port of the worker node itself. After you expose Pods using the worker node, you'll be able to reach the Pods if you know the IP address of your worker node and the port of the NodePort service in this format:

```
<WORKER_NODE_IP_ADDRESS>:<PORT_DECLARED_IN_NODE_PORT_SVC>.
```

Let's illustrate all of this by declaring some Kubernetes objects.

Most of the time, the NodePort service is used as an entry point to your Kubernetes cluster. In the following example, I will create two Pods based on the containous/whoami Docker image available on the Docker Hub, which is a very nice Docker image that will simply print the container hostname.

I'll create two Pods so that I get two containers with different hostnames, and I'll expose them using a NodePort service.

Creating two containous/whoami Pods

Let's start by creating two Pods without forgetting about adding one or more labels because we will need labels to tell the service which Pods it's going to expose.

I'm also going to open the port on the Pod side. That won't make it exposed on its own, but it will open a port the service will be able to reach. The code is illustrated here:

```
$ kubectl run whoami1 -image=containous/whoami -port 80 -
labels="app=whoami"


pod/whoami1 created
```

```
$ kubectl run whoami1 –image=containous/whoami -port 80 –
labels="app=whoami"
```

```
pod/whoami2 created
```

Now, I can run a `kubectl get pods` command in order to verify that our two Pods are correctly running. I can also add the `--show-labels` parameter in order to display the labels as part of the command output, as follows:

```
$ kubectl get pods --show-labels
NAME        READY    STATUS      RESTARTS    AGE      LABELS
whoami1     1/1      Running     0           5m4s     app=whoami
whoami2     1/1      Running     0           4m59s    app=whoami
```

Everything seems to be okay! Now that we have two Pods created with a label set for each of them, we will be able to expose them using a service. We're now going to discover the YAML manifest file that will create the `NodePort` service to expose these two Pods.

Understanding NodePort YAML definition

Unlike most Kubernetes clusters, it is impossible to create services of the `NodePort` type using the declarative way. And since services are quite complex resources, it is better to create them using a YAML file rather than direct command input.

Here is the YAML file that will expose the `whoami1` and `whoamo2` Pods using a `NodePort` service:

```
# ~/nodeport-whoami.yaml
apiVersion: v1
kind: Service
metadata:
  name: nodeport-whoami
spec:
  type: NodePort
  selector:
    app: whoami
  ports:
  - nodePort: 30001
    port: 80
    targetPort: 80
```

This YAML can be difficult to understand because it refers to three different ports as well as a `selector` block.

Before explaining the YAML file, let's apply it and check if the service was correctly   ·
created afterward, as follows:

```
$ kubectl create -f nodeport-whoami.yaml
service/nodeport-whoami created


$ kubectl get services


NAME                    TYPE        CLUSTER-IP      EXTERNAL-IP
PORT(S)            AGE


nodeport-whoami    NodePort    10.98.159.88    <none>
80:30001/TCP    77s
```

The previous `kubectl get services` command indicated that the service was properly created!

Now, let's discuss the `selector` block, which is an important one: it basically instructs the `NodePort` service which Pods it will expose. `selector` stands for label selector. Without it, your service won't do anything. Here, we're telling the service to expose all the Pods that have a label with an key app containing the `whoami` value, which makes our two `whoami1` and `whoami2` Pods exposed through this service.

Then, we have under the key `type` as a child key of `spec`. This is where we specify the type of our service. When we create `ClusterIP` or `LoadBalancer` services, we will have to update this line. Here, we're creating a `NodePort` service, so that's fine for us.

The last thing that is quite hard to understand is that `ports` block. Here, we define a map of multiple port combinations. We indicated three ports, as follows:

- `nodePort`
- `port`
- `targetPort`

The first one is the easiest one to understand. It is basically the port on the host machine/worker node you want this `NodePort` service to be accessible from. Here, we're specifying port `30001`, which makes this `NodePort` service accessible from port `30001` on the IP address of the worker node. You'll be reaching this `NodePort` service and the Pods it exposes by calling the following address: `<WORKER_NODE_IP_ADDRESS>:30001`.

This `NodePort` setting cannot be set arbitrarily. Indeed, on default Kubernetes installation, it can be a port from the `30000 - 32767` range.

The second port specified in YAML is just called `port`: this setting indicates the port of the `NodePort` service itself. It can be hard to understand, but `NodePort` services do have a port of their own too, and this is where you specify it. You can put whatever you want here if it is a valid port.

The third and last one is called `targetPort`. As you can imagine, `targetPort` is the port of the targeted Pods. It is where the application runs: the port where the `NodePort` will forward traffic to the Pod found by the selector mentioned previously.

Here is a quick diagram to sum all of this up:

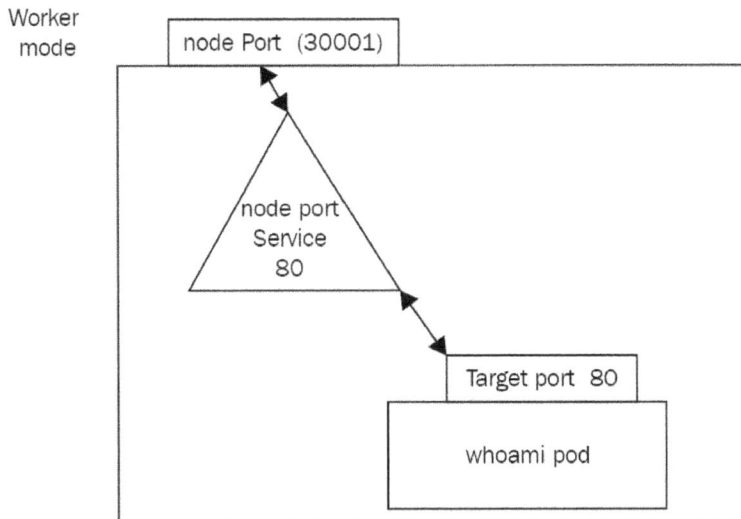

Figure 7.5 – The three ports involved for NodePort setup—nodePort is on the worker node, port is on the service itself, whereas targetPort is on the top

That setup is complex because of the naming. Pay attention to where the three open ports sit to make sure to not confuse the three of them.

For convenience and to reduce complexity, the `NodePort` service port and target port (the Pods' port) are often defined to the same value.

Making sure NodePort works as expected

To try out your `NodePort` setup, the first thing to do is to retrieve the public IP of your machine running it. In my example, I'm running a single-machine Kubernetes setup with `minikube` locally. On AWS, GCP, or Azure, your worker node might have a public IP address or a private one if you access your node with a **virtual private network** (**VPN**).

Based on your setup, keep in mind that the IP address you use must be the one of the worker nodes directly.

On `minikube`, the easiest way to retrieve the IP address is to issue the following command:

```
$ minikube ip
192.168.64.2
```

Now that I have the IP address of my worker node, I must retrieve the worker node's port used by the `NodePort` service.

Now that I have all the information, I can open a web browser and enter the following URL to access the `NodePort` service and the Pods running. You should see the round-robin algorithm in place and reaching `whoami1` and then `whoami2`, and so on. The `NodePort` service is doing its job!

Is this setup production-ready?

There's no answer to this question because it depends on your setup.

`NodePort` provides a way to expose Pods to the outside world by exposing them on a worker node port. With the current setup, you have no HA: if our two Pods were to fail, you have no way to relaunch them automatically, so your service won't be able to forward traffic to anything, resulting in a poor experience for your end user.

Another problem is the fact that the choice of port is limited. Indeed, by default, you are just forced to use a port in the `30000-32767` range, and as it's forced, it will be inconvenient for a lot of people. Indeed, if you want to expose an HTTP application, you'll want to use port `80` or `443` of your frontal machine and not a port for `30000-32767`, because all web browsers are configured with port `80` and `443` as standard HTTP and **HTTP Secure** (**HTTPS**) ports.

The solution to this consists of using a tiered architecture. Indeed, a lot of Kubernetes architects tend to not expose a `NodePort` service as the first layer in an architecture but to put the Kubernetes cluster behind a reverse proxy, such as the AWS Application Load Balancer, and so on. Two other concepts of Kubernetes are `Ingress` and `IngressController` objects: these two objects allow you to configure a reverse proxy such as nginx or HAProxy directly from Kubernetes objects and help you in making your application publicly accessible as the first layer of entry to Kubernetes. But this is way beyond the scope of Kubernetes services.

Listing NodePort services

Listing `NodePort` services is achieved through the usage of the `kubectl` command-line tool. You must simply issue a `kubectl get services` command to fetch the services created within your cluster.

Please note that the plural form `services`, as well as the shorter alias `svc`, works too for all operations related to services. Unfortunately, there is currently no command that allows us to fetch services based on their type.

That being said, let's now discover how we can update `NodePort` services to have them do what we want.

Adding more Pods to NodePort services

If you want to add a Pod to the pool served by your Services, things are very easy. In fact, you just need to add a new Pod that matches the label selector defined on the Service—Kubernetes will take care of the rest. The Pod will be part of the pool served by the Service. In case you delete a Pod, it will be deleted from the pool of services as soon as it enters the `Terminating` state.

Kubernetes handles service traffic based on Pods' availability—for example, if you have three replicas of a web server and one goes down, creating an additional replica that matches the label selector on the service will be enough. You'll discover later in *Chapter 11, Deployment – Deploying Stateless Applications,* that this behavior can be entirely automated.

Describing NodePort services

Describing `NodePort` services is super easy and is achieved with the `kubectl describe` command, just as with any other Kubernetes object.

Deleting NodePort services

Deleting a service, whether it is a `NodePort` service or not, should not be done recurrently. Indeed, whereas Pods are supposed to be easy to delete and recreate, services are supposed to be for the long term. Indeed, they provide a consistent way to expose your Pod, and deleting them will impact how your applications can be reached.

Therefore, you should be careful when deleting services: it won't delete the Pods behind it, but they won't be accessible anymore!

Here is the command to delete the service created to expose the `whoami1` and `whoami2` Pods:

```
$ kubectl delete svc/nodeport-whoami
deleted
```

You can run a `kubectl get svc` command now to check that the service was properly destroyed, and then access it once more through the web browser by refreshing it. You'll notice that the application is not reachable anymore, but the Pods will remain on the cluster. Pods and services have completely independent life cycles. If you want to delete Pods, then you'll need to delete them separately.

You probably remember the `kubectl port-forward` command we already used when we created an nginx Pod and tested it to display the home page. You may think `NodePort` and `kubectl port-forward` are the same thing, but they are not. Let's explain quickly the difference between the two.

NodePort or kubectl port-forward?

It might be tempting to compare `NodePort` services with the `kubectl port-forward` command because so far, we used these two methods to access a running Pod in our cluster using a web browser. But still, these two methods are completely different things—they have nothing to do one with another and shouldn't be confused.

The `kubectl port-forward` command is a testing tool, whereas `NodePort` services are for real use cases and are a production-ready feature.

Keep in mind that `kubectl port-forward` must be kept open in your terminal session for it to work. As soon as the command is killed, the port forwarding is stopped too, and your application becomes inaccessible from outside the cluster once more. It is only a testing tool meant to be used by the `kubectl` user and is just one of the useful tools bundled into the `kubectl` CLI.

`NodePort`, on the other hand, is really meant for production use and is a long-term production-ready solution. It doesn't require `kubectl` to work and makes your application accessible to anyone calling the service, provided the service is properly configured and Pods correctly labeled.

Simply put: if you just need to test your app, go for `kubectl port-forward`. If you need to expose your Pod to the outside world for real, go for `NodePort`. Don't create `NodePort` for testing, and don't try to use `kubectl port-forward` for production! Stick with one tool for each use case!

Now, we will discover another type of Kubernetes service called `ClusterIP`. This one is probably the most used of them all, even more than the `NodePort` type!

The ClusterIP service

We're now going to discover another type of service called `ClusterIP`.

`ClusterIP` is, in fact, the simplest type of service Kubernetes provides. With a `ClusterIP` service, you can expose your Pod so that other Pods in Kubernetes can communicate with it via its IP address or DNS name.

Why do you need ClusterIP services?

The `ClusterIP` service type greatly resembles the `NodePort` service type, but they have one big difference: `NodePort` services are meant to expose Pods to the outside world, whereas `ClusterIP` services are meant to expose Pods to other Pods inside the Kubernetes cluster.

Indeed, `ClusterIP` services are the services that allow different Pods in the same cluster to communicate with each other through a static interface: the `ClusterIP Service` object itself.

`ClusterIP` answers the exact same need for static DNS name or IP address we had with the `NodePort` service: if a Pod fails, is recreated, deleted, relaunched, and so on, then Kubernetes will assign it another IP address. `ClusterIP` services are here to remediate this issue, by providing an internal DNS name only accessible from within your cluster that will resolve to the Pods defined by the label selector.

As the name `ClusterIP` suggests, this service grants a static IP within the cluster! Let's now discover how to expose our Pods using `ClusterIP`! Keep in mind that `ClusterIP` services are not accessible from outside the cluster—they are only meant for inter-Pod communication.

How do I know if I need NodePort or ClusterIP services to expose my Pods?

Choosing between the two types of services is extremely simple, basically because they are not meant for the same thing.

If you need your app to be accessible from outside the cluster, then you'll need a NodePort service, but if you need your app to be accessible from inside the cluster, then you'll need a ClusterIP service. It's as simple as that! ClusterIP services are also good for stateless applications that can be scaled, destroyed, recreated, and so on. The reason is that the ClusterIP service will maintain a static entry point to a whole pool of Pods without being constrained by a port on the worker node, unlike the NodePort service.

Contrary to NodePort services, ClusterIP services will not take one port of the worker node, and thus it is impossible to reach it from outside the Kubernetes cluster.

Keep in mind that nothing prevents you from using both types of services for the same pool of Pods. Indeed, if you have an app, that should be publicly accessible, but also privately exposed to other Pods, then you can simply create two services, one NodePort service and one ClusterIP service.

In this specific use case, you'll simply have to name the two services differently so that they won't conflict when creating them against kube-apiserver. Nothing else prevents you from doing so!

Listing ClusterIP services

Listing ClusterIP services is easy. It's basically the same command as the one used for NodePort services. Here is the command to run:

```
$ kubectl get svc
```

As always, this command lists the services with their type added to the output.

Creating ClusterIP services using the imperative way

Creating ClusterIP services can be achieved with a lot of different methods. Since it is an extremely used feature, there are lots of ways to create these, as follows:

- Using the --expose parameter (the imperative way)
- Using a YAML manifest file (the declarative way)

The imperative way consists of using the –expose method. This will create a
ClusterIP service directly from a kubectl run command, for example. In the
following example, I will create an nginx-clusterip Pod as well as a ClusterIP Pod
to expose them both at the same time. Using the –expose parameter will also require
defining a ClusterIP port. ClusterIP will listen to make the Pod reachable. The code
is illustrated here:

```
$ kubectl run  nginx-clusterip --image nginx --expose --port 80
service/nginx-clusterip created
pod/nginx-clusterip created
```

As you can see, I get both a Pod and a service to expose it. Let's describe the service.

Describing ClusterIP services

Describing ClusterIP services is the same process as describing any type of object in
Kubernetes and is achieved using the kubectl describe command. You just need to
know the name of the service to describe to achieve that.

Here, I'm going to the ClusterIP service created previously:

```
$ kubectl describe svc/nginx-clusterip
Name:              nginx-clusterip
Namespace:         default
Labels:            <none>
Annotations:       <none>
Selector:          run=nginx-clusterip
Type:              ClusterIP
IP Family Policy:  SingleStack
IP Families:       IPv4
IP:                10.96.89.255
IPs:               10.96.89.255
Port:              <unset>  80/TCP
TargetPort:        80/TCP
Endpoints:         10.244.120.79:80
Session Affinity:  None
Events:            <none>
```

What I really like here is that the output of this command shows us the `Selector` block, which shows that the `ClusterIP` service was created by the `--expose` parameter with the proper label configured. This label matches the `nginx-clusterip` Pod we created at the same time. To be sure about that, let's display the labels of the said Pod, as follows:

```
$ kubectl get pods/nginx-clusterip --show-labels
NAME                   READY   STATUS    RESTARTS   AGE    LABELS
nginx-clusterip   1/1      Running   0                   4m1s
run=nginx-clusterip
```

As you can see, the selector on the service matches the labels defined on the Pod. Communication is thus established between the two. I'll now call the `ClusterIP` service directly from another Pod on my cluster.

Since my `ClusterIP` service is named `nginx-clusterip`, I know that it is reachable at this address: `nginx-clusterip.default.svc.cluster.local`.

Let's reuse the `dnsutils` container, as follows:

```
$ kubectl exec -ti dnsutils -- wget nginx.default.svc.cluster.
local
Connecting to nginx.default.svc.cluster.local
(10.98.191.187:80)
index.html              100% |*****************************|
615    0:00:00 ETA
```

Using the `wget` command, we have downloaded the `index.html` file properly. I can now display its content using `cat`, as follows:

```
$ kubectl exec -ti dnsutils -- cat index.html
<html>
<head>
<title>Welcome to nginx!</title>
<style>
html { color-scheme: light dark; }
body { width: 35em; margin: 0 auto;
font-family: Tahoma, Verdana, Arial, sans-serif; }
</style>
</head>
<body>
<h1>Welcome to nginx!</h1>
```

Looks good! The `ClusterIP` service correctly forwarded my request to the nginx Pod, and I do have the nginx default home page. The service is working!

We did not use `containous/whoami` as a web service this time, but keep in mind that the `ClusterIP` service is also doing load balancing internally following the round robin algorithm. If you have 10 Pods behind a `ClusterIP` service and your service received 1,000 requests, then each Pod is going to receive 100 requests.

Let's now discover how to create a `ClusterIP` service using YAML.

Creating ClusterIP services using the declarative way

`ClusterIP` services can also be created using the declarative way by applying YAML configuration files against `kube-apiserver`.

Here's a YAML manifest file we can use to create the exact same `ClusterIP` service we created before using the imperative way:

```
# ~/clusterip-service.yaml
apiVersion: v1
kind: Service
metadata:
  name: nginx-clusterip
spec:
  type: ClusterIP # Indicates that the service is a ClusterIP
  ports:
  - port: 80 # The port exposed by the service
    protocol: TCP
    targetPort: 80 # The destination port on the pods
  selector:
    run: nginx-clusterip
```

Take some time to read the comments in the YAML, especially the `port` and `targetPort` ones.

Indeed, `ClusterIP` services have their own port independent of the one exposed on the Pod side. You reach the `ClusterIP` service by calling its DNS name and its port, and the traffic is going to be forwarded to the destination port on the Pods matching the labels and selectors.

Keep in mind that no worker node port is involved here. The ports we are mentioning when it comes to `ClusterIP` scenarios have absolutely nothing to do with the host machine!

Deleting ClusterIP services

Deleting `ClusterIP` services is the same process as deleting any type of object in Kubernetes. You just need to know the name of the service and to pass it to the `kubectl delete svc` command, as follows:

```
$ kubectl delete svc/my-service
service/my-service deleted
```

This way, the service gets deleted from the Kubernetes cluster. Keep in mind that deleting the cluster won't delete the Pods exposed by it. It is a different process; you'll need to delete Pods separately. We will now discover one additional resource related to `ClusterIP` services, which are headless services.

Understanding headless services

Headless services are derived from the `ClusterIP` service. They are not technically a dedicated type of service (such as `NodePort`), but they are an option from `ClusterIP`.

Headless services can be configured by setting the `.spec.clusterIP` option to `None` in a YAML configuration file for the `ClusterIP` service. Here is an example derived from our YAML file previously:

```
# ~/clusterip-headless.yaml
apiVersion: v1
kind: Service
metadata:
  name: nginx-clusterip-headless
spec:
  clusterIP: None
  type: ClusterIP # Indicates that the service is a ClusterIP
  ports:
```

```
  - port: 80 # The port exposed by the service
    protocol: TCP
    targetPort: 80 # The destination port on the pods
  selector:
    run: nginx-clusterip
```

A headless service roughly consists of a `ClusterIP` service without load balancing and without a pre-allocated `ClusterIP` address. Thus, the load-balancing logic and the interfacing with the Pod are not defined by Kubernetes.

Since a headless service has no IP, you are going to reach the Pod behind it directly, without the proxying and the load-balancing logic. What the headless service does is return you the DNS names of the Pods behind it so that you can reach them directly. There is still a little load-balancing logic here, but it is implemented at the DNS level, not as a Kubernetes logic.

When you use a normal `ClusterIP` service, you'll always reach one static IP address allocated to the service and this is going to be your proxy to communicate with the Pod behind it. With a headless service, the `ClusterIP` service will just return the DNS names of the Pods behind it and the client will have the responsibility to establish a connection with the DNS name of its choosing.

It can be hard to think of a concrete use case for this, and honestly, headless services are not used as much as normal `ClusterIP` services. They are still helpful when you want to build connectivity with clustered stateful services such as **Lightweight Directory Access Protocol (LDAP)**. In that case, you may want to use an LDAP client that will have access to the different DNS names of the Pods hosting the LDAP server, and this can't be done with a normal `ClusterIP` service since it will bring both a static IP and Kubernetes' implementation of load balancing. That's the kind of scenario when headless services are helpful.

Let's now briefly introduce another type of service called `LoadBalancer`.

The LoadBalancer service

`LoadBalancer` services are a very interesting service to explain because this service relies on the cloud platform where the Kubernetes cluster is provisioned. For it to work, it is thus required to use Kubernetes on a cloud platform that supports the `LoadBalancer` service type.

Explaining the LoadBalancer services

Not all cloud providers support the `LoadBalancer` service type, but we can name a few that do support it, as follows:

- AWS
- GCP
- Azure
- OpenStack

The list is not exhaustive, but it's good to know that all three major public cloud providers are supported.

If your cloud provider is supported, keep in mind that the load-balancing logic will be the one implemented by the cloud provider: you cannot control how the traffic will be routed to your Pods from Kubernetes; you will have to know how the load-balancer component of your cloud provider works. Consider it as a third-party component implemented as a Kubernetes kind.

Should I use the LoadBalancer service type?

This question is difficult to answer but a lot of people tend to not use a `LoadBalancer` service type for a few reasons.

The main reason is that `LoadBalancer` services are near impossible to configure from Kubernetes. Indeed, if you must use a cloud provider, it is better to configure it from the tooling provided by the provider rather than from Kubernetes.

The `LoadBalancer` service type is a generic way to provision a `LoadBalancer` service but does not expose all the advanced features that the cloud provider may provide. Let's take the example of AWS. This cloud provider has three `LoadBalancer` components, as follows:

- Classic Load Balancer
- Application Load Balancer
- Network Load Balancer

All these three offerings differ. Their usage, configurations, and behaviors are not the same from one to another. The `LoadBalancer` service type, when used on an AWS-backed Kubernetes cluster, will create a Classic Load Balancer: you have no way to create an Application Load Balancer or a Network Load Balancer from the Kubernetes `Service` object.

> **Just to Say**
>
> A lot of configurations are hidden, and you won't have access to all the features your cloud provider offers. That's why you should avoid using the `LoadBalancer` service and stick to different tooling such as Terraform or the CLI.

Implementing ReadinessProbe

`ReadinessProbe`, along with `LivenessProbe`, is an important aspect to master if you want to provide the best possible experience to your end user. We will first discover how to implement `ReadinessProbe` and how it can help you to ensure your containers are fully ready to serve traffic.

Why do you need ReadinessProbe?

Readiness probes are technically not part of services, but I think it is important to discover this feature alongside Kubernetes services.

Just as with everything in Kubernetes, `ReadinessProbe` was implemented to bring a solution to a problem. This problem is this: how to ensure a Pod is fully ready before it can receive traffic, possibly from a service?

Indeed, Services obey a simple rule: they serve traffic to every Pod that matches their label selector. As soon as a Pod gets provisioned, if this pod's labels match the selector of a service in your cluster, then this service will immediately start forwarding traffic to it. This can lead to a simple problem: if the app is not fully launched, because it has a slow launch process or requires some configuration from a remote API, and so on, then it might receive traffic from services before being ready for it. The result would be a poor **user experience (UX)**.

To make sure this scenario never happens, we can use what we call a readiness probe, which is an additional configuration to add to Pods' configuration.

When a Pod is configured with a readiness probe, it can send a signal to the control plane that it is not ready to receive traffic, and when a Pod is not ready, services won't forward any traffic to it. Let's see how we can implement a readiness probe.

Implementing ReadinessProbe

ReadinessProbe implementation is achieved by adding some configuration data to a Pod YAML manifest. It has nothing to do with the Service object itself. By adding some configuration to the container spec in the Pod object, you can basically tell Kubernetes to wait for the Pod to be fully ready before it can receive traffic from services.

ReadinessProbe can be of three different types, as outlined here:

- **Command**—You issue a command that should complete with exit code 0, indicating the Pod is ready.

- **HTTP**—You issue an HTTP request that should complete with a response code >= 200 and < 400, which indicates the Pod is ready.

- **TCP**—You issue a TCP connection attempt. If the connection is established, the Pod is ready.

Here is a YAML file configuring a nginx Pod with a readiness probe of type HTTP:

```
# ~/nginx-pod-with-readiness-http.yaml # ~/nginx-pod-with-
readiness-http.yaml
apiVersion: v1
kind: Pod
metadata:
  name: nginx-pod-with-readiness-http
spec:
  containers:
  - name: nginx-pod-with-readiness-http
    image: nginx
    readinessProbe:
      initialDelaySeconds: 5
      periodSeconds: 5
      httpGet:
        path: /ready
        port: 80
```

As you can see, we have two important inputs under the `readinessProbe` key, as follows:

- `initialDelaySeconds`, which indicates the number of seconds the probe will wait before running the first health check.
- `periodSeconds`, which indicates the number of seconds the probe will wait between two consecutive health checks.

Indeed, the readiness probe will be replayed regularly, and the interval between two checks will be defined by the `periodSeconds` parameter.

In our case, our `ReadinessProbe` will run an HTTP call against the `/ready` path. If this request receives an HTTP response code >= `200` and < `400`, then the probe will be a success and the Pod will be considered healthy.

`ReadinessProbe` is important. In our example, the endpoint being called should really test that the application is really in such a state that it can receive traffic. So, try to call an endpoint that is relevant to the state of the actual application. For example, you can try to call a page that will open a MySQL connection internally to make sure the application is capable of communicating with its database if it is using one, and so on. If you're a developer, do not hesitate to create a dedicated endpoint that will just open connections to the different backends to be fully sure that the application is ready for real.

The Pod will then join the pool being served by the service and will start receiving traffic. `ReadinessProbe` can also be configured as TCP and commands, but we will keep these examples for `LivenessProbe`. Let's discover them now!

What is LivenessProbe and why do you need it?

`LivenessProbe` resembles `ReadinessProbe` a lot. In fact, if you already used some cloud providers before, you might already have heard about something called health checks. So, in other words, `LivenessProbe` is basically a health check.

Basically, liveness probes are used to determine whether a Pod is in a broken state or not, and the usage of `LivenessProbe` is especially suited for long-running processes such as web services. Indeed, imagine a situation where you have a service forwarding traffic to three Pods and one of them is broken. Services cannot detect that on their own, and they will just continue to serve traffic to the three Pods, including the broken one. In such situations, 33% of your request will inevitably lead to an error response, resulting in a poor UX, as illustrated in the following screenshot:

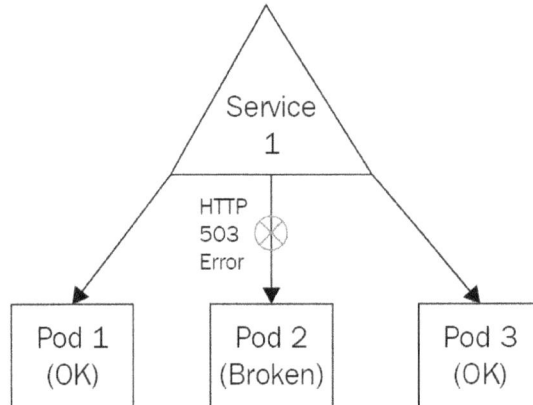

Figure 7.6 – One of the Pods is broken but the service will still forward traffic to it

You want to avoid such situations, and to do that, you need a way to detect situations where Pods are broken, plus a way to kill such a container so that it goes out of the pool of Pods being targeted by the service.

`LivenessProbe` is the solution to this problem and is implemented at the Pod level. Be careful because `LivenessProbe` cannot repair a Pod: it can only detect that a Pod is not healthy and command its termination. Let's see how we can implement a Pod with `LivenessProbe`.

Implementing LivenessProbe

`LivenessProbe` is a health check that will be executed on a regular schedule to keep track of the application state in the long run. These health checks are executed by the `kubelet` component and can be of three types, as outlined here:

- **Command**, where you issue a command in the container and its result will tell whether the Pod is healthy or not (exit code = 0 means healthy)

- **HTTP**, where you run an HTTP request against the Pod, and its result tells whether the Pod is healthy or not (HTTP response code >= 200 and < 400 means the Pod is healthy)

- **TCP**, where you define a TCP call (a successful connection means the Pod is healthy)

Each of these three liveness probes will require you to input a parameter called `periodSeconds`, which must be an integer. This will tell the `kubelet` component the number of seconds to wait before performing a new health check. You can also use another parameter called `initialDelaySeconds`, which will indicate the number of seconds to wait before performing the very first health check. Indeed, in some common situations, a health check might lead to flagging an application as unhealthy just because it was performed too early. That's why it might be a good idea to wait a little bit before performing the first health check, and that parameter is here to help.

`LivenessProbe` configuration is achieved at the Pod YAML configuration manifest, not at the service one. Each container in the Pod can have its own `livenessProbe`.

Here is a configuration file that checks if a Pod is healthy by running an HTTP call against a `/healthcheck` endpoint in an nginx container:

```
# ~/nginx-pod-with-liveness-http.yaml
apiVersion: v1
kind: Pod
metadata:
  name: nginx-pod-with-liveness-http
spec:
  containers:
  - name: nginx-pod-with-liveness-http
    image: nginx
    livenessProbe:
      initialDelaySeconds: 5
      periodSeconds: 5
      httpGet:
        path: /healthcheck
        port: 80
        httpHeaders:
        - name: My-Custom-Header
          value: My-Custom-Header-Value
```

Please pay attention to all sections after the `livenessProbe` blocks. If you understand this well, you can see that we will wait 5 seconds before performing the first health check, and then, we will run one HTTP call against the `/healthcheck` path on port `80` every 5 seconds. One custom HTTP header was added. Adding such a header will be useful to identify our health checks in the access logs. Be careful because the `/healthcheck` path probably won't exist in our nginx container and thus, this container will never be considered healthy because the liveness probe will result in a `404` HTTP response. Keep in mind that for an HTTP health check to succeed, it must answer an HTTP >= `200` and < `400`. `404` being out of this range, the answer Pod won't be healthy.

You can also use a command to check if a Pod is healthy or not. Let's grab the same YAML configuration, but now, we will use a command instead of an HTTP call in the liveness probe, as follows:

```
# ~/nginx-pod-with-liveness-command.yaml
apiVersion: v1
kind: Pod
metadata:
  name: nginx-pod-with-liveness-command
spec:
  containers:
  - name: nginx-pod-with-liveness-command
    image: nginx
    livenessProbe:
      initialDelaySeconds: 5
      periodSeconds: 5
      exec:
        command:
        - cat
        - /hello/world
```

If you check this example, you can see that this one is much simpler than the HTTP one. Here, we are basically running a `cat /hello/world` command every 5 seconds. If the file exists and the `cat` command completes with an exit code equal to `0`, then the health check will succeed. Otherwise, if the file is not present, the health check will fail, and the Pod will never be considered healthy and will be terminated.

We will now complete this section by discovering one last example with a TCP `livenessProbe`. In this situation, we will attempt a connection to a TCP socket on port `80`. If the connection is successfully established, then the health check will pass, and the container will be considered ready. Otherwise, the health check will fail, and the Pod will be terminated eventually. The code is illustrated in the following snippet:

```
# ~/nginx-pod-with-liveness-tcp.yaml
apiVersion: v1
kind: Pod
metadata:
  name: nginx-pod-with-liveness-tcp
spec:
  containers:
  - name: nginx-pod-with-liveness-tcp
    image: nginx
    livenessProbe:
      initialDelaySeconds: 5
      periodSeconds: 5
      tcpSocket:
        port: 80
```

Using TCP health checks greatly resembles HTTP ones since HTTP is based on TCP anyway. But having TCP as a liveness probe is nice especially if you want to keep track of an application that is not based on using HTTP as protocol and using that command is irrelevant to you, such as health-checking an LDAP connection, for example.

Using ReadinessProbe and LivenessProbe together

You can use `ReadinessProbe` and `LivenessProbe` together in the same Pod. There is nothing wrong with that since `ReadinessProbe` is here for you to prevent a Pod from joining a service before being ready to serve traffic, and `LivenessProbe` is here to kill a Pod that has become unhealthy.

So, they are configured almost the same way—they don't have the exact same purpose, and it is fine to use them together. Please note that both probes share these parameters:

- `initialDelaySeconds`: The number of seconds to wait before the first probe execution
- `periodSeocnds`: The number of seconds between two probes
- `timeoutSeconds`: The number of seconds to wait before timeout

- `successThreshold`: The number of successful attempts to consider a Pod is ready (for `ReadinessProbe`) or healthy (for `LivenessProbe`)
- `failureThreshold`: The number of failed attempts to consider a Pod is not ready (for `ReadinessProbe`) or ready to be killed (for `LivenessProbe`)

We now have discovered `ReadinessProbe` and `LivenessProbe`, and we can now move on to our final section, which introduces the control of network flow in Kubernetes using the `NetworkPolicy` object. You will see that you can build a kind of network firewall directly in Kubernetes so that you can prevent Pods from being able to reach one another.

Securing your Pods using the NetworkPolicy object

The `NetworkPolicy` object is the last resource kind we need to discover as part of this chapter to have an overview of services in this chapter. `NetworkPolicy` will allow you to define network firewalls directly implemented in your cluster.

Why do you need NetworkPolicy?

When you have to manage a real Kubernetes workload in production, you'll have to deploy more and more applications onto it, and it is possible that these applications will have to communicate with each other.

Achieving communication between applications is really one of the fundamental objectives of a microservice architecture. Most of this communication will be done through the network, and the network is forcibly something that you want to secure by using firewalls.

Kubernetes has its own implementation of network firewalls called `NetworkPolicy`. This is a new resource kind we are going to discover. Say that you want one nginx resource to be accessible on port `80` from a particular IP address and to block any other traffic that doesn't match these requirements. To do that, you'll need to use `NetworkPolicy` and attach it to that Pod.

`NetworkPolicy` brings three benefits, as follows:

- You can build egress/ingress rules based on **Classless Inter-Domain Routing (CIDR)** blocks.
- You can build egress/ingress rules based on Pods labels and selectors (just as we've seen before with services' and Pods' association).

- You can build egress/ingress rules based on namespaces (a notion we will discover in the next chapter).

Lastly, keep in mind that for `NetworkPolicy` to work, you'll need to have a Kubernetes cluster with a CNI plugin installed. CNI plugins are generally not installed by default on Kubernetes. If you're using `minikube`, the good news is that it has an integration with Calico, which is a CNI plugin with `NetworkPolicy` support implemented out of the box. You just need to relaunch `minikube` this way:

```
$ minikube start --network-plugin=cni --cni=calico
```

If you're using Kubernetes on top of a cloud platform, I suggest you read the documentation of your cloud provider in order to verify which CNI options your cloud platform offers and whether it implements `NetworkPolicy` support.

Understanding Pods are not isolated by default

This is something extremely important to understand. By default, in Kubernetes, Pods are not isolated. It means that any Pod can be reached by any other Pod without any constraint.

All Pods you have deployed so far can be reached by any other Pod in the cluster. If you don't use `NetworkPolicy`, Pods will remain just like that: accessible by everything without any constraint. Once you attach the `NetworkPolicy` to a Pod, the rules described on the `NetworkPolicy` will be applied to the Pod.

To establish communication between two Pods associated with network policies, both sides must be open. It means Pod *A* must have an egress rule to Pod *B*, and Pod *B* must have an ingress rule from Pod *A*; otherwise, the traffic will be denied. The following screenshot illustrates this:

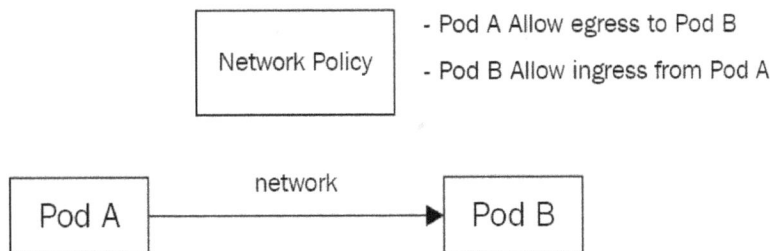

Figure 7.7 – One of the Pods is broken but the service will still forward traffic to it

Keep that in mind that you'll have to troubleshoot `NetworkPolicy` because it can be the root cause of a lot of issues. Let's now configure a `NetworkPolicy` between two Pods by using labels and selectors.

Configuring NetworkPolicy with labels and selectors

First, let's create two nginx Pods to demonstrate our example. I'll create the two Pods with two distinct labels so that they become easier to target with the NetworkPolicy, as follows:

```
$ kubectl run nginx-1 --image nginx   --labels 'app=nginx-1'
pod/nginx-1 created


$ kubectl run nginx-2 --image nginx --labels 'app=nginx-2'
pod/nginx-2 created


$ kubectl get pods -o wide
nginx-1                                                      1/1
Running    0            30s    172.17.0.12    minikube
nginx-2                                                      1/1
Running    0            30s    172.17.0.13    minikube
```

Now that the two Pods are created with distinct labels. I could use the -o wide flag to get the IP address of both Pods. If you follow along, your IP addresses will be different. Now, I can run a curl command from the nginx-1 Pod to reach the nginx-2 Pod, to confirm that by default, network traffic is allowed because no NetworkPolicy is created at this point. The code is illustrated here:

```
$ kubectl exec   nginx-1 -- curl 172.17.0.13
<html>
<head>
<title>Welcome to nginx!</title>
<style>
html { color-scheme: light dark; }
body { width: 35em; margin: 0 auto;
font-family: Tahoma, Verdana, Arial, sans-serif; }
</style>
</head>
<body>
<h1>Welcome to nginx!</h1>
...
```

As you can see, I correctly receive the nginx home page from the `nginx-2` Pod. Now, I'll add `NetworkPolicy` to `nginx-2` to explicitly allow traffic coming from `nginx-1`. Here is how to proceed with the YAML code:

```yaml
# ~/nginx-2-networkpolicy.yaml
apiVersion: networking.k8s.io/v1
kind: NetworkPolicy
metadata:
  name: nginx-2-networkpolicy
spec:
  podSelector:
   matchLabels:
      app: nginx-2 # Applies to which pod
  policyTypes:
   - Ingress
  ingress:
  - from:
    - podSelector:
        matchLabels:
          app: nginx-1 # Allows calls from which pod
    ports:
    - protocol: TCP
      port: 80
```

Let's apply this `NetworkPolicy`, as follows:

```
$ kubectl create -f nginx-2-networkpolicy.yaml
networkpolicy.networking.k8s.io/nginx-2-networkpolicy created
```

Now, let's run the same `curl` command we did before, as follows:

```
$ kubectl exec  nginx-1 -- curl 172.17.0.13
<html>
<head>
<title>Welcome to nginx!</title>
<style>
html { color-scheme: light dark; }
body { width: 35em; margin: 0 auto;
font-family: Tahoma, Verdana, Arial, sans-serif; }
```

```
</style>
</head>
<body>
<h1>Welcome to nginx!</h1>
...
```

As you can see, it works just like it did before. Why? For the following two reasons:

- nginx-2 now explicitly allows ingress traffic on port 80 from nginx-1; everything else is denied.

- nginx-1 has no NetworkPolicy, and thus, egress traffic to everything is allowed for it.

Keep in mind that if no NetworkPolicy is set on the Pod, the default behavior applies—everything is allowed for the Pod. Now, let's update the NetworkPolicy attached to the Pod to change the allowed port from 80 to 8080 by updating the file and applying it against the cluster, as follows:

```
$ kubectl apply -f nginx-2-networkpolicy.yaml
networkpolicy.networking.k8s.io/nginx-2-networkpolicy
configured
```

Now, let's run the curl command from the nginx-1 Pod once more, as follows:

```
$ kubectl exec nginx-1 -- curl 172.17.0.13:80
curl: (28) Failed to connect to 10.244.120.77 port 80:
Connection timed out
command terminated with exit code 28
```

As you can see, the curl command failed with a timeout. The reason is that the traffic is blocked by the NetworkPolicy, and thus, the curl command never reaches the nginx-2 Pod.

Again, there are two reasons for that, as outlined here:

- There is still no NetworkPolicy set to the nginx-1 Pod, so egress is opened.

- We have a NetworkPolicy on the nginx-2 Pod, allowing ingress traffic on port 8080 from Pod nginx-1.

Since `curl` is trying to reach the `nginx-2` Pod on port `80`, the traffic is blocked. Let's now try to run `curl` to make a call against port `8080`, which is the one allowed by the `NetworkPolicy`, to see what happens. Here's the code we need to execute:

```
$ kubectl exec nginx-1 -- curl 172.17.0.13:8080
curl: (7) Failed to connect to 172.17.0.13 port 8080:
Connection refused
command terminated with exit code 7
```

As you can see, no timeout this time. The traffic reaches the `nginx-2` Pod, but since nothing is listening on port `8080` on `nginx-2`, then we have a `Connection refused` error.

I strongly encourage you to take the habit of using `NetworkPolicy` along with your Pod. Lastly, please be aware that `NetworkPolicy` can also be used to build firewalls based on CIDR blocks. It might be useful especially if your Pods are called from outside the cluster. Otherwise, when you need to configure firewalls between Pods, it is recommended to proceed with labels and selectors as you already did with the services' configuration.

Summary

This chapter was dense and contained a huge amount of information on networking in general when applied to Kubernetes. Services are just like Pods: they are the foundation of Kubernetes, and mastering them is crucial to being successful with the orchestrator.

Overall, in this chapter, we discovered that Pods have dynamic IP assignment, and they get a unique IP address when they're created. To establish a reliable way to connect to your Pods, you need a proxy called `Service` in Kubernetes. We've also discovered that Kubernetes services can be of multiple types and that each type of service is designed to address a specific need. We've also discovered what `ReadinessProbe` and `LivenessProbe` are and how they can help you in designing health checks to ensure your pods gets traffic when they are ready and live.

Lastly, we discovered how to control traffic flow between Pods by using an additional object called `NetworkPolicy` that behaves like a networking firewall within the cluster.

We've now discovered most of the Kubernetes fundamentals, but there are still some missing pieces that you need to discover. In the next chapter, we're going to cover another important aspect of Kubernetes called namespaces. We've already discussed them a little bit during this chapter, and if you've read it carefully, you should be aware that up to now, we have somehow used a namespace called `default`.

8

Managing Namespaces in Kubernetes

So far, we've learned about Kubernetes' key concepts by launching objects into our clusters and observing their behavior. As you deployed all of these objects, such as Pods, ConfigMaps, Secrets, and more, you may have noticed that, in the long run, it would be difficult to maintain a cleanly organized cluster. As your clusters grow, it's going to become more and more difficult to maintain the ever-increasing number of resources that are managed in your cluster. That's when Kubernetes namespaces come into play.

In this chapter, we will learn about **namespaces**. They help us keep our clusters well organized by grouping our resources by application or environment. Kubernetes namespaces are another key aspect of Kubernetes management and it's really important to master them!

In this chapter, we're going to cover the following main topics:

- Introduction to Kubernetes namespaces
- How namespaces impact your resources and services
- Configuring ResourceQuota at the namespace level

Technical requirements

For this chapter, you will need the following:

- A working Kubernetes cluster (local or cloud-based, but this is not important)
- A working Kubectl CLI configured to communicate with the cluster

If you do not have these technical requirements, please read *Chapter 2, Kubernetes Architecture – from Docker Images to Running Pods*, and *Chapter 3, Installing Your First Kubernetes Cluster*, to get them.

Introduction to Kubernetes namespaces

The more applications you deploy on your Kubernetes clusters, the more you will feel the need to keep your cluster organized. You can already use labels and annotations to tidy up the objects inside your cluster, but you can go even further by using namespaces, which are presented by Kubernetes as virtual clusters within your cluster.

Using namespaces in Kubernetes allows you to split your cluster into smaller parts that will be logically isolated. Once namespaces have been created, you can launch Kubernetes objects such as Pods, which will only exist in that namespace. So, all the operations that are run against the cluster with kubectl will be scoped to that individual namespace: you can perform as many operations as possible while eliminating the risk of impacting resources that are in another namespace.

Lastly, namespaces allow you to define resources limits and quotas for the Pods running in them.

Let's start by finding out what exactly namespaces are and why they were created.

Why do you need namespaces?

As we mentioned previously, namespaces in Kubernetes are a way to help the cluster administrator keep everything clean and organized, while providing resource isolation when running kubectl commands.

The biggest Kubernetes clusters can run hundreds or even thousands of applications in Pods. Very often, those Pods come with services, ConfigMaps, volumes, and more. When everything is deployed in the same namespace, it can become very complex to know which particular resource belongs to which application.

If, by misfortune, you update the wrong resource using `kubectl`, you might end up breaking an app running in your cluster. To resolve that, you can use labels and selectors but even then, as the number of resources grows, managing the cluster will quickly become chaotic if you don't start using namespaces.

Let's learn how namespaces are used to split Kubernetes clusters.

How namespaces are used to split resources into chunks

Right after you've installed Kubernetes, when your cluster is brand new, it is created with namespaces. So, even if you didn't notice previously, you are already using namespaces.

You can freely add new namespaces: the whole idea is to deploy your Pods and other objects in Kubernetes, making sure to always define a namespace of your choosing. This way, you will maintain a clean and well-organized cluster. We will discover right after that by default, Kubernetes is created with a namespace, and it is the one that is used if you do not specify another explicitly.

A key architecting rule is that an application should not be aware of the Kubernetes namespace that it is running in. This means that you should think of Kubernetes namespaces as objects for cluster administrators, not developers or application users. It shouldn't have any impact on how the application works, and an application should be able to deploy on any namespace. If an app is running on namespace A and is then redeployed on namespace B, it should not have any impact on any of its features.

More generally, Kubernetes namespaces will help you achieve the following as an administrator:

- Cluster partitioning to ease resource organization
- Scoping resource names
- Hardware sharing and consumption limitation
- Permissions access

I recommend that you create one namespace per microservice and deploy all the resources that belong to a microservice within its namespace. However, be aware that Kubernetes does not impose any specific rules on you. For example, you could decide to use namespaces in these ways:

- **Differentiate environments**

  For example, one namespace for a production environment and another one for a development environment.

- **Differentiate the tiers**

  One namespace for databases, one for application Pods, and another for middleware deployment.

- **Just using the default namespaces**

  For the smallest clusters that only deploy a few resources, you can go for the simplest setup and just use one big default namespace and deploy everything into it.

Either way, keep in mind that even though two Pods are deployed in different namespaces and exposed through services, they can still interact and communicate with each other. Even though Kubernetes services are created in a given namespace, they'll receive a **fully qualified domain name (FQDN)** that will be accessible on the whole cluster. So, even if an application running on namespace A needs to interact with an application in namespace B, it will have to call the service exposing app B by its FQDN. You don't need to worry about cross-namespace communication.

Now, let's learn about the `default` namespace.

Understanding default namespaces

Most Kubernetes clusters are created with a few namespaces by default. You can list your namespaces using `kubectl get namespaces`, as follows:

```
$ kubectl get namespaces # Or kubectl get ns
# You can use "ns" as an alias for namespaces
NAME              STATUS    AGE
default           Active    5m31s
kube-public       Active    5m33s
kube-system       Active    5m33s
```

For instance, I'm using a Minikube cluster. By reading this command's output, we can see that the cluster I'm currently using was set up with three namespaces from the start:

- `default`
- `kube-public`
- `kube-system`

These namespaces are the ones provided by default in Kubernetes.

The most important of them is `default`: it is in this namespace that, up until now, all your resources have been created, and it is also the namespace that's used by default when no other one is specified.

The other default namespaces are also interesting, especially the one called `kube-system`. It is in this namespace that Kubernetes deploys the objects it needs to function. In most Kubernetes distributions, components such as `kube-scheduler` are deployed in the form of Pods and are themselves by the Kubernetes cluster that they are running for. They are then launched in this namespace. So, it is better to leave this namespace alone and not touch it because it is a very important namespace, and changing something could damage your cluster.

A third namespace is usually created by default when the cluster is set up, and this is called `kube-public`. It is a namespace where public objects are accessible. It is not used at the moment, so you can leave it alone.

> **Important Note**
> Depending on which Kubernetes distribution you use, the default namespaces can change. But most of the time, these three different ones will be created by default.

If your cluster is small, you can continue to use the default namespace – this is not a problem. However, let's leave this namespace aside for now because we are going to get to the heart of the matter a little more and start creating namespaces. We will look at the impacts that these can have on your Pods, particularly at the level of the DNS resolution of your services.

How namespaces impact your resources and services

Previously, we learned what namespaces are and we will continue to discover them. In this section, we'll learn how to create, update, and delete namespaces, as well as the impacts that namespaces have on services and Pods.

We will also learn how to create ConfigMaps and Pods by specifying a custom namespace so that we don't rely on the default one.

Listing namespaces inside your cluster

We saw this in the previous section but here is how to list the namespaces that have been created in your Kubernetes cluster:

```
$ kubectl get namespaces #kubectl get ns works too
NAME              STATUS    AGE
default           Active    5m31s
kube-public       Active    5m33s
kube-system       Active    5m33s
```

Keep in mind that all the commands that make use of the `namespaces` resource `Kind` can also use the `ns` alias to benefit from a shorter format.

Retrieving the data of a specific namespace

Retrieving the data of a specific namespace can be achieved using the `kubectl describe` command. Here is how to grab the data from that command:

```
$ kubectl get namespaces default #kubectl get ns works too
NAME      STATUS    AGE
default   Active    13m
```

You can also use the `get` command and redirect the YAML format to a file to get the data from a specific namespace. By reading the output of the `describe` command, you can see a field called `STATUS`. Please note that a namespace can be in one of two statuses:

- `Active`: The namespace is active; it can be used to place new objects into it.
- `Terminating`: The namespace is being deleted, along with all its objects. It can't be used to host new objects while in this status.

Now, let's learn how to create a new namespace imperatively.

Creating a namespace using imperative syntax

You can create a namespace with imperative or declarative syntax. To imperatively create a namespace, you can use the `kubectl create namespaces` command by specifying the name of the namespace to create. Here, I'm going to create a new namespace called `custom-ns`. Please notice that all the operations related to namespaces in `kubectl` can be written with the shorter ns alias:

```
$ kubectl create ns custom-ns #kubectl get ns works too
namespace/custom-ns created
```

The new namespace, called `custom-ns`, should now be created in your cluster. You can check it by running the `kubectl get` command once more:

```
$ kubectl get ns custom-ns

NAME          STATUS    AGE
custom-ns     Active    35s
```

As you can see, the namespace has been created and is in the `Active` state. We can now place resources in it.

> **Important Note**
>
> Please do avoid naming your cluster with a name starting with the `kube-` prefix as this is the terminology for Kubernetes' objects and system namespaces.

Now, let's learn how to create another namespace using declarative syntax.

Creating a namespace using declarative syntax

Let's discover how to create a namespace using declarative syntax. As always, you must use a YAML (or JSON) file. Here is a basic YAML file for creating a new namespace in your cluster. Please pay attention to `kind: Namespace` in the file. `Namespace` is a resource kind, just like `Pod`:

```
# ~/custom-ns-2.yaml
apiVersion: v1
kind: Namespace
metadata:
  name: custom-ns-2
```

Once this file has been created, you can apply it using the `kubectl create` command by defining the YAML file path:

```
$ kubectl create -f custom-ns.yaml
namespace/custom-ns-2 created
```

With that, we have created two custom namespaces. The first one, which was created imperatively, is called `custom-ns`, while the second one, which was created declaratively, is called `custom-ns-2`.

Now, let's remove these two namespaces using `kubectl`.

Deleting a namespace

You can delete a namespace using `kubectl delete`, as follows:

```
$ kubectl delete namespaces custom-ns
namespace "custom-ns" deleted
```

Please note this can also be achieved using declarative syntax. Let's delete the `custom-ns-2` namespace that was created using the previous YAML file:

```
$ kubectl delete -f custom-ns-2.yaml
namespace "custom-ns-2" deleted
```

Running this command will take the namespace out of the `Active` status: it will enter the `Terminating` status. Right after the command, the namespace will be unable to host new objects, and after a few moments, it should completely disappear from the cluster.

However, I have to warn you about using this command as it is extremely dangerous. Deleting a namespace is permanent and definitive. All the resources that were created in the namespace will be destroyed. If you need to use this command, be sure to have YAML files to recreate the destroyed resources and even the destroyed namespace.

Now, let's discover how to create resources inside a specific namespace.

Creating a resource inside a namespace with the -n option

Creating a resource within a namespace is very easy. The following code shows how to create an NGINX Pod by specifying a custom namespace. Here, we are going to recreate a new `custom-ns` namespace and launch an NGINX Pod in it:

```
$ kubectl create ns custom-ns
$ kubectl run nginx --image nginx:latest -n custom-ns
Pod/nginx created
```

Pay attention to the `-n` option, which, in its long form, is the `--namespace` option. This is used to enter the name of the namespace where you want to create the resource. This option is supported by all the kind resources that can be scoped in a namespace.

Here is another command to demonstrate this. The following command will create a new `configmap` in the `custom-ns` namespace:

```
$ kubectl create configmap configmap-custom-ns --from-
literal=Lorem=Ipsum -n custom-ns
configmap/configmap-custom-ns created
```

You can also specify a namespace when using declarative syntax. Here is how to create a Pod in a specific namespace with declarative syntax:

```
# ~/Pod-in-namespace.yaml
apiVersion: v1
kind: Pod
metadata:
  name: nginx2
  namespace: custom-ns
spec:
  containers:
  - name: nginx
    image: nginx:latest
```

Please note the `namespace` key under the `metadata` section, just under the Pod's name. Now, we can apply this file using `kubectl`:

```
$ kubectl create -f Pod-in-namespace.yaml
Pod/nginx2 created
```

At this point, we should have a namespace called `custom-ns` that contains two `nginx` Pods, as well as a `configmap` called `configmap-custom-ns`.

> **Important Note**
>
> When you're using namespaces, you should always specify the `-n` flag to target the specific namespace of your choosing. Otherwise, you might end up running operations in the wrong namespace. This is a common mistake.

Now that we have created some resources in the `custom-ns` namespace, we will learn how to list the resources in a specific namespace using `kubectl get Pods`. We can do this by specifying the namespace using the `-n` option:

```
$ kubectl get Pods -n custom-ns
NAME      READY   STATUS    RESTARTS   AGE
nginx     1/1     Running   0          66m
nginx2    1/1     Running   0          36s
```

Listing resources inside a specific namespace

To be able to list the resources within a namespace, you must add the `-n` option, just like when creating a resource. Use the following command to list the Pods in the `custom-ns` namespace:

```
$ kubectl get Pods -n custom-ns
NAME      READY   STATUS    RESTARTS   AGE
nginx     1/1     Running   0          66m
nginx2    1/1     Running   0          36s
```

Here, you can see that the `nginx` Pod that we created earlier is present in the namespace. From now on, all the commands that target this particular Pod should contain the `-n custom-ns` option.

The reason for this is that the Pod does not exist in the default namespace, and if you omit passing the `-n` option, then the default namespace will be requested. Let's try to remove `-n custom-ns` from the `get` command. We will see that the `nginx` Pod is not here anymore:

```
$ kubectl get Pods -n custom-ns
NAME      READY   STATUS    RESTARTS   AGE
nginx     1/1     Running   0          9m17s
```

```
$ kubectl get Pods
No resources found in default namespace.
```

Now, we can also run the get configmap command to check whether configmap is listed in the output. As you can see, the behavior is the same as when trying to list Pods. If you omit the -n option, the list operation will run against the default namespace that does not contain configmap:

```
$ kubectl get cm
No resources found in default namespace.


$ kubectl get cm -n custom-ns
NAME                   DATA    AGE
configmap-custom-ns    1       76m
```

With all that has been said so far, the idea is to never forget to add the -n option when working on a cluster that has multiple namespaces. This little carelessness could waste your time because if you forget it, everything you do will be done on the default namespace.

You can also decide to remove the default namespace. So, if you forget to specify the namespace, an error will occur because if you try to work on a namespace that does not exist, Kubernetes will give you an explicit error.

Now, let's discover how to list all the resources inside a specific namespace.

Listing all the resources inside a specific namespace

If you want to list all the resources in a specific namespace, there is a very useful command that you can use called kubectl get all -n custom-ns:

```
$ kubectl get all -n custom-ns
```

As you can see, this command can help you retrieve all the resources that are created in the namespace specified in the -n flag.

Understanding that names are scoped to a namespace

You should know that namespaces bring an additional benefit: they provide scope to the names of the resources they contain.

Take the example of Pod names. When you work without namespaces, you are interacting with the default namespace: when you create two Pods with the same name, you get an error because Kubernetes uses the names of the Pods as their unique identifiers to distinguish them.

Let's try to create two Pods in the default namespace. Both will be called `nginx`. Here, we can simply run the same command twice in a row:

```
$ kubectl run nginx --image nginx:latest
Pod/nginx created
$ kubectl run nginx --image nginx:latest
Error from server (AlreadyExists): Pods "nginx" already exists
```

The second command produces an error, saying that the Pod already exists, which it does. If we run `kubectl get Pods`, we can see that only one Pod exists:

```
$ kubectl get Pods
NAME     READY     STATUS      RESTARTS     AGE
nginx    1/1       Running     0            64s
```

Now, let's try to list the Pods again, but this time in the `custom-ns` namespace:

```
$ kubectl get Pods --namespace custom-ns
NAME      READY     STATUS      RESTARTS     AGE
nginx     1/1       Running     0            89m
nginx2    1/1       Running     0            23m
```

As you can see, this namespace also has a Pod called `nginx`, and it's not the same Pod that is contained in the default namespace. This is one of the major advantages of namespaces. By using them, your Kubernetes cluster can now define multiple resources with the same names, so long as they are in different namespaces. You can easily duplicate microservices or applications by using this element.

Also, note that you can override the key to the namespaces of the resources that you create declaratively. By adding the `-n` option to the `kubectl create` command, you force a namespace as the context for your command: `kubectl` will take the namespace that was passed in the command into account, not the one present in the YAML file. By doing this, it becomes very easy to duplicate your resources between different namespaces; for example, a production environment in a `production` namespace and a test environment in a `test` namespace. The possibilities are endless!

Understanding that not all resources are in a namespace

In Kubernetes, not all objects belong to a namespace. This is the case, for example, with nodes, which are represented at the cluster level by an entry of the Node kind but that does not belong to any particular namespace. You can list resources that do not belong to a namespace using the following command:

```
$ kubectl api-resources --namespaced false

NAME                        SHORTNAMES    APIGROUP
NAMESPACED    KIND
bindings
true          Binding
configmaps                  cm
true          ConfigMap
endpoints                   ep
true          Endpoints
events                      ev
true          Event
limitranges                 limits
true          LimitRange
...
```

You can also list all the resources that belong to a namespace by passing --namespaced to true:

```
$ kubectl api-resources --namespaced true
```

Now, let's learn how namespaces affect services DNSes.

Resolving a service using namespaces

As we discovered in *Chapter 7*, *Exposing Your Pods with Services*, Pods can be exposed through a type of object called Services. When created, services are assigned a DNS record that allows Pods in the cluster to access them.

However, when a Pod tries to call a service through DNS, it can only reach it if the service is in the same namespace as the Pod, which is limiting. Namespaces have a solution to this problem. When a service is created in a particular namespace, the name of its service is added to its DNS:

```
<service_name>.<namespace_name>.svc.cluster.local
```

By querying this domain name, you can easily query any service that is in any namespace in your Kubernetes cluster. So, you are not limited to that level. Pods are still capable of achieving inter-communication, even if they are not running in the same namespace.

Switching between namespaces with kubectl

When using kubectl to use Kubernetes, you saw that you have to use the -n or ---namespace option with your kubectl command to tell it the namespace you're targeting. But it is also possible to define a namespace at the configuration level. This way, you'll have kubectl always running requests against that namespace, even if you omit the -n parameter.

Let's create a few more namespaces. Then, we will demonstrate how to switch between them without the -n flag:

```
$ kubectl create ns another-ns
namespace/another-ns created
```

At the moment, we know that we have three namespaces in our cluster. You can use the following command to switch to another namespace:

```
$ kubectl config set-context $(kubectl config current-context)
--namespace=another-ns
Context "minikube" modified.
```

Running this command will update the current configuration that kubectl is using to make it point to the specified namespace. At the moment, any command you run will be executed against the another-ns namespace.

Important Note

Running the kubectl config command and sub-commands will only trigger modification or read operations against the ~/.kube/config file, which is the configuration file that kubectl is using.

When you're using the kubectl config set-context command, you're just updating that file to make it point to another namespace.

Knowing how to switch between namespaces with kubectl is important, but before you run any *write operations* such as kubectl delete or kubectl create, make sure that you are in the correct namespace. Otherwise, you should continue to use the -n flag. As this switching operation might be executed a lot of times, Kubernetes users tend to create Linux aliases to make them easier to use. Do not hesitate to define a Linux alias if you think it can be useful to you.

Now, let's learn how to retrieve the current namespace we're in.

Displaying the current namespace with kubectl

There is no easy way to display the current namespace besides displaying the ~/.kube/config file where this information is stored, as we mentioned previously. You should avoid running cat ~/.kube/config and just go for the kubectl config view ~/.kube/config command.

However, the problem is that this command displays a lot of data we don't need because we're just searching for the current namespace that is in use. That's why I recommend that you simply run the grep command against the output of kubectl config view to retrieve just the namespace:

```
$ kubectl config view | grep -i "namespace"
namespace: another-ns
```

Let's launch an NGINX Pod in the another-ns namespace. My ~/kube/config is pointing to it, so I don't have to use the -n flag anymore:

```
$ kubectl run nginx-Pod --image nginx # in another-ns
Pod/nginx created
```

Here, the command is telling me that I'm currently in the another-ns namespace. If I run kubectl get Pods, I should get no Pods at all:

```
$ kubectl get Pods
NAME     READY   STATUS    RESTARTS   AGE
nginx    1/1     Running   0          10s
```

Let's switch back to the `default` namespace. Immediately after, run `grep` against the `kubectl config view` output to see that the context in `~/kube/config` was updated successfully to the `default` namespace:

```
$ kubectl config set-context $(kubectl config current-context)
--namespace=default
Context "minikube" modified.

$ kubectl config view | grep -i "namespace"
namespace: default
```

Everything seems good. Now, let's run the `kubectl get Pods` command once more:

```
$ kubectl get Pods
No resources found in another-ns namespace.
```

Everything is good! I can't see `nginx-Pod` in the default namespace because it was created when the context was set to `another-ns`! Again, I strongly recommend that you define Linux aliases to switch and display namespaces quickly because you'll probably have to run these commands very often!

With that, we're done with the basics of namespaces in Kubernetes. We have learned what namespaces are, how to create and delete them, how to use them to keep a cluster clean and organized, and how to update the `kubeconfig` context to make `kubectl` point to a specific namespace. Now, we'll look at a few more advanced options related to namespaces. I think it is a good moment to introduce ResourceQuota and `Limit`, which you can use to limit the computing resources an application deployed on Kubernetes can access!

Configuring ResourceQuota and Limit at the namespace level

In this section, we're going to discover that namespaces can not only be used to sort resources in a cluster but also to limit the computing resources that Pods can access.

Using ResourceQuota and `Limits` with namespaces, you can create limits regarding the computing resources your Pods can access. We're going to learn how to proceed and exactly how to use these new concepts. In general, defining ResourceQuota and `Limits` is considered good practice for production clusters – that's why you should use them wisely.

Understanding why you should set ResourceQuota

Just like applications or systems, Kubernetes Pods will require a certain amount of computing resources to work properly. In Kubernetes, you can configure two types of computing resources:

- CPU
- Memory

All your worker nodes work together to provide CPU and memory, and in Kubernetes, adding more CPU and memory simply consists of adding more worker nodes to make room for more Pods. Depending on whether your Kubernetes cluster is based on-premises or in the cloud, adding more worker nodes can be achieved by purchasing the hardware and setup to do so or simply calling an API to create additional virtual machines.

Understanding how Pods consume these resources

When you launch a Pod on Kubernetes, a control plane component, known as kube-scheduler, will elect a worker node and assign the Pods to it. Then, the kubelet on the elected worker node will attempt to launch the containers defined in the Pod.

This process of worker node election is called Pod scheduling in Kubernetes.

When a Pod gets scheduled and launched on a worker node, it has, by default, access to all the resources the worker node has. Nothing is preventing it from accessing more and more CPU and memory as the application is used and ultimately, if the Pods run out of memory or CPU resources to work properly, then it simply crashes.

This can become a real problem because worker nodes can be used to run multiple applications – and so multiple Pods – at the same time. So, if 10 Pods are launched on the same worker node but one of them is consuming all the computing resources, then this will have an impact on all 10 Pods running on the worker node.

This problem means that you have two things you must consider:

- Each Pod should be able to require some computing resource to work.
- The cluster should be able to restrict the Pod's consumption so that it doesn't take all the resources available and share them with other Pods too.

It is possible to address these two problems in Kubernetes and we will discover how to use two options that are exposed to the Pod object. The first one is called **resources**, which is the option that's used to let a Pod indicate what amount of computing resources it needs, while the other one is called **limit** and will be used to indicate the maximum computing resources the Pod will have access to.

Let's discover these options now.

Understanding how Pods can require computing resources

The `request` and `limit` options will be declared within the YAML definition file of a Pod resource. Here, we're going to focus on the `request` option.

`request` is not a resource kind on its own – it is simply an option you can directly specify within your YAML definition file. request is simply the minimal amount of computing resource a Kubernetes Pod will need to work properly, and it is a really good practice to always define a request option for your Pods, at least for those that are meant to run in production.

Let's say that you want to launch an NGINX Pod on your Kubernetes cluster. By filling in the request option, you can tell Kubernetes that your NGINX Pod will need, at the bare minimum, 512 MiB of memory and 25% of a CPU core to work properly.

Here is the YAML definition file that will create this Pod:

```
# ~/Pod-in-namespace-with-request.yaml
apiVersion: v1
kind: Pod
metadata:
  name: nginx-with-request
  namespace: custom-ns
spec:
  containers:
  - name: nginx
    image: nginx:latest
    resources:
      requests:
        memory: "512Mi"
        cpu: "250m"
```

As you can see, you can define `request` at the container level and set different ones for each container.

There are three things to notice about this Pod:

- It is created inside the `custom-ns` namespace.
- It requires `512Mi` of memory.
- It requires `250m` of CPU.

But what do these metrics mean?

Memory is expressed in bytes, whereas CPU is expressed in **millicores** and allows fractional values. If you want your Pod to consume one entire CPU core, you can set the `cpu` key to `1000m`. If you want two cores, you must set it to `2000m`; for half of a core it will be `500m` or `0.5`, and so on. So, to explain, the preceding YAML definition is saying that the NGINX Pod we will create will forcibly need 512 MiB of memory since memory is expressed in bytes, and one-quarter of a CPU core of the underlying worker node. There is nothing related to the CPU or memory frequency here.

When you apply this YAML definition file to your cluster, the scheduler will look for a worker node that is capable of launching your Pods. This means that you need a worker node where there is enough room in terms of available CPU and memory to meet your Pods' requests.

But what if no worker node is capable of fulfilling these requirements? Here, the Pod will never be scheduled and never be launched. Unless you remove some running Pods to make room for this one, or unless you add a worker node that is capable of launching this Pod, it won't ever be launched.

> **Important Note**
>
> Keep in mind that Pods cannot span multiple nodes. So, if you set 8,000 m, which represents eight CPU cores, but your cluster is made up of two worker nodes with four cores each, then no worker node will be able to fulfill the request and your Pod won't be scheduled. That's why it's not good to set requests with really high values.

So, use the `request` option wisely – consider it as the minimum amount of resources your Pod will need to work. You have the risk that your Pod will never be scheduled if you set too high a request, but on the other hand, if your Pod is scheduled and launched successfully, this amount of resources is guaranteed.

Understanding how you can limit resource consumption

When you write a YAML definition file, you can define resource limits regarding what the Pod will be able to consume.

Setting a request won't suffice to do things properly. You should set a limit each time you set a resource. Setting a limit will tell Kubernetes to let the Pod consume resources up to that limit, and never above. This way, you're ensuring that your Pod won't take all the resources of the worker for itself.

Be careful, though – Kubernetes won't behave the same, depending on what kind of limit is reached. If the Pod reaches its CPU limit, it is going to be throttled and you'll notice performance degradation. But if your Pod reaches its memory limit, then it might be terminated. The reason for this is that memory is not something that can be throttled and Kubernetes still needs to ensure that other applications are not impacted and remain stable. So, be aware of that.

Without a limit, the Pod will be able to consume all the resources of the worker node. Here is an updated YAML file corresponding to the NGINX Pod we saw earlier, but now, it has been updated to define a limit on memory and CPU.

Here, the Pod will be able to consume up to 1 GiB of memory and up to 1 entire CPU core of the underlying worker node:

```
# ~/Pod-in-namespace-with-request-and-limit.yaml
apiVersion: v1
kind: Pod
metadata:
  name: nginx-with-request-and-limit
  namespace: custom-ns
spec:
  containers:
  - name: nginx
    imaage: nginx:latest
    resources:
```

```
    requests:
        memory: "512Mi"
        cpu: "250m"
    limits:
        memory: "1Gi"
        cpu: "1000m"
```

So, when you set a request, set a limit too. Now that you are aware of this request and limit consideration, don't forget to add it to your Pods!

Understanding why you need ResourceQuota

The good news is that you can entirely manage your Pod's consumptions by relying entirely on the Pod's request and limits options. All the applications in Kubernetes are just Pods, so setting these two options provides you with a strong and reliable way to manage resource consumption on your cluster, given that you never forget to set these.

It is super easy to forget these two options and deploy a Pod on your cluster that won't define any request or limit. Maybe it will be you, or maybe a member of your team, but the risk of deploying such a Pod is high because everyone can forget about these two options. And if you do so, the risk of application instability is high because a Pod without a limit can eat all the resources on the worker node is it launched on.

Kubernetes provides a way to mitigate this issue thanks to two objects called ResourceQuota and LimitRange. These two objects are extremely useful because they can enforce these constraints at the namespace level.

ResourceQuota is another resource kind, just like Pod or ConfigMap. The workflow is quite simple and consists of two steps:

1. First, you must create a new namespace.
2. Then, you must create a ResourceQuota and a LimitRange object inside that namespace.

Then, all the Pods that are launched in that namespace will be constrained by these two objects.

These quotas are used, for example, to ensure that all the containers that are accumulated in a namespace do not consume more than 4 GiB of RAM.

Therefore, it is possible and even recommended to set restrictions on what can and cannot run within Pods. It is strongly recommended that you always define a ResourceQuota and a LimitRange object for each namespace you create in your cluster!

Without these quotas, the deployed resources could consume as much CPU or RAM as they want, which could ultimately make your cluster and all the applications running on it unstable, given that the Pods don't hold requests and limits as part of their respective configuration.

In general, ResourceQuota is used to do the following:

- Limit CPU consumption within a namespace
- Limit memory consumption within a namespace
- Limit the absolute number of Pods operating within a namespace

There are a lot of use cases and you can discover all of them directly in the Kubernetes documentation. Now, let's learn how to define ResourceQuota in a namespace.

Creating a ResourceQuota

To demonstrate the usefulness of ResourceQuota, we are going to create one for a namespace we are going to call `custom-ns-with-resource-quota`. This ResourceQuota will be used to create requests and limits that all the Pods within this namespace combined will be able to use. Here is the YAML file that will create ResourceQuota; please note the resource kind:

```
# ~/resourcequota.yaml
apiVersion: v1
kind: ResourceQuota
metadata:
  name: my-resourcequota
spec:
  hard:
    requests.cpu: "1000m"
    requests.memory: "1Gi"
    limits.cpu: "2000m"
    limits.memory "2Gi"
```

ResourceQuota is enforcing some requests and limits on all the Pods that will be launched in the namespace where it is created. Keep in mind that the ResourceQuota object is scoped to one namespace.

This one is stating that in this namespace, the following will occur:

- All the Pods combined won't be able to request more than one CPU core.

- All the Pods combined won't be able to request more than 1 GiB of memory.

- All the Pods combined won't be able to consume more than two CPU cores.

- All the Pods combined won't be able to consume more than 2 GiB of memory.

You can have as many Pods and containers in the namespace, so long as they are respecting these constraints. Most of the time, ResourceQuotas are used to enforce constraints on requests and limits, but they can also be used to enforce these limits per namespace.

In the following example, the preceding ResourceQuota has been updated to specify that the namespace where it is created cannot hold more than 10 ConfigMaps and 5 services, which is pointless but a good example to demonstrate the different possibilities with ResourceQuota:

```
# ~/resourcequota-with-object-count.yaml
apiVersion: v1
kind: ResourceQuota
metadata:
  name: my-resourcequota
spec:
  hard:
    requests.cpu: "1000m"
    requests.memory: "1Gi"
    limits.cpu: "2000m"
    limits.memory "2Gi"
    configmaps: "10"
    services: "5"
```

Lastly, when you apply a ResourceQuota YAML definition file, don't forget to add the --namespace setting to set the namespace where ResourceQuota will be applied:

```
$ kubectl create -f ~/resourcequota-with-object-count.yaml
--namespace=custom-ns
```

Now, let's learn how to list ResourceQuotas.

Listing ResourceQuota

ResourceQuota objects can be accessed through `kubectl` using the quota's resource name option. The `kubectl get` command will do this for us:

```
$ kubectl get quotas -n custom-ns
```

Now, let's learn how to delete `ResourceQuota` from a Kubernetes cluster.

Deleting ResourceQuota

To remove a ResourceQuota object from your cluster, use the `kubectl delete` command:

```
$ kubectl delete -f quotas/resourcequota-with-object-count -n custom-ns
```

Now, let's introduce the notion of LimitRanges.

Introducing LimitRange

LimitRange is another object that is similar to ResourceQuota as it is created at the namespace level. The LimitRange object is used to enforce default requests and limit values to individual containers. Even by using the ResourceQuota object, you could create one object that consumes all the available resources in the namespace, so the LimitRange object is here to prevent you from creating too small or too large containers within a namespace.

Here is a YAML file that will create LimitRange:

```
# ~/limitrange.yaml
apiVersion: v1
kind: LimitRange
metadata:
  name: my-limitrange
spec:
  limits:
  - default:
      memory: 256Mi
      cpu: 500m
    defaultRequest:
```

```
      memory: 128Mi
      cpu: 250Mib
    max:
      memory: 1000Mi
      cpu: 1Gib
    min:
      memory: 128Mi
      cpu: 250m
    type: Container
```

As you can see, the LimitRange object consists of four important keys that all contain memory and cpu configuration. These keys are as follows:

- default: This helps you enforce default values for the memory and cpu limits of containers if you forget to apply them at the Pod level. Each container that is set up without limits will inherit these default ones from the LimitRange object.

- defaultRequest: This is the same as default but for the request option. If you don't set a request option to one of your containers in a Pod, the ones from this key in the LimitRange object will be automatically used by default.

- max: This value indicates the maximum limit (not request) a Pod can set. You cannot configure a Pod with a limit value that is higher than this one. It is the same as the default value in that it cannot be higher than the one defined here.

- min: This value works like max but for requests. It is the minimum amount of computing resources a Pod can request, and the defaultRequest option cannot be lower than this one.

Finally, note that if you omit the default and defaultRequest keys, then the max key will be used as the default key, and the min key will be used as the default key.

Defining LimitRange is a good idea if you want to protect yourself from forgetting to set requests and limits on your Pods. At least with LimitRange, these objects will have default limits and requests!

Just like the ResourceQuota object, don't forget to set the -n option to create LimitRange inside a namespace:

```
$ kubectl create -f ~/limitrange.yaml --namespace=custom-ns
```

Now, let's learn how to list LimitRanges.

Listing LimitRange

The `kubectl` command line will help you list your LimitRanges. Don't forget to add the `-n` flag to scope your request to a specific namespace:

```
$ kubectl get limit -n custom-ns
```

Now, let's learn how to delete LimitRange from a namespace.

Deleting LimitRange

Deleting LimitRange can be achieved using the `kubectl` command-line tool. Here is how to proceed:

```
$ kubectl delete limit/my-limitrange -n custom-ns
```

As always, don't forget to add the `-n` flag to scope your request to a specific namespace; otherwise, you may target the wrong one!

Summary

This chapter introduced you to namespaces, which are extremely important in Kubernetes. You cannot manage your cluster effectively without using namespaces because they provide logical resource isolation in your cluster. Most people use production and development namespaces, for example, or one namespace for each application. It is generally not rare to see clusters where dozens of namespaces are created.

We discovered that most Kubernetes resources are scoped to a namespace, though some are not. Keep in mind that, by default, Kubernetes is set up with a few default namespaces, such as `kube-system`, and that it is generally a bad idea to change the things that run in these namespaces, especially if you do not know what you are doing.

We also discovered that namespaces can be used to set quotas and limit the resources that Pods can consume, and it is a really good practice to set these quotas and limits at the namespace level using the ResourceQuota and LimitRange objects to prevent your Pods from consuming too many computing resources. Overall, you'll improve the stability of all the applications running on your cluster by setting these options and using them wisely.

In the next chapter, we'll continue to discover the basics of Kubernetes by discovering the concepts of `PersistentVolume` and `PersistentVolumeClaims`, which are the methods Kubernetes uses to deal with persistent data. It is going to be a very interesting chapter if you want to build and provision stateful applications on your Kubernetes clusters, such as database or file storage solutions.

9
Persistent Storage in Kubernetes

So far, we've learned about Kubernetes' key concepts, and this chapter is going to be the last one about Kubernetes' core concepts. So far, you've understood that Kubernetes is all about creating an object in the `etcd` datastore that will be converted into actual computing resources on the Nodes that are part of your cluster.

This chapter will focus on a concept called `PersistentVolume`. This is going to be another object that you will need to master in order to get persistent storage on your cluster. Persistent storage is achieved in Kubernetes by using the `PersistentVolume` resource kind, which has its own mechanics. Honestly, these can be relatively difficult to approach at first, but we are going to discover all of that!

In this chapter, we're going to cover the following main topics:

- Why you would want to use `PersistentVolume`
- Understanding how to mount `PersistentVolume` to your Pod claims
- Understanding the life cycle of `PersistentVolume` in Kubernetes
- Static and dynamic `PersistentVolume` provisioning

Technical requirements

- A working Kubernetes cluster (either local or cloud-based)

- A working `kubectl` CLI configured to communicate with the cluster

If you do not meet these technical requirements, you can follow *Chapter 2*, *Kubernetes Architecture – From Docker Images to Running Pods*, and *Chapter 3*, *Installing Your Kubernetes Cluster*, to get these two prerequisites.

Why you would want to use PersistentVolume

When you're creating your Pods, you have the opportunity to create volumes in order to share files between the containers created by them. However, these volumes can represent a massive problem: they are bound to the life cycle of the Pod that created them.

That is why Kubernetes offers another object called `PersistentVolume`, which is a way to create storage in Kubernetes that will not be bound to the life cycle of a Pod.

Introducing PersistentVolumes

Just like the Pod of `ConfigMap`, `PersistentVolume` is a resource kind that is exposed through `kube-apiserver`: you can create, update, and delete persistent volumes using YAML and `kubectl` just like any other Kubernetes object.

The following command will demonstrate how to list the `PersistentVolume` resource kind currently provisioned within your Kubernetes cluster:

```
$ kubectl get persistentvolume
No resource found
```

The `persistentvolume` object is also accessible with the plural form of `persistentvolumes` along with the alias of `pv`. The following three commands are essentially the same:

```
$ kubectl get persistentvolume
No resource found
$ kubectl get persistentvolumes
No resource found

$ kubectl get pv
No resource found
```

You'll find that the `pv` alias is very commonly used in the Kubernetes world, and a lot of people refer to persistent volumes as simply `pv`, so be aware of that. As of now, no `PersistentVolume` object has been created within my Kubernetes cluster, and that is why I don't see any resource listed in the output of the preceding command.

`PersistentVolume` is the object and, essentially, represents a piece of storage that you can attach to your Pod. That piece of storage is referred to as a *Persistent* one because it is not supposed to be tied with the life cycle of a Pod.

Indeed, as mentioned in *Chapter 5*, *Using Multi-Container Pods and Design Patterns*, Kubernetes Pods uses the notion of volumes. Additionally, we discovered the `emptyDir` and `hostPath` volumes, which, respectively, initiate an empty directory that your Pods can share. It also defines a path within the worker Node filesystem that will be exposed to your Pods. Both of these volumes were supposed to be attached to the life cycle of the Pod. This means that once the Pod is destroyed, the data stored within the volumes will be destroyed as well.

However, sometimes, you don't want the volume to be destroyed. You just want it to have its life cycle to keep both the volume and its data alive even if the Pod fails. That's where `PersistentVolumes` comes into play: essentially, they are volumes that are not tied to the life cycle of a Pod. Since they are a resource kind just like the Pods themselves, they can live on their own!

> **Important Note**
>
> Bear in mind that `PersistentVolumes` objects are just entries within the `etcd` datastore, and they are not an actual disk on their own.
>
> `PersistentVolume` is just a kind of pointer within Kubernetes to an external piece of storage, such as an NFS, a disk, an Amazon EBS volume, and more. This is so that you can access these technologies from within Kubernetes and in a Kubernetes way.

Simply put, `PersistentVolume` is essentially made up of two different things:

- A backend technology called a `PersistentVolume` type
- An access mode, such as `ReadWriteOnce`

You need to master both concepts in order to understand how to use `PersistentVolumes`. Let's begin by explaining what `PersistentVolume` types are.

Introducing PersistentVolume types

Kubernetes is supposed to be able to run on as much infrastructure as possible, and even though it started as a Google project, it can be used on many platforms, whether they are public clouds or private solutions.

As you already know, the simplest Kubernetes setup consists of a simple `minikube` installation, whereas the most complex Kubernetes setup can be made of dozens of servers on massively scalable infrastructure. All of these different setups will forcibly have different ways in which to manage persistent storage. For example, the three biggest public cloud providers have a lot of different solutions. Let's name a few, as follows:

- Amazon AWS EBS volumes
- Amazon AWS EFS filesystems
- Google GCE **Persistent Disk (PD)**
- Microsoft Azure disks

These solutions have their own design and set of principles along with their own logic and mechanics. Kubernetes was built with the principle that all of these setups should be abstracted using just one object to abstract all of the different technologies. And that single object is the `PersistentVolume` resource kind. The `PersistentVolume` resource kind is the object that is going to be attached to a running Pod. Indeed, a Pod is a Kubernetes resource and does not know what an EBS or a PD is; Kubernetes Pods only play well with `PersistentVolumes`, which is also a Kubernetes resource.

Whether your Kubernetes cluster is running on Google GKE, Amazon EKS, or whether it is a single `Minkube` cluster on your local machine has no importance. When you wish to manage persistent storage, you are going to create, use, and deploy `PersistentVolumes` objects, and then bind them to your Pods!

Here are some of the backend technologies supported by Kubernetes out of the box:

- `awsElasticBlockStore`: **Amazon EBS volumes**
- `gcePersistentDisk`: **Google Cloud PD**
- `azureDisk`: **Azure Disk**
- `azureFile`: **Azure File**
- `cephfs`: **Ceph-based filesystems**
- `csi`: **Container storage interface**

- `glusterfs`: **GlusterFS-based filesystems**
- `nfs`: **Regular network file storage**

The preceding list is not exhaustive: Kubernetes is extremely versatile and can be used with many storage solutions that can be abstracted as `PersistentVolume` objects in your cluster.

When you create a `PersistentVolume` object, essentially, you are creating a `YAML` file. However, this YAML file is going to have a different key/value configuration based on the backend technology used by the `PersistentVolume` objects.

The benefits brought by PersistentVolume

There are three major benefits of `PersistentVolume`:

- `PersistentVolume` is not bound to the life cycle of a Pod. If you want to remove a Pod that is attached to a `PersistentVolume` object, then the volume will survive.

- The preceding statement is also valid when a Pod crashes: the `PersistentVolume` object will survive the fault and not be removed from the cluster.

- `PersistentVolume` is cluster-wide; this means that it can be attached to any Pod running on any Node.

Bear in mind that these three statements are not always 100% valid. Indeed, sometimes, a `PersistentVolume` object can be affected by its underlying technology.

To demonstrate this, let's consider a `PersistentVolume` object that is, for example, a pointer to an Amazon EBS volume created on your AWS cloud. In this case, the worker Node will be an Amazon EC2 instance. In such a setup, `PersistentVolume` won't be available to any Node.

The reason is that AWS has some limitations around EBS volumes, which relates to the fact that an EBS volume can only be attached to one instance at a time, and that instance must be provisioned in the same availability zones as the EBS volume. From a Kubernetes perspective, this would only make `PersistentVolume` (EBS volumes) accessible from EC2 instances (that is, worker Nodes) in the same AWS availability zone, and several Pods running on different Nodes (EC2 instances) won't be able to access the `PersistentVolume` object at the same time.

However, if you take another example, such as an NFS setup, it wouldn't be the same. Indeed, you can access an NFS from multiple machines at once; therefore, a PersistentVolume object that is backed by an NFS would be accessible from several different Pods running on different Nodes without much problem. To understand how to make a PersistentVolume object on several different Nodes at a time, we need to consider the concept of access modes.

Introducing access modes

As the name suggests, access modes are an option you can set when you create a PersistentVolume type that will tell Kubernetes how the volume should be mounted.

PersistentVolumes supports three access modes, as follows:

- ReadWriteOnce: This volume allows read/write by only one Node at the same time.

- ReadOnlyMany: This volume allows read-only mode by many Nodes at the same time.

- ReadWriteMany: This volume allows read/write by multiple Nodes at the same time.

It is necessary to set at least one access mode to a PersistentVolume type, even if said volume supports multiple access modes. Indeed, not all PersistentVolume types will support all access modes.

Understanding that not all access modes are available to all PersistentVolume types

As mentioned earlier, PersistentVolume types are only a pointer to an external piece of storage. And that piece of storage is constrained by the backend technology that is providing it.

As mentioned earlier, one good example that we can use to explain this is the Amazon EBS volume technology that is accessible within the AWS cloud. When you create a PersistentVolume in Kubernetes, which is a pointer to an Amazon EBS volume, then that PersistentVolume will only support the ReadWriteOnce access mode, whereas NFS supports all three. This is because of the hard limitation mentioned earlier: an EBS volume can only be attached to one Amazon EC2 instance at a time, and it is a hard limit set by AWS. So, in the Kubernetes world, it can only be represented by a PersistentVolume type with an access mode set to ReadWriteOnce.

Simply put, these `PersistentVolume` types, and the concepts surrounding them, are simply Kubernetes concepts that are only valid within the Kubernetes scope and have absolutely no meaning outside of Kubernetes.

Some `PersistentVolume` objects will be permissive, while others will have a lot of constraints. And all of this is determined by the underlying technology they are pointing to. No matter what you do with `PersistentVolume`, you'll have to deal with the restrictions set by your cloud provider or underlying infrastructure.

Now, let's create our first `PersistentVolume` object.

Creating our first PersistentVolume

So, let's create a `PersistentVolume` on your Kubernetes cluster using the declarative way. Since this is a kind of complex resource, I heavily recommend that you try not to use the imperative way to create such resources:

```
# ~/pv-hostpath.yaml
apiVersion: v1
kind: PersistentVolume
metadata:
  name: pv-hostpath
  spec:
    volumeMode: Filesystem
    accessModes:
      - ReadWriteOnce
    capacity:
      storage: 1Gi
```

This is the simplest form of `PersistentVolume`. Essentially, this YAML file creates a `PersistentVolume` entry within the Kubernetes cluster. So, this `PersistentVolume` will be a `hostPath` type.

It could be a more complex volume such as a cloud-based disk, or an NFS, but in its simplest form, a `PersistentVolume` can simply be a `hostPath` type on the worker Node running your Pod.

How does Kubernetes PersistentVolumes handle cloud-based storage?

A bare `PersistentVolume` entry in our cluster can do nothing on its own and must be seen as a layer of abstraction on the Kubernetes level: outside Kubernetes, the `PersistentVolume` resource kind has no meaning.

That being said, the `PersistentVolume` resource kind is a pointer to something else, and that something else can be, for example, a disk, an NFS drive, a Google Cloud PD, or an Amazon EBS volume. All of these different technologies are managed differently. However, fortunately for us, in Kubernetes, they are all represented by the `PersistentVolume` object.

Simply put, the YAML file to build a `PersistentVolume` will be a little bit different depending on the backend technology that the `PersistentVolume` is backed by. For example, if you want your `PersistentVolume` to be a pointer to an Amazon EBS volume, you have to meet the following two conditions:

- The Amazon EBS volume must already be provisioned in your AWS cloud.
- The YAML file for your `PersistentVolume` must include the ID of the EBS volume, as it will be displayed in the AWS console.

And the same logic goes for everything else. For a `PersistentVolume` to work properly, it needs to forcibly be able to make the link between Kubernetes and the actual storage. So, you need to create a piece of storage or provision it outside of Kubernetes and then create the `PersistentVolume` entry by including the unique ID of the disk, or the volume, that is backed by a storage technology that is external to Kubernetes. Next, let's take a closer look at some examples of `PersistentVolume` YAML files.

Amazon EBS PersistentVolume YAML

This example displays a `PersistentVolume` object that is pointing to an Amazon EBS volume on AWS:

```
# ~/persistent-volume-ebs.yaml
apiVersion: v1
kind: PersistentVolume
metdata:
  name: persistent-volume-ebs
spec:
  capacity:
```

```
    storage: 2Gi
  accessModes:
    - ReadWriteOnce
  awsElasticBlockStore:
    volumeId: vol-xxxx
    fsType: ext4
```

As you can see, in this YAML file, awsElasticBlockStore is indicating that this PersistentVolume object is pointing to a volume on my AWS account. The exact Amazon EBS volume is identified by the volumeId key. And that's pretty much it. With this YAML file, Kubernetes is capable of finding the proper EBS volume and maintaining a pointer to it thanks to this PersistentVolume entry.

Of course, since EBS volumes are pure AWS, they can only be mounted on EC2 instances, which means this volume will never work if you attempt to attach it to something else. Now, let's examine a very similar YAML file; however, this time, it's going to point to a GCE PD.

GCE PersistentDisk PersistentVolume YAML

Here is the YAML file that is creating a PersistentVolume object that is pointing to an existing GCE PD:

```
# ~/persistent-volume-pd.yaml
apiVersion: v1
kind: PersistentVolume
metdata:
  name: persistent-volume-pd
spec:
  capacity:
    storage: 2Gi
  accessModes:
    - ReadWriteOnce
  gcePersistentDisk:
    pdName: xxxx
    fsType: ext4
```

Once again, please note that it is the same `kind: PersistentVolume` object as the one used by the Amazon EBS `PersistentVolume` object. In fact, it is the same object and the same interface from the Kubernetes side. The only difference is the configuration under `gcePersistentDisk`, which, this time, points to a PD created on Google Cloud. Kubernetes is so versatile that it can fetch and use different cloud storage solutions just like that.

Next, let's explore one last example in YAML, this time using NFS.

NFS PersistentVolume YAML

Here is an example YAML file that can create a `PersistentVolume` object that is backed by an NFS drive:

```
# ~/persistent-volume-nfs.yaml
apiVersion: v1
kind: PersistentVolume
metdata:
  name: persistent-volume-nfs
spec:
  capacity:
    storage: 2Gi
  accessModes:
    - ReadWriteMany
  nfs:
    path: /opt/nfs
    server: nfsxxxx
    fsType: ext4
```

Again, note that this time, we're still using the `kind: PersistentVolume` entry. Additionally, we are now specifying an `nfs` path configuration with the path as well as the server address. Now, let's discuss a little bit about the provisioning of the storage resources.

Can Kubernetes handle the provisioning or creation of the resource itself?

The fact that you need to create the actual storage resource separately and then create a `PersistentVolume` in Kubernetes might be tedious.

Fortunately for us, Kubernetes is also capable of communicating with the **APIs** of your cloud provider in order to create the volumes or disk on the fly. There is something called **dynamic provisioning** that you can use when it comes to managing `PersistentVolume`. It makes things a lot simpler when dealing with `PersistentVolume` provisioning, but it only works on supported cloud providers.

However, this is an advanced topic, so we will discuss it, in more detail, later in this chapter.

Now that we know how to provision `PersistentVolume` objects inside our cluster, we can try to mount them. Indeed, in Kubernetes, once you create a `PersistentVolume`, you need to mount it to a Pod so that it becomes in use. Things will get slightly more advanced and conceptual here; this Kubernetes uses an intermediate object in order to mount a `PersistentVolume` to Pods. And this intermediate object is called `PersistentVolumeClaim`. Let's focus on it next.

Understanding how to mount a PersistentVolume to your Pod claims

So far, we've learned that Kubernetes makes use of two objects to deal with persistent storage technologies. The first one is `PersistentVolumes`, which represents a piece of storage, and we quoted Google Cloud PD and Amazon EBS volumes as possible backends for `PersistentVolume`. Additionally, we discovered that depending on the technology that `PersistentVolume` is relying on, it is going to be exposed to one or more Pods using access modes.

That being said, we can now try to mount a `PersistentVolume` object to a Pod. To do that, we will need to use another object, which is the second object we need to explore in this chapter, called `PersistentVolumeClaim`.

Introducing PersistentVolumeClaim

Just like `PersistentVolume` and `ConfigMap`, `PersistentVolumeClaim` is another independent resource kind living within your Kubernetes cluster and is the second resource kind that we're going to examine in this chapter.

This object can appear to be a little bit more complex to understand compared to the others. First, bear in mind that even if both names are almost the same, `PersistentVolume` and `PersistentVolumeClaim` are two distinct resources that represent two different things.

You can list the `PersistentVolumeClaim` resource kind created within your cluster using `kubectl`, as follows:

```
$ kubectl get persistentvolumeclaims
No resources found in default namespace.
```

The following output is telling me that I don't have any `PersistentVolumeClaim` resources created within my cluster. Please note that the `pvc` alias works, too:

```
$ kubectl get pvc
No resources found in default namespace.
```

You'll quickly find that a lot of people working with Kubernetes refer to the `PersistentVolumeClaim` resources simply with `pvc`. So, don't be surprised if you see the term `pvc` here and there while working with Kubernetes. That being said, let's explain what `PersistentVolumeClaim` resources are in Kubernetes.

Splitting storage creation and storage consumption

The key to understanding the difference between `PersistentVolume` and `PersistentVolumeClaims` is to understand that one is meant to represent the storage itself, whereas the other one represents the request for storage that a Pod makes to get the actual storage.

The reason is that Kubernetes is supposed to be used by two types of people:

- **Kubernetes administrator**: This person is supposed to maintain the cluster, operate it, and also add computation resources and persistent storage.

- **Kubernetes application developer**: This person is supposed to develop and deploy an application, so put simply, consume the computation resource and storage offered by the administrator.

In fact, there is no problem if you handle both roles in your organization; however, this information is crucial to understand the workflow to mount `PersistentVolume` to Pods.

Kubernetes was built with the idea that a `PersistentVolume` object should belong to the cluster administrator scope, whereas `PersistentVolumeClaims` objects belong to the application developer scope. It is up to the cluster administrator to add `PersistentVolumes` since they might be hardware resources, whereas developers have a better understanding of what amount of storage and what kind of storage is needed, and that's why the `PersistentVolumeClaim` object was built.

Essentially, a Pod cannot mount a `PersistentVolume` object directly. It needs to explicitly ask for it. And that *asking* action is achieved by creating a `PersistentVolumeClaim` object and attaching it to the Pod that needs a `PersistentVolume` object.

This is the only reason why this additional layer of abstractions exists.

The summarized PersistentVolume workflow

Once the developer has built the application, it is their responsibility to ask for a `PersistentVolume` object if needed. To do that, the developer will write two YAML manifest files:

- One file is for the Pod application.
- The other file is for `PersistentVolumeClaim`.

The Pod application must be written so that the `PersistentVolumeClaim` object is mounted as a `volumeMount` configuration key in the YAML file. Please note that for it to work, the `PersistentVolumeClaim` object needs to be in the same namespace as the application Pod that is mounting it. The `PersistentVolume` object is never mounted directly to the Pod.

When both YAML files are applied and both resources are created in the cluster, the `PersistentVolumeClaim` object will look for a `PersistentVolume` object that matches the criteria required in the claim. Supposing that a `PersistentVolume` object capable of fulfilling the claim is created and ready in the Kubernetes cluster, the `PersistentVolume` object will be attached to the `PersistentVolumeClaim` object.

If everything is okay, the claim is considered fulfilled, and the volume is correctly mounted to the Pod: if you understand this workflow, essentially, you understand everything related to `PersistentVolume` usage.

Let's summarize this as follows:

1. A Kubernetes administrator created a `PersistentVolume` object.
2. A Kubernetes developer requests a `PersistentVolume` object for their application using the `PersistentVolumeClaim` object.
3. The developer writes its YAML file so that the `PersisentVolumeClaim` object is configured as a volume mount to the Pod.

4. Once the Pod and its `PersisentVolumeClaim` are created, Kubernetes fetches a `PersisentVolume` object that answers what is requested in the PVC.

5. Then, the `PersisentVolume` object is accessible from the Pod and is ready to receive read or write operations based on the `PersisentVolume` access mode.

This setup might seem complex to understand at first, but you will quickly become used to it.

Creating a Pod with a PersistentVolumeClaim object

In this section, I will create a Pod that mounts `PersisentVolume` within a `minikube` cluster. This is going to be a kind of `PersisentVolume` object, but this time, it will not be bound to the life cycle of the Pod. Indeed, since it will be managed as a real `PersisentVolume` object, the `hostPath` type will get its life cycle independent of the Pod.

The very first thing to do is to create the `PersisentVolume` object that will be a `hostPath` type. Here is the YAML file to do that. Please note that I created this `PersisentVolume` object with some arbitrary labels in the `metadata` section. This is so that it will be easier to fetch it from the `PersistentVolumeClaim` object later:

```
# ~/pv.yaml
apiVersion: v1
kind: PersistentVolume
metadata:
  name: my-hostpath-pv
  labels:
    type: hostpath
    env: prod
spec:
  capacity:
    storage: 1Gi
  accessModes:
    - ReadWriteOnce
  hostPath:
    path: "/Users/me/test"
```

We can now list the `PersisentVolume` entries available in our cluster, and we should observe that this one exists. Please note that the pv alias works, too:

```
$ kubectl create -f pv.yaml
persistentvolume/my-hostpath-pv created
$ kubectl get pv


NAME        CAPACITY    ACCESS MODES    RECLAIM POLICY    STATUS
CLAIM    STORAGECLASS    REASON    AGE
my-hostpath-pv    1Gi           RWO              Retain
Available                                          49s
```

We can see that the `PersisentVolume` was successfully created.

Now, we need to create two things in order to mount the `PersisentVolume` object:

- A `PersistentVolumeClaim` object that targets this specific `PersisentVolume` object

- A Pod to use the `PersistentVolumeClaim` object

Let's proceed, in order, with the creation of the `PersistentVolumeClaim` object:

```
# ~/pvc.yaml
apiVersion: v1
kind: PersistentVolumeClaim
metadata:
  name: my-hostpath-pvc
spec:
  resources:
    requests:
      storage: 1Gi
  selector:
    matchLabels:
        type: hostpath
        env: prod
```

The important aspect of this `PersisentVolumeClaim` object is that it is going to fetch the proper volume by using its labels, using the `selector` key. Let's create it and check that it was successfully created in the cluster. Please note that the `pvc` alias also works here:

```
$ kubectl create -f pvc.yaml
persistentvolumeclaim/my-hostpath-pvc created
$ kubectl get pvc
```

```
NAME                STATUS      VOLUME      CAPACITY    ACCESS MODES
STORAGECLASS    AGE
my-hostpath-pvc     Pending
standard        53s
```

Now that the `PersisentVolume` object and the `PersistentVolumeClaim` object exist, I can create a Pod that will mount the PV using the PVC.

Let's create an NGINX Pod that will do the job:

```yaml
# ~/Pod.yaml
apiVersion: v1
kind: Pod
metadata:
  name: nginx
spec:
  containers
  - image: nginx
    name: nginx
    volumeMounts:
      - mountPath: "/var/www/html"
        name: mypersistentvolume
    volumes:
      - name: mypersistentvolume
        persistentVolumeClaim:
          claimName: my-hostpath-pvc
```

As you can see, in the `volumeMounts` section, the `PersistentVolumeClaim` object is referenced as a volume, and we reference the PVC by its name. Note that the PVC must live in the same namespace as the Pod that mounts it. This is because PVCs are namespace-scoped resources, whereas PVs are not. There are no labels and selectors for this one; to bind a PVC to a Pod, you simply need to use the PVC name.

That way, the Pod will become attached to the `PersistentVolumeClaim` object, which will find the corresponding `PersisentVolume` object. This, in the end, will make the host path available and mounted on my NGINX Pod.

Now we can create the three objects in the following order:

1. The `PersisentVolume` object
2. The `PersistentVolumeClaim` object
3. The `Pod` object

Note that before you go any further, you need to make sure that the `/Users/me/test` directory exists on your host machine or your worker Node. This is because this is the path specified in the PV definition.

You can achieve that using the following commands if you have not already created these resources in your cluster:

```
$ kubectl create -f pvc.yaml
persistentvolumeclaim/my-hostpath-pvc created
$ kubectl create -f Pod.yaml
Pod/nginx created
```

Now, let's check that everything is okay by checking the status of our `PersistentVolumeClaim` object:

```
$ kubectl get pvc
NAME                  STATUS     VOLUME
CAPACITY     ACCESS MODES     STORAGECLASS      AGE
my-hostpath-pvc       Bound      pvc-f7e322d1-6ac8-45b6-b9b2-
636323e38b55    1Gi           RWO                standard          3m44s
```

Everything seems to be okay! We have just demonstrated a typical workflow. No matter what kind of storage you need, it's always going to be the same:

1. First, the storage must be provisioned.

2. Second, you create a PV entry in order to have a pointer to it in Kubernetes.

3. Third, you provision a PVC capable of fetching this PV.

4. Fourth, you mount the PVC (not the PV directly) as a volume mount to a Pod.

And that's it!

So far, we have learned what `PersistentVolume` and `PersistentVolumeClaim` objects are and how to use them to mount persistent storage on your Pods.

Next, we must continue our exploration of the `PersistentVolume` and `PersistentVolumeClaim` mechanics by explaining the life cycle of these two objects. Because they are independent of the Pods, their life cycles have some dedicated behaviors that you need to be aware of.

Understanding the life cycle of a PersistentVolume object in Kubernetes

`PersistentVolume` objects are good if you want to keep the state of your app without being constrained by the life cycle of the Pods or containers that are running them.

However, since `PersistentVolume` objects get their very own life cycle, they have some very specific mechanics that you need to be aware of when you're using them. We'll take a closer look at them next.

Understanding that PersistentVolume objects are not bound to namespaces

The first thing to be aware of when you're using `PersistentVolume` objects is that they are not `namespaced` resources, but `PersistentVolumeClaims` objects are.

That's something very important to know. This is because when a Pod is using a `PersistentVolume` object, it is only exposed to the `PersistentVolumeClaims` object. So, its one requirement is that it is created in the same `namespace` as the Pod that is using it.

That being said, PersistentVolume objects are constrained by namespaces, unlike PersistentVolumeClaim objects. Indeed, they are created cluster-wide. So, do bear that in mind: PersistentVolumeClaim needs to be created in the same namespace resource as the Pods using them, but they are able to fetch PersistentVolume resources that are not in any namespace resource at all.

To figure this out, I invite you to create the following PersistentVolume object using the following YAML file, which will create a PersistentVolume called new-pv-hostpath:

```
# ~/new-pv-hostpath.yaml
apiVersion: v1
kind: PersistentVolume
metadata:
  name: new-pv-hostpath
  spec:
    accessModes:
      - ReadWriteOnce
    capacity:
      storage: 1Gi
    hostPath:
      path: "/home/user/mydirectory"
```

Once the file has been created, we can apply it against our cluster using the kubectl create -f new-pv-hostpath.yaml command:

```
$ kubectl create -f new-pv-hostpath
persistentvolume/new-pv-hostpath created
```

Then, we will run the kubectl get pv/new-hostpath-pv -o yaml | grep -i "namespace" command, which will output nothing. This means the PersistentVolume object is not a namespace:

```
$ kubectl get api-resources -namespaced=false #kubectl
Olala...
```

As you can see, the PersistentVolume object appears in the output of the command, which means it is not living in any namespace!

Now, let's examine another important aspect of `PersistentVolume` known as reclaiming a policy. This is something that is going to be important when you want to unmount a PVC from a running Pod.

Reclaiming a PersistentVolume object

When it comes to `PersistentVolume`, there is a very important option that you need to understand, which is the reclaim policy. But what does this option do?

This option will tell Kubernetes what treatment it should give to your `PersistentVolume` object when you delete the corresponding `PersistentVolumeClaim` object that was attaching it to the Pods.

Indeed, deleting a `PersistentVolumeClaim` object consists of deleting the link between the Pod(s) and your `PersistentVolume` object, so it's like you unmount the volume and then the volume becomes available again for another application to use. However, in some cases, you don't want that behavior; instead, you want your `PersistentVolume` object to be automatically removed when its corresponding `PersistentVolumeClaim` object has been deleted. That's why the reclaim policy option exists and it is what you should configure.

The reclaim policy can be set to three statuses, as follows:

* Delete
* Retain
* Recycle

Let's explain these three reclaim policies.

The delete one is the simplest of the three. When you set your reclaim policy to delete, the `PersistentVolume` object will be wiped out and the `PersistentVolume` entry will be removed from the Kubernetes cluster when the corresponding `PersistentVolumeClaim` object is deleted. That's the behavior for sensitive data. So, use this when you want your data to be deleted and not used by any other application. Bear in mind that this is a permanent option, so you might want to build a backup strategy with your underlying storage provider if you need to recover anything.

The retain policy is the second policy and is contrary to the delete policy. If you set this reclaim policy, the `PersistentVolume` object won't be deleted if you delete its corresponding `PersistentVolumeClaim` object. Instead, the `PersistentVolume` object will enter the released status, which means it is still available in the cluster, and all of its data can be manually retrieved by the cluster administrator.

The third policy is the recycle reclaim policy, which is a kind of combination of the previous two policies. First, the volume is wiped of all its data, such as a basic `rm -rf volume/*` volume. However, the volume itself will remain accessible in the cluster, so you can mount it again on your application.

The reclaim policy can be set in your cluster directly in the YAML definition file at the `PersistentVolume` level.

Updating a reclaim policy

The good news with a reclaim policy is that you can change it after the `PersistentVolume` object has been created; it is a mutable setting. To do that, you can simply list the PVs in your cluster and then issue a `kubectl patch` command to update the PV of your choice:

```
$ kubectl get pv
NAME                                         CAPACITY
ACCESS MODES    RECLAIM POLICY    STATUS     CLAIM
STORAGECLASS    REASON     AGE
my-hostpath-pv                               1Gi
RWO             Retain            Available
24m
```

As you can see, this PV has a `Retain` reclaim policy. I'll now update it to `Delete` using the `kubectl` patch command against the `my-hostpath-pv` PersistentVolume:

```
$ kubectl patch pv/my-hostpath-pv -p
'{"spec":{"persistentVolumeReclaimPolicy":"Delete"}}'
persistentvolume/my-hostpath-pv patched
```

```
$ kubectl get pv
NAME                                         CAPACITY
ACCESS MODES    RECLAIM POLICY    STATUS     CLAIM
STORAGECLASS    REASON     AGE
my-hostpath-pv                               1Gi
RWO             Delete            Available
3h13m
```

We can observe that the reclaim policy was updated from `Retain` to `Delete`!

Now, let's discuss the different statuses that PVs and PVCs can have.

Understanding PersistentVolume and PersistentVolumeClaims statuses

Just like Pods can be in a different state, such as `Pending`, `ContainerCreating`, `Running`, and more, `PersistentVolume` and `PersistentVolumeClaim` can also hold different states. You can identify their state by issuing the `kubectl get pv` and `kubectl get pvc` commands.

`PersistentVolume` has the following different states that you need to be aware of:

- `Available`

- `Bound`

- `Terminating`

On their side, `PersistentVolumeClaim` can hold one additional status, which is the `Terminating` one.

Let's explain these different states in more detail.

The `Available` status indicates that the `PersistentVolume` object is created and ready to be mounted by a `PersistentVolumeClaim` object. There's nothing wrong with it, and the PV is just ready to be used.

The `Bound` status indicates that the `PersistentVolume` object is currently mounted to one or several Pods. The `PersistentVolume` is bound to a `PersistentVolumeClaim` object. Essentially, it indicates that the volume is currently in use. When this status is applied to a `PersistentVolumeClaim` object, this indicates that the PVC is currently in use: that is, a Pod is using it and has access to a PV through it.

The `Terminating` status applies to a `PersistentVolumeClaim` object. This is the status the PVC enters after you issue a `kubectl delete pvc` command. It is during this phase that the PV the PVC is bound to is destroyed and wiped out. This happens if its reclaim policy is set to `Retain` and is destroyed when it is then set to `Delete`.

We now have all the basics relating to `PersistentVolume` and `PersistentVolumeClaim` that should be enough to start using persistent storage in Kubernetes. However, there's still something important to know about this topic, and it is called dynamic provisioning. This is a very impressive aspect of Kubernetes that makes it able to communicate with cloud provider APIs to create persistent storage on the cloud. Additionally, it can make this storage available on the cluster by dynamically creating PV objects. In the next section, we will compare static and dynamic provisioning.

Static and dynamic PersistentVolume provisioning

So far, we've only provisioned `PersistentVolume` by doing static provisioning. Now we're going to discover dynamic `PersistentVolume` provisioning, which enables `PersistentVolume` provisioning directly from the Kubernetes cluster.

Static versus dynamic provisioning

So far, when using static provisioning, you have learned that you have to follow this workflow:

1. You create the piece of storage against the cloud provider or the backend technology.

2. Then, you create the `PersistentVolume` object to serve as a Kubernetes pointer to this actual storage.

3. Following this, you create a Pod and a PVC to bind the PV to the Pod.

That is called static provisioning. It is static because you have to create the piece of storage before creating the PV and the PVC in Kubernetes. It works well; however, at scale, it can become more and more difficult to manage, especially if you are managing dozens of PV and PVC. Let's say you are creating an Amazon EBS volume to mount it as a `PersistentVolume` object, and you would do it like this with static provisioning:

1. Authenticate against the AWS console.

2. Create an EBS volume.

3. Copy/paste its unique ID to a `PersistentVolume` YAML definition file.

4. Create the PV using your YAML file.

5. Create a PVC to fetch this PV.

6. Mount the PVC to the Pod object.

Again, it should work, but it would become extremely time-consuming to do at scale, with possibly dozens and dozens of PVs and PVCs.

That's why Kubernetes developers decided that it would be better if Kubernetes was capable of provisioning the piece of actual storage on your behalf along with the `PersistentVolume` object to serve as a pointer to it. This is known as dynamic provisioning.

Introducing dynamic provisioning

When using dynamic provisioning, you configure your Kubernetes cluster so that it authenticates for you on your AWS account. Then, you issue a command to provision an EBS volume and automatically create a `PersistentVolume` claim to bind it to a Pod.

That way, you can save a huge amount of time by getting things automated. Dynamic provisioning is so useful because Kubernetes supports a wide range of storage technologies. We already introduced a few of them earlier in this chapter, when we mentioned NFS, Google PD, Amazon EBS volumes, and more.

But how does Kubernetes achieve this versatility? Well, the answer is that it makes use of a third resource kind, which we're going to discover in this chapter, that is the `StorageClass` object.

Introducing StorageClasses

`StorageClass` is another resource kind exposed by `kube-apiserver`. This resource kind is the one that grants Kubernetes the ability to deal with several underlying technologies transparently.

You can access and list the `storageclasse` resources created within your Kubernetes cluster by using `kubectl`. Here is the command to list the storage classes:

```
$ kubectl get storageclass
NAME                     PROVISIONER                     RECLAIMPOLICY
VOLUMEBINDINGMODE        ALLOWVOLUMEEXPANSION    AGE
standard (default)       k8s.io/minikube-hostpath    Delete
Immediate                false                   24d
```

Additionally, you can use the plural form of `storageclasses` along with the `sc` alias. The following three commands are essentially the same:

```
$ kubectl get storageclass
$ kubectl get storageclasses
$ kubectl get sc
```

Note that I haven't included the output of the command for simplicity, but it is essentially the same for the three commands. There are two fields within the command output that are important to us:

- NAME: This is the name and the unique identifier of the `storageclass` object.
- PROVISIONNER: This is the name of the underlying storage technology: this is basically a piece of code the Kubernetes cluster uses to interact with the underlying technology.

> **Important Note**
>
> Note that you can create multiple `StorageClass` objects that use the same `provisioner`.

As I'm currently using a `minikube` cluster, I have a `storageclass` resource called standard that is using the `k8s.io/minikube-hostpath` provisioner.

This provider deals with my host filesystem to automatically create provisioned host path volumes for my Pods, but it could be the same for Amazon EBS volumes or Google PDs.

Here is the same output for a Kubernetes cluster based on Google GKE:

```
$ kubectl get sc
```

And here is the same output for a Kubernetes cluster based on Amazon EKS:

```
$ kubectl get sc
```

As you might have gathered, by default, we get different storage because all of these clusters need to access different kinds of storage. In GKE, Google built a storage class with a provisioner that was capable of interacting with the Google PD's API, which is a pure Google Cloud feature, In contrast, in AWS, we have a `storageclass` object with a provisioner that is capable of dealing with EBS volume APIs. These provisioners are just libraries that interact with the APIs of these different cloud providers.

The `storageclass` objects are the reason why Kubernetes is capable of dealing with so many different storage technologies. From a Pod perspective, no matter if it is an EBS volume, NFS drive, or GKE volume, the Pod will only see a `PersistentVolume` object. All the underlying logic dealing with the actual storage technology is implemented by the provisioner the `storageclass` object uses.

The good news is that you can add as many `storageclass` objects with their provisioner as you want to your Kubernetes cluster in a plugin-like fashion. As of writing, the following is a list of `PersistentVolume` types that are supported in Kubernetes:

- `awsElasticBlockStore`: Amazon EBS volumes
- `gcePersistentDisk`: Google Cloud PD
- `azureDisk`: Azure Disk
- `azureFile`: Azure File
- `cephfs`: Ceph-based filesystems
- `csi`: Container storage interface
- `glusterfs`: GlusterFS-based filesystems
- `nfs`: Regular network file storage

By the way, nothing is preventing you from expanding your cluster by adding `storageclasses` to your cluster. You'll simply add the ability to deal with different storage technologies from your cluster. For example, I can add an Amazon EBS `storageclass` object to my `minikube` cluster. But while it is possible, it's going to be completely useless. Indeed, since my `minikube` setup is not running on an EC2 instance but my local machine, I won't be able to attach an EBS.

Understanding the role of PersistentVolumeClaim for dynamic storage provisioning

When using dynamic storage provisioning, the `PersistentVolumeClaim` object will get an entirely new role. Since `PersistentVolume` is gone in this use case, the only object that will be left for you to manage is the `PersistentVolumeClaim` one.

Let's demonstrate this by creating an NGINX Pod that will mount a `hostPath` type dynamically. In this example, the administrator won't have to provision a `PersistentVolume` object at all. This is because the `PersistentVolumeClaim` object and the `StorageClass` object will be able to create and provision it together.

Let's start by creating a new namespace, called `dynamicstorage`, where we will run our examples:

```
$ kubectl create ns dynamicstorage
namespace/dynamicstorage created
```

Now, let's run a kubectl get sc command to check that we have a storage class that is capable of dealing with the hostPath that is provisioned in our cluster.

For this specific storageclass object in this specific Kubernetes setup (minikube) we don't have to do anything to get the storageclass object as it is created by default at cluster installation. However, this might not be the case depending on your Kubernetes distribution.

Bear that in mind because it is very important: clusters that have been set up on GKE might have default storage classes that are capable of dealing with Google's storage offerings, whereas an AWS-based cluster might have storageclass to communicate with Amazon's storage offerings and more. With minikube, we have at least one default storageclass object that is capable of dealing with a hostPath-based PersistentVolume object. If you understand that, you should understand that the output of the kubectl get sc command will be different depending on where your cluster has been set up:

```
$ kubectl get sc
NAME                    PROVISIONER                 RECLAIMPOLICY
VOLUMEBINDINGMODE       ALLOWVOLUMEEXPANSION    AGE
standard (default)      k8s.io/minikube-hostpath    Delete
Immediate               false                   11h
```

As you can see, we do have a storage class called standard on our cluster that is capable of dealing with hostPath.

> **Important Note**
>
> Some complex clusters spanning across multiple clouds and or on-premises might be provisioned with a lot of different storageclass objects to be able to communicate with a lot of different storage technologies. Bear in mind that Kubernetes is not tied to any cloud provider and, therefore, does not force or limit you in your usage of backing storage solutions.

Now, we will create a PersistentVolumeClaim object that will dynamically create a hostPath type. Here is the YAML file to create the PVC. Please note that storageClassName is set to standard:

```
# ~/pvc-dynamic.yaml
apiVersion: v1
kind: PersistentVolumeClaim
metadata:
```

```
    name: my-dynamic-hostpath-pvc
spec:
  accessModes:
    - ReadWriteOnce
  storageClassName: standard # VERY IMPORTANT !
  resources:
    requests:
      storage: 1Gi
  selector:
    matchLabels:
        type: hostpath
        env: prod
```

Following this, we can create it in the proper namespace:

```
$ kubectl create -f pvc-dynamic.yaml -n dynamicstorage
persistentvolumeclaim/my-dynamic-hostpath-pvc created
```

Now that this PVC has been created, we can add a new Pod that will mount this `PersistentVolumeClaim` object. As soon as this claim has been mounted, it will create a `PersistentVolume` object using the provisioner and then bind to it.

That's how dynamic provisioning works, and it is the same behavior no matter if it is on-premise or in the cloud. Here is a YAML definition file of a Pod that will mount the `PersistentVolumeClaim` object created earlier:

```
# ~/pvc-dynamic.yaml
apiVersion: v1
kind: Pod
metadata:
  name: nginx-dynamic-storage
spec:
  containers
  - image: nginx
    name: nginx
    volumeMounts:
      - mountPath: "/var/www/html"
        name: mypersistentvolume
    volumes:
```

```
    - name: mypersistentvolume
      persistentVolumeClaim:
        claimName: my-dynamic-hostpath-pvc
```

Now let's create it in the correct namespace:

```
$ kubectl create -f Pod-dynamic.yaml -n dynamicstorage
Pod/nginx-dynamic-storage created
```

Next, let's list the `PersistentVolume` object. If everything worked, we should get a brand new `PersistentVolume` object that has been dynamically created and is in the bound state:

```
$ kubectl get pv
NAME                                        CAPACITY
ACCESS MODES    RECLAIM POLICY    STATUS    CLAIM
STORAGECLASS    REASON    AGE
pvc-56b79a65-86f6-4db5-b800-2ec415156097    1Gi         RWO
Delete          Bound             dynamicstorage/my-dynamic-
hostpath-pvc    standard                 7m19s
```

Everything is OK! We're finally done with dynamic provisioning! Please note, by default, the reclaim policy will be set to `Delete` so that the PV is removed when the PVC that created it is removed, too. Don't hesitate to change the reclaim policy if you need to retain sensitive data.

Summary

We have arrived at the end of this chapter, which taught you how to manage persistent storage on Kubernetes. You discovered that `PersistentVolume` is a resource kind that acts as a point to an underlying resource technology, such as `hostPath` and NFS, along with cloud-based solutions such as Amazon EBS and Google PDs.

Additionally, you discovered that you cannot use your `PersistentVolume` object without `PersistentVolumeClaim`, and that `PersistentVolumeClaim` acts as an object to fetch and mount `PersistentVolume` to your Pods. You learned that `PersistentVolume` can hold different reclaim policies, which makes it possible to remove, recycle, or retain them when their corresponding `PersistentVolumeClaim` object gets removed.

Finally, we discovered what dynamic provisioning is and how it can help us. Bear in mind that you need to be aware of this feature because if you create and retain too many volumes, it can have a negative impact on your cloud bill at the end of the month.

We're now done with the basics of Kubernetes, and this chapter is also the end of this section. In the next section, you're going to discover Kubernetes controllers, which are objects designed to automate certain tasks in Kubernetes, such as maintaining a number of replicas of your Pods, either using the Deployment resource kind or the StatefulSet resource kind. There are still a lot of things to learn!

Section 3: Using Managed Pods with Controllers

At this point in the book, you know how to use Kubernetes. But now, we want to familiarize you with cloud excellence. The header was introduced as part of Kubernetes basics in the previous chapters; we now want it to be able to run Kubernetes workloads in production for real. You will be taught the concepts and objectives of high availability, fault tolerance, elasticity, and scaling as well as application life cycle management.

This part of the book comprises the following chapters:

- *Chapter 10, Running Production-Grade Kubernetes Workloads*
- *Chapter 11, Deployment – Deploying Stateless Applications*
- *Chapter 12, StatefulSet – Deploying Stateful Applications*
- *Chapter 13, DaemonSet – Maintaining Pod Singletons on Nodes*

10
Running Production-Grade Kubernetes Workloads

In the previous chapters, we have focused on containerization concepts and the fundamental Kubernetes building blocks, such as Pods, Jobs, and ConfigMaps. Our journey so far has covered mostly single-machine scenarios, where the application requires only one container host or Kubernetes Node. For **production-grade** Kubernetes container workloads, you have to consider different aspects, such as scalability, **high availability (HA)**, and load balancing, and this always requires **orchestrating** containers running on multiple hosts.

Briefly, **container orchestration** is a way of managing multiple containers' life cycles in large, dynamic environments—this can range from provisioning and deploying containers to managing networks, providing redundancy and HA of containers, automatically scaling up and down container instances, automated health checks, and telemetry (log and metrics) gathering. Solving the problem of efficient container orchestration at cloud scale is not straightforward—this is why Kubernetes exists!

In this chapter, we will cover the following topics:

- Ensuring HA and **fault tolerance** (**FT**) on Kubernetes
- What is ReplicationController?
- What is ReplicaSet and how does it differ from ReplicationController?

Technical requirements

For this chapter, you will need the following:

- A Kubernetes cluster deployed. You can use either a local or a cloud-based cluster, but in order to fully understand the concepts, we recommend using a multi-node, cloud-based Kubernetes cluster.
- The Kubernetes **command-line interface** (**CLI**) (`kubectl`) installed on your local machine and configured to manage your Kubernetes cluster.

Kubernetes cluster deployment (local and cloud-based) and `kubectl` installation have been covered in *Chapter 3, Installing Your First Kubernetes Cluster*.

You can download the latest code samples for this chapter from the official GitHub repository at `https://github.com/PacktPublishing/The-Kubernetes-Bible/tree/master/Chapter10`.

Ensuring HA and FT on Kubernetes

First, let's quickly recap on how we define HA and FT and how they differ. These are key concepts in cloud applications that describe the ability of a system or a solution to be continuously operational for a desirably long length of time. From a system end user perspective, the aspect of availability, alongside data consistency, is usually the most important requirement.

High Availability

In short, the term *availability* in systems engineering describes the percentage of time when the system is fully functional and operation for the end user. In other words, it is a measure of system uptime divided by the sum of uptime and downtime (which is basically total time). For example, if, in the last 30 days (720 hours), your cloud application had 1 hour of unplanned maintenance and was not available to the end user, it means that the availability measure of your application is $\frac{719h}{720h} = 99.861\%$. Usually, to

simplify the notation when designing systems, the availability will be expressed in so-called **nines**: for example, if we say that a system has availability of *five nines*, it means it is available at least 99.999% of the total time. To put this into perspective, such a system can have only up to 26 seconds per month of downtime! These measures are often the base indicators for defining **service-level agreements** (**SLAs**) for billed cloud services.

The definition of HA, based on that, is relatively straightforward, although not precise—a system is highly available if it is operational (available) without interruption for long periods of time. Usually, we can say that *five nines* availability is considered the gold standard of HA.

Achieving HA in your system usually involves one or a combination of the following techniques:

- **Eliminating single points of failure (SPOF) in the system**: Usually achieved by components' redundancy.

- **Failover setup**, which is a mechanism that can automatically switch from the currently active (possibly unhealthy) component to a redundant one.

- **Load balancing**, which means managing traffic coming into the system and routing it to redundant components that can serve the traffic. This will, in most cases, involve proper failover setup and component monitoring and telemetry.

Let's introduce a related concept of FT, which is also important in distributed systems such as applications running on Kubernetes.

Fault Tolerance

Now, FT can be presented as a complement to the HA concept: a system is fault-tolerant if it can continue to be functional and operating in the event of the failure of one or more of its components. Achieving full FT means achieving 100% HA, which in many cases requires complex solutions actively detecting faults and remediating the issues in the components without interruptions. Depending on the implementation, the fault may result in a graceful degradation of performance that is proportional to the severity of the fault. This means that a small fault in the system will have a small impact on the overall performance of the system while serving requests from the end user.

HA and FT for Kubernetes applications

In the previous chapters, you have learned about the concept of Pods and how you can expose them to external traffic using Services (*Chapter 7*, *Exposing Your Pods with Services*). Services are Kubernetes objects that expose a set of healthy Pods using a single network address that remains fixed and stable for the lifetime of the Service. Internally, inside the Kubernetes cluster, the Service will make its Pods addressable using a **cluster Internet Protocol (IP) address**. These cluster-internal IP addresses are virtual IPs managed by a `kube-proxy` component on each node as a set of `iptables` rules on Linux Nodes or **Host Networking Service** (**HNS**) policies on Windows Nodes. Externally, there are multiple approaches offered by Kubernetes that are described in *Chapter 7*, *Exposing Your Pods with Services*, but in cloud environments, you will most often expose a service behind a cloud **load balancer** (**LB**). An external LB integrates with your cluster using a plugin implementation specific to each cloud service provider, in the Kubernetes `cloud-controller-manager` component. In this way, the microservices or workloads running on Kubernetes can achieve request load balancing to healthy Pods, which is a necessary building block to HA.

Services are required for load balancing requests to Pods, but we haven't yet covered how to maintain multiple replicas of the same Pod object definition that are possibly redundant and allocated on different nodes. Kubernetes offers multiple building blocks to achieve this goal, outlined as follows:

- A ReplicationController object—the original form of defining Pod replication in Kubernetes.

- A ReplicaSet object—the successor to ReplicationController. The main difference is that ReplicaSet has support for set-based requirement selectors for Pods.

- A Deployment object—another level of abstraction on top of ReplicaSet. Provides *declarative* updates for Pods and ReplicaSets, including rollouts and rollbacks. Used for managing *stateless* microservices and workloads.

- A StatefulSet object—similar to Deployment but used to manage *stateful* microservices and workloads in the cluster. Managing state inside a cluster is usually the toughest challenge to solve in distributed systems design.

- A DaemonSet object—used for running a singleton copy of a Pod on all (or some) of the nodes in the cluster. These objects are usually used for managing internal services for log aggregation or Node monitoring.

In the next sections, we will cover ReplicationController and ReplicaSets. The more advanced objects, such as Deployment, StatefulSet, and DaemonSet, will be covered in the next chapters.

> **Important note**
> This chapter covers HA and FT for Kubernetes workloads and applications. If you are interested in how to ensure HA and FT for Kubernetes itself, please refer to the official documentation at `https://kubernetes.io/docs/tasks/administer-cluster/highly-available-master/`. Please note that in managed Kubernetes offerings in the cloud, such as **Azure Kubernetes Service (AKS)**, Amazon **Elastic Kubernetes Service (EKS)**, or **Google Kubernetes Engine (GKE)**, you are provided with highly available clusters and you do not need to manage the master nodes yourself.

What is ReplicationController?

Achieving HA and FT requires providing redundancy of components and proper load balancing of incoming traffic between the replicas of components. Let's take a look at the first Kubernetes object that allows you to create and maintain multiple replicas of the Pods in your cluster: ReplicationController. Please note that we are discussing ReplicationController mainly for historical reasons as it was the initial way of creating multiple Pod replicas in Kubernetes. We advise you to use ReplicaSet whenever possible, which is basically the next generation of ReplicationController with an extended specification API.

> **Tip**
> The Controller objects in Kubernetes have one main goal: to observe the *current* and the *desired* cluster state that is exposed by the Kubernetes API server and command changes that attempt to change the *current* state to the *desired* one. They serve as **continuous feedback loops**, doing all they can to bring clusters to the desired state described by your object templates.

ReplicationController has a straightforward task—it needs to ensure that a specified number of Pod replicas (defined by a template) are running and healthy in a cluster at any time. This means that if ReplicationController is configured to maintain three replicas of a given Pod, it will try to keep exactly three Pods by creating and terminating Pods when needed. For example, right after you create a ReplicationController object, it will create three new Pods based on its template definition. If, for some reason, there are four such Pods in the cluster, ReplicationController will terminate one Pod, and if by any chance a Pod gets deleted or becomes unhealthy, it will be replaced by a new, hopefully healthy, one.

> **Tip**
>
> You can think of ReplicationController as a container analog of process or service supervisors from operating systems, such as `systemd` or `supervisor` on Linux systems. You just define how many Pod replicas you want to have in the cluster, and let the ReplicationController object do the job of keeping them running in all circumstances!

Creating a ReplicationController object

In order to create a ReplicationController object, we will need to use *declarative* management of objects using the `kubectl` command. Imperative commands for creating ReplicationController objects are not available and even if they were, it would be not advised to use them—declarative management using YAML files (**manifests**) for object configuration is much easier for more complex Kubernetes objects. Additionally, it perfectly fits all recommended paradigms such as **Infrastructure-as-Code (IaC)**.

Let's first create an example YAML manifest file for our example ReplicationController object. This ReplicationController object will have a simple task: maintain three replicas of an `nginx` Pod with specific metadata labels. The `nginx-replicationcontroller.yaml` example file should have the following content:

```
apiVersion: v1
kind: ReplicationController
metadata:
  name: nginx-replicationcontroller-example
spec:
  replicas: 3
  selector:
    app: nginx
    environment: test
```

```
template:
  metadata:
    labels:
      app: nginx
      environment: test
  spec:
    containers:
    - name: nginx
      image: nginx:1.17
      ports:
      - containerPort: 80
```

There are three main components of the ReplicationController specification, as outlined here:

- `replicas`: Defines the number of Pod replicas that should run using the given `template` and matching label `selector`. Pods may be created or deleted in order to maintain the required number.

- `selector`: A label selector, which defines how to identify Pods that the ReplicationController object owns or acquires. Note that this may have a consequence of acquiring existing bare Pods by ReplicationController if they match the selector! By bare Pods, we mean any Pods in the cluster that were created directly using the `kubectl run` command.

- `template`: Defines the template for Pod creation—this has exactly the same structure as you have already learned about in *Chapter 4*, *Running Your Docker Containers*. Labels used in `metadata` must match `selector`.

Now, let's apply the ReplicationController manifest to the cluster using the `kubectl apply` command, as follows:

```
$ kubectl apply -f ./nginx-replicationcontroller.yaml
```

You can immediately observe the status of your new ReplicationController object named `nginx-replicationcontroller-example` using the following command:

```
$ kubectl get replicationcontroller/nginx-
replicationcontroller-example
NAME                                        DESIRED   CURRENT   READY
AGE
nginx-replicationcontroller-example   3         3         2
8s
```

> **Tip**
>
> When using the `kubectl` commands, you can use the `rc` abbreviation instead of typing `replicationcontroller`.

As you can see, the ReplicationController object has already created three Pods, and two of them are already in a `Ready` state.

Similarly, you can describe the state of your ReplicationController object to get more details by using the `kubectl describe` command, as follows:

```
$ kubectl describe rc/nginx-replicationcontroller-example
```

And if you would like to get the Pods in your cluster, use the following command:

```
$ kubectl get pods
NAME                                          READY   STATUS
. . .
nginx-replicationcontroller-example-btz5t    1/1     Running
. . .
nginx-replicationcontroller-example-c6sl6    1/1     Running
. . .
nginx-replicationcontroller-example-xxl7f    1/1     Running
. . .
```

You will see that the ReplicationController object has indeed created three new Pods. If you are interested, you can use the `kubectl describe pod <podId>` command in order to inspect the labels of the Pods and also see that it contains a `Controlled By:  ReplicationController/nginx-replicationcontroller-example` property that identifies our example ReplicationController object.

> **Tip**
> If you would like to know how the Pods are distributed among the Kubernetes Nodes, you can use the `kubectl get pods -o wide` command.

Testing the behavior of ReplicationController

To demonstrate the agility of our ReplicationController object, let's now delete one of the Pods that are owned by the example ReplicationController object. You can find the names using the usual `kubectl get pods` command. In our case, we will be deleting the `nginx-replicationcontroller-example-btz5t` Pod using the following `kubectl delete` command:

```
$ kubectl delete pod/nginx-replicationcontroller-example-btz5t
```

Now, if you are quick enough, you will be able to see from using the `kubectl get pods` command that the `nginx-replicationcontroller-example-btz5t` Pod is being terminated and ReplicationController is immediately creating a new one in order to match the target number of replicas!

If you want to see more details about **events** that happened in relation to our example ReplicationController object, you can use the `kubectl describe` command, as illustrated in the following code snippet:

```
$ kubectl describe rc/nginx-replicationcontroller-example
...
  Type     Reason           Age    From
Message
  ----     ------           ----   ----
-------
  Normal   SuccessfulCreate 33m    replication-controller
Created pod: nginx-replicationcontroller-example-c6sl6
  Normal   SuccessfulCreate 33m    replication-controller
Created pod: nginx-replicationcontroller-example-btz5t
  Normal   SuccessfulCreate 33m    replication-controller
Created pod: nginx-replicationcontroller-example-xxl7f
  Normal   SuccessfulCreate 112s   replication-controller
Created pod: nginx-replicationcontroller-example-gf27c
```

The `nginx-replicationcontroller-example-gf27c` Pod is a new pod that was created in place of the terminated `nginx-replicationcontroller-example-btz5t` Pod.

Scaling ReplicationController

It is possible to modify an existing ReplicationController object and change the `replicas` property in the specification to a different number. This process is called **scaling**—if you increase the number of replicas, you are *scaling up* (or, more precisely, *scaling out*), and if you decrease the number of replicas, you are *scaling down*. This is a basic building block of **rolling upgrades** and **rollbacks**—you will learn more about this in the next chapters. More generally, this is how **horizontal scaling** of your component (packaged as a pod) can be performed.

To illustrate how horizontal scaling of Pods works at a high level, please take a look at the following diagram:

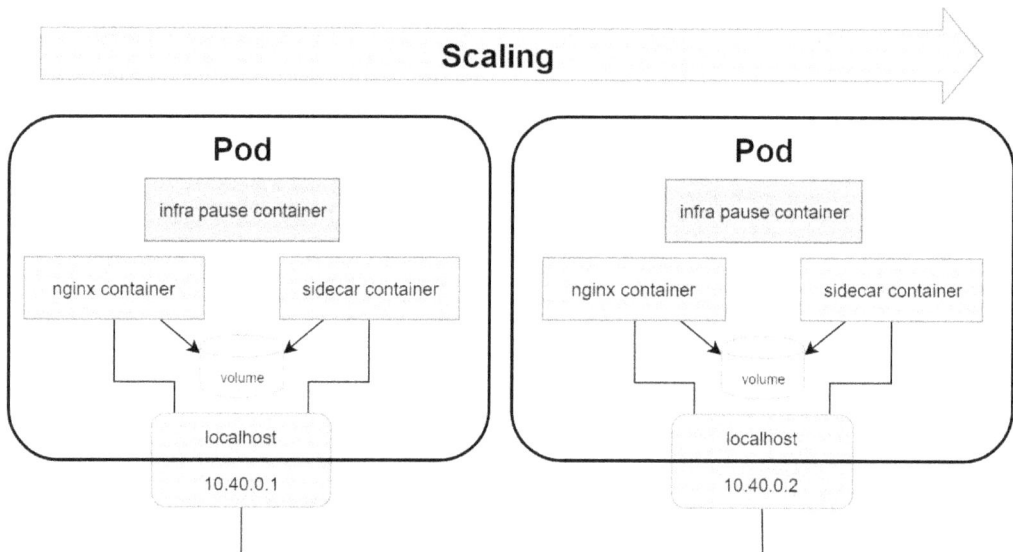

Figure 10.1 – Horizontal scaling of Kubernetes Pods

Let's first scale up our example ReplicationController object. Open the `nginx-replicationcontroller.yaml` file and modify the `replicas` property to 5, as follows:

```
...
spec:
  replicas: 5
...
```

Now, we need to declaratively apply the changes to the cluster state. Use the following `kubectl apply` command to do this:

```
$ kubectl apply -f ./nginx-replicationcontroller.yaml
```

To see that the number of Pods controlled by the ReplicationController object has changed, you can use the `kubectl get pods` or `kubectl describe rc/nginx-replicationcontroller-example` commands.

> **Tip**
>
> You can achieve similar results using the `kubectl scale rc/nginx-replicationcontroller-example --replicas=5` imperative command. In general, such imperative commands are recommended only for development or learning scenarios.

Similarly, if you want to scale down, you need to open the `nginx-replicationcontroller.yaml` file and modify the `replicas` property to 2, as follows:

```
...
spec:
  replicas: 2
...
```

Again, declaratively apply the changes to the cluster state. Use the following `kubectl apply` command to do this:

```
$ kubectl apply -f ./nginx-replicationcontroller.yaml
```

At this point, you can use the `kubectl get pods` or `kubectl describe rc/nginx-replicationcontroller-example` commands to verify that the number of Pods has been reduced to just 2.

> **Important note**
>
> In real-world scenarios, if you would like to roll out a new version of your component (Pod), you would need to have two ReplicationController objects (one for the *old* version of the component, and another one for the *new* version) and scale them up and down in a **coordinated** manner. As this process is quite complex, Kubernetes provides more advanced abstractions, such as Deployment and StatefulSet objects.

Deleting ReplicationController

Lastly, let's take a look at how you can delete a ReplicationController object. There are two possibilities, outlined as follows:

1. Delete ReplicationController together with Pods that it owns—this is performed by first scaling down automatically.

2. Delete ReplicationController and leave the Pods unaffected.

To delete the ReplicationController object together with Pods, you can use the regular `kubectl delete` command, as follows:

```
$ kubectl delete rc/nginx-replicationcontroller-example
```

You will see that the Pods will first get terminated, and then the ReplicationController object is deleted.

Now, if you would like to delete just the ReplicationController object, you need to use the `--cascade=orphan` option for `kubectl delete`, as follows:

```
$ kubectl delete rc/nginx-replicationcontroller-example
--cascade=orphan
```

After this command, if you inspect which Pods are in the cluster, you will still see all the Pods that were owned by the `nginx-replicationcontroller-example` replication controller. These Pods can now, for example, be acquired by another ReplicationController object that has a matching label selector.

Congratulations — you have successfully created your first ReplicationController object in a Kubernetes cluster! Now, it is time to introduce a close relative of ReplicationController, called ReplicaSet.

What is ReplicaSet and how does it differ from ReplicationController?

Let's introduce another Kubernetes object: ReplicaSet. This is very closely related to ReplicationController, which we have just discussed. In fact, this is a **successor** to ReplicationController, which has a very similar specification API and capabilities. The purpose of ReplicaSet is also the same—it aims to maintain a fixed number of healthy, identical Pods (replicas) that fulfill certain conditions. So, again, you just specify a template for your Pod, along with appropriate label selectors and the desired number of replicas, and Kubernetes ReplicaSetController (this is the actual name of the controller responsible for maintaining ReplicaSet objects) will carry out the necessary actions to keep the Pods running.

Now, what are the differences between ReplicaSet and ReplicationController? We have summarized these here:

- Most importantly, ReplicaSet allows more advanced, *set-based* (inclusion, exclusion) label selectors. For example, you can easily define a selector so that it matches when a Pod has `environment=test` or `environment=dev` labels, but exclude those that have an `environment=prod` label. For ReplicationController, it is only possible to use simple *equality-based* label selectors.

- ReplicaSet is a powerful building block of other Kubernetes objects such as Deployment and HorizontalPodAutoscaler (**HPA**) objects. Deployment objects provide a way of *declarative* management of ReplicaSets, doing staged rollouts and rollbacks.

- For ReplicationController, you could achieve a similar Pod update rollout using the `kubectl rolling-update` *imperative* command. However, this command is now deprecated.

- In general, you can expect that in the future, ReplicationController will be eventually **deprecated**.

The bottom line—always choose ReplicaSet over ReplicationController. However, you should also remember that using bare ReplicaSets is generally not useful in production clusters, and you should use higher-level abstractions such as Deployment objects for managing ReplicaSets. We will introduce this concept in the next chapter.

Creating a ReplicaSet object

First, let's take a look at the structure of an `nginx-replicaset.yaml` example YAML manifest file that maintains three replicas of an `nginx` Pod, as follows:

```yaml
apiVersion: apps/v1
kind: ReplicaSet
metadata:
  name: nginx-replicaset-example
spec:
  replicas: 3
  selector:
    matchLabels:
      app: nginx
      environment: test
  template:
    metadata:
      labels:
        app: nginx
        environment: test
    spec:
      containers:
      - name: nginx
        image: nginx:1.17
        ports:
        - containerPort: 80
```

There are three main components of the ReplicaSet specification, as follows:

- `replicas`: Defines the number of Pod replicas that should run using the given `template` and matching label `selector`. Pods may be created or deleted in order to maintain the required number.

- `selector`: A label selector that defines how to identify Pods that the ReplicaSet object owns or acquires. Again, similar to the case of ReplicationController, please take note that this may have a consequence of existing bare Pods being acquired by ReplicaSet if they match the selector!

- `template`: Defines a template for Pod creation. Labels used in `metadata` must match the `selector` label query.

These concepts have been visualized in the following diagram:

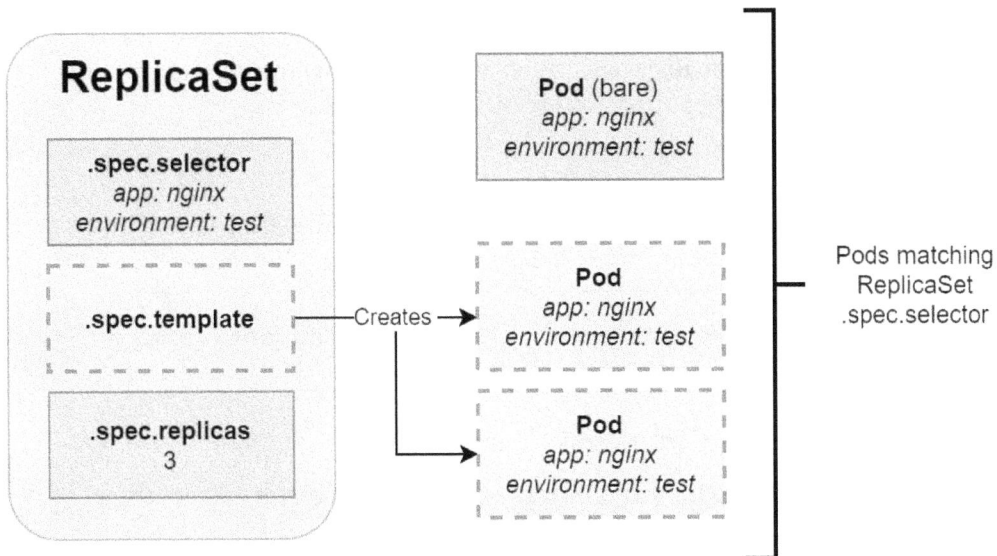

Figure 10.2 – Kubernetes ReplicaSet

As you can see, the ReplicaSet object uses `.spec.template` in order to create Pods. These Pods must match the label selector configured in `.spec.selector`. Please note that it is also possible to acquire existing bare Pods that have labels matching the ReplicaSet object. In the case shown in *Figure 10.2*, the ReplicaSet object only creates two new Pods, whereas the third Pod is a bare Pod that was acquired.

In the preceding example, we have used a simple, **equality-based** selector specified by `spec.selector.matchLabels`. A more advanced, **set-based** selector can be defined using `spec.selector.matchExpressions`—for example, like this:

```
...
spec:
  replicas: 3
  selector:
    matchLabels:
      app: nginx
    matchExpressions:
    - key: environment
      operator: In
      values:
      - test
```

```
        - dev
  . . .
```

This specification would make ReplicaSet still match only Pods with `app=nginx`, and `environment=test` or `environment=dev`.

> **Important note**
> When defining ReplicaSet, `.spec.template.metadata.labels` must match `spec.selector`, or it will be rejected by the API.

Now, let's apply the ReplicaSet manifest to the cluster using the `kubectl apply` command, as follows:

```
$ kubectl apply -f ./nginx-replicaset.yaml
```

You can immediately observe the status of your new ReplicaSet object named `nginx-replicaset-example` using the following command:

```
$ kubectl describe replicaset/nginx-replicaset-example
. . .
Replicas:      3 current / 3 desired
Pods Status:   3 Running / 0 Waiting / 0 Succeeded / 0 Failed
. . .
```

Similarly, as in the case of ReplicationController, you can use the `kubectl get pods` command to observe the Pods that are managed by the ReplicaSet object. If you are interested, you can use the `kubectl describe pod <podId>` command in order to inspect the labels of the Pods and also see that it contains a `Controlled By: ReplicaSet/nginx-replicaset-example` property that identifies our example ReplicaSet object.

> **Tip**
> When using `kubectl` commands, you can use an `rs` abbreviation instead of typing `replicaset`.

Testing the behavior of ReplicaSet

Just as in the previous section about ReplicationController, you can experiment with the `kubectl delete pod` command to observe how ReplicaSet reacts to the sudden termination of a Pod that matches its label selector.

Now, let's try something different and create a bare Pod that matches the label selector of our ReplicaSet object. You can expect that the number of Pods that match the ReplicaSet will be four, so ReplicaSet is going to terminate one of the Pods to bring the replica count back to three.

First, let's create a simple bare Pod manifest file named `nginx-pod-bare.yaml`, as follows:

```
apiVersion: v1
kind: Pod
metadata:
  name: nginx-pod-bare-example
  labels:
    app: nginx
    environment: test
spec:
  containers:
  - name: nginx
    image: nginx:1.17
    ports:
    - containerPort: 80
```

The metadata of `Pod` must have labels matching the ReplicaSet selector. Now, apply the manifest to your cluster using the following command:

```
$ kubectl apply -f nginx-pod-bare.yaml
```

Immediately after that, check the events for our example ReplicaSet object using the `kubectl describe` command, as follows:

```
$ kubectl describe rs/nginx-replicaset-example
...
Events:
  Type    Reason            Age   From
Message
  ----    ------            ----  ----
-------
...
  Normal  SuccessfulDelete  29s   replicaset-controller
Deleted pod: nginx-pod-bare-example
```

As you can see, the ReplicaSet object has immediately detected that there is a new Pod created matching its label selector and has terminated the Pod.

> **Tip**
> Similarly, it is possible to remove Pods from a ReplicaSet object by modifying their labels so that they no longer match the selector. This is useful in various debugging or incident-investigation scenarios.

Scaling ReplicaSet

For ReplicaSet, we can do a similar scaling operation as for ReplicationController in the previous section. In general, you will not perform manual scaling of ReplicaSets in usual scenarios. Instead, the size of the ReplicaSet object will be managed by another, higher-level object such as Deployment.

Let's first scale up our example ReplicaSet object. Open the `nginx-replicaset.yaml` file and modify the `replicas` property to 5, as follows:

```
...
spec:
  replicas: 5
...
```

Now, we need to declaratively apply the changes to the cluster state. Use the following `kubectl apply` command to do this:

```
$ kubectl apply -f ./nginx-replicaset.yaml
```

To see that the number of Pods controlled by the ReplicaSet object has changed, you can use the `kubectl get pods` or `kubectl describe rs/nginx-replicaset-example` commands.

> **Tip**
> Exactly as in the case of ReplicationController, you can achieve similar results using the `kubectl scale rs/nginx-replicaset-example --replicas=5` imperative command. In general, such imperative commands are recommended only for development or learning scenarios.

Similarly, if you would like to scale down, you need to open the `nginx-replicaset.yaml` file and modify the `replicas` property to 2, as follows:

```
...
spec:
  replicas: 2
...
```

Again, declaratively apply the changes to the cluster state. Use the following `kubectl apply` command to do this:

```
$ kubectl apply -f ./nginx-replicaset.yaml
```

At this point, you can use the `kubectl get pods` or `kubectl describe rs/nginx-replicaset-example` commands to verify that the number of Pods has been reduced to just 2.

Using Pod liveness probes together with ReplicaSet

Sometimes, you may want to consider a Pod unhealthy and requiring a container restart, even if the main process in the container has not crashed. For such cases, Kubernetes offers different types of **probes** for containers to determine whether they need to be restarted or whether they can serve the incoming traffic. We will quickly demonstrate how you can use **liveness probes** together with `ReplicaSet` to achieve even greater resilience to failures of containerized components.

> Tip
> Kubernetes provides more container probes, such as **readiness probes** and **startup probes**. If you want to learn more about them and when to use them, please refer to the official documentation at `https://kubernetes.io/docs/concepts/workloads/pods/pod-lifecycle/#container-probes`.

In our example, we will create a ReplicaSet object that runs `nginx` Pods with an additional liveness probe on the main container, which checks whether an `HTTP GET` request to the `/` path responds with a *successful* HTTP status code. You can imagine that, in general, your `nginx` process running in the container will always be healthy (until it crashes), but it doesn't mean that the Pod can be considered healthy. If the web server is not able to successfully provide content, it means that the web server process is running but something else might have gone wrong, and this Pod should no longer be used. We will simulate this situation by simply deleting the `/index.html` file in the container, which will cause the liveness probe to fail.

First, let's create a YAML manifest file named `nginx-replicaset-livenessprobe.yaml` for our new `nginx-replicaset-livenessprobe-example` ReplicaSet object with the following content:

```
apiVersion: apps/v1
kind: ReplicaSet
metadata:
  name: nginx-replicaset-livenessprobe-example
spec:
  replicas: 3
  selector:
    matchLabels:
      app: nginx
      environment: test
  template:
    metadata:
      labels:
        app: nginx
        environment: test
    spec:
      containers:
      - name: nginx
        image: nginx:1.17
        ports:
        - containerPort: 80
        livenessProbe:
          httpGet:
            path: /
```

```
     port: 80
     initialDelaySeconds: 2
     periodSeconds: 2
```

The highlighted part of the preceding code block contains the liveness probe definition and is the only difference between our earlier ReplicaSets examples. The liveness probe is configured to execute an HTTP GET request to the / path at port 80 for the container every 2 seconds (`periodSeconds`). The first probe will start after 2 seconds (`initialDelaySeconds`) from the container start.

> **Tip**
>
> If you are modifying an existing ReplicaSet object, you need to first delete it and recreate it in order to apply changes to the Pod template.

Now, apply the manifest file to the cluster using the following command:

```
$ kubectl apply -f ./nginx-replicaset-livenessprobe.yaml
```

Verify that the Pods have been successfully started using the following command:

```
$ kubectl get pods
```

Now, you need to choose one of the ReplicaSet Pods in order to simulate the failure inside the container that will cause the liveness probe to fail. In the case of our example, we will be using a Pod with the name `nginx-replicaset-livenessprobe-example-2qbhk`.

To simulate the failure, run the following command. This command will remove the `index.html` file served by the `nginx` web server and will cause the HTTP GET request to fail with a non-successful HTTP status code:

```
$ kubectl exec -it nginx-replicaset-livenessprobe-example-2qbhk
-- rm /usr/share/nginx/html/index.html
```

Inspect the events for this Pod using the `kubectl describe` command, as follows:

```
$ kubectl describe pod/nginx-replicaset-livenessprobe-example-
2qbhk
...
Events:
   Type      Reason     Age               From
Message
```

```
  ----        ------      ----                        ----
-------
  . . .
  Normal    Created    20s (x2 over 3m53s)    kubelet
Created container nginx
  Warning   Unhealthy  20s (x3 over 24s)      kubelet
Liveness probe failed: HTTP probe failed with statuscode: 403
  Normal    Killing    20s                    kubelet
Container nginx failed liveness probe, will be restarted
  Normal    Started    19s (x2 over 3m53s)    kubelet
Started container nginx
```

As you can see, the liveness probe has correctly detected that the web server became unhealthy and restarted the container inside the Pod.

However, please note that the ReplicaSet object itself did not take part in the restart in any way—the action was performed at a Pod level. This demonstrates how individual Kubernetes objects provide different features that can work together to achieve improved FT. Without the liveness probe, the end user could be served by a replica that is not able to provide content, and this would go undetected!

Deleting a ReplicaSet object

Lastly, let's take a look at how you can delete a ReplicaSet object. There are two possibilities, outlined as follows:

1. Delete the ReplicaSet object together with the Pods that it owns—this is performed by first scaling down automatically.
2. Delete the ReplicaSet object and leave the Pods unaffected.

To delete the ReplicaSet object together with the Pods, you can use the regular `kubectl delete` command, as follows:

```
$ kubectl delete rs/nginx-replicaset-livenessprobe-example
```

You will see that the Pods will first get terminated and then the ReplicaSet object is deleted.

Now, if you would like to delete just the ReplicaSet object, you need to use the `--cascade=orphan` option for `kubectl delete`, as follows:

```
$ kubectl delete rs/nginx-replicaset-livenessprobe-example
--cascade=orphan
```

After this command, if you inspect which Pods are in the cluster, you will still see all the Pods that were owned by the `nginx-replicaset-livenessprobe-example` ReplicaSet object. These Pods can now, for example, be acquired by another ReplicaSet object that has a matching label selector.

Summary

In this chapter, you have learned about the key building blocks for providing **High Availability** (**HA**) and **Fault Tolerance** (**FT**) for applications running in Kubernetes clusters. First, we have explained why HA and FT are important. Next, you have learned more details about providing component replication and failover using ReplicationController and ReplicaSet, which are used in Kubernetes in order to provide multiple copies (replicas) of identical Pods. We have demonstrated the differences between ReplicationController and ReplicaSet and eventually explained why using ReplicaSet is currently the recommended way to provide multiple replicas of Pods.

The next chapters in this part of the book will give you an overview of how to use Kubernetes to orchestrate your container applications and workloads. You will familiarize yourself with concepts of the most important Kubernetes objects, such as ReplicaSet, Deployment, StatefulSet, or DaemonSet, and in the next chapter, we will focus on the next level of abstraction over ReplicaSets: Deployment objects. You will learn how to deploy and easily manage rollouts and rollbacks of new versions of your application.

Further reading

For more information regarding ReplicationController and ReplicaSet, please refer to the following *Packt Publishing* books:

- *The Complete Kubernetes Guide*, by *Jonathan Baier, Gigi Sayfan*, and *Jesse White* (`https://www.packtpub.com/virtualization-and-cloud/complete-kubernetes-guide`)

- *Getting Started with Kubernetes – Third Edition*, by *Jonathan Baier* and *Jesse White* (`https://www.packtpub.com/virtualization-and-cloud/getting-started-kubernetes-third-edition`)

- *Kubernetes for Developers*, by *Joseph Heck* (https://www.packtpub.com/virtualization-and-cloud/kubernetes-developers)

- *Hands-On Kubernetes on Windows*, by *Piotr Tylenda* (https://www.packtpub.com/product/hands-on-kubernetes-on-windows/9781838821562)

You can also refer to the excellent official Kubernetes documentation (https://kubernetes.io/docs/home/), which is always the most up-to-date source of knowledge about Kubernetes in general.

11

Deployment – Deploying Stateless Applications

The previous chapter introduced two important Kubernetes objects: ReplicationController and ReplicaSet. At this point, you already know that they serve similar purposes in terms of maintaining identical, healthy replicas (copies) of Pods. In fact, ReplicaSet is a successor of ReplicationController and, in the most recent versions of Kubernetes, ReplicaSet should be used in favor of ReplicationController.

Now, it is time to introduce the Deployment object, which provides easy scalability, rolling updates, and versioned rollbacks for your stateless Kubernetes applications and services. Deployment objects are built on top of ReplicaSets and they provide a declarative way of managing them – just describe the desired state in the Deployment manifest and Kubernetes will take care of orchestrating the underlying ReplicaSets in a controlled, predictable manner. Alongside StatefulSet, which will be covered in the next chapter, it is the most important workload management object in Kubernetes. This will be the bread and butter of your development and operations on Kubernetes! The goal of this chapter is to make sure that you have all the tools and knowledge you need to deploy your stateless application components using Deployment objects, as well as to safely release new versions of your components using rolling updates of deployments.

This chapter will cover the following topics:

- Introducing the Deployment object
- How does a Deployment object manage revisions and version rollout?
- Deployment object best practices

Technical requirements

For this chapter, you will need the following:

- A Kubernetes cluster that's been deployed. You can use either a local or cloud-based cluster, but to fully understand the concepts shown in this chapter, we recommend using a multi-node, cloud-based Kubernetes cluster.
- The Kubernetes CLI (`kubectl`) must be installed on your local machine and configured to manage your Kubernetes cluster.

Kubernetes cluster deployment (local and cloud-based) and `kubectl` installation were covered in *Chapter 3, Installing Your First Kubernetes Cluster*.

You can download the latest code samples for this chapter from the official GitHub repository: `https://github.com/PacktPublishing/The-Kubernetes-Bible/tree/master/Chapter11`.

Introducing the Deployment object

Kubernetes gives you out-of-the-box flexibility when it comes to running different types of workloads, depending on your use cases. Let's have a brief look at the supported workloads to understand where the Deployment object fits, as well as its purpose. When implementing cloud-based applications, you will generally need the following types of workloads:

- **Stateless**: Applications and services that are stateless, by definition, do not have any modifiable client data (state) that is needed for further operations or sessions. In the container-based world, what we mean by stateless containers or Pods is that they do not store any application state inside the container or attached volume. Imagine that you have two different `nginx` containers serving the same purpose: the first one stores some data coming from user requests in a JSON file inside the container, while the second one stores data in MongoDB that's running in a separate container. Both serve the same purpose, but the first one is stateless and the second one is not stateless (in other words, it's stateful). However, our small application, as a whole, is not stateless because some of its components are stateful. In the first case, the `nginx` container is the stateful component, while in the second case, the MongoDB container is the stateful component. Stateless workloads are easy to orchestrate, scale up and down, and, in general, are simple to manage. In Kubernetes, Deployment objects are used to manage stateless workloads.

- **Stateful**: In the case of containers and Pods, we call them stateful if they store any modifiable data inside themselves. A good example of such a Pod is a MySQL or `MongoDB` Pod that reads and writes the data to a persistent volume. Stateful workloads are much harder to manage – you need to carefully manage sticky sessions or data partitions during rollouts, rollbacks, and when scaling. As a rule of thumb, try to keep stateful workloads outside your Kubernetes cluster if possible, such as by using cloud-based Software-as-a-Service database offerings. In Kubernetes, StatefulSet objects are used to manage stateful workloads. *Chapter 12, StatefulSet – Deploying Stateful Applications*, provides more details about these types of objects.

- **Batch**: This type of workload is anything that performs job or task processing, either scheduled or on demand. Depending on the type of application, batch workloads may require thousands of containers and a lot of nodes – this can be anything that happens *in the background*. Containers that are used for batch processing should also be stateless, to make it easier to resume interrupted jobs. In Kubernetes, Job and CronJob objects are used to manage batch workloads. *Chapter 4, Running Your Docker Containers*, provides more details about these types of objects.

- **Node-local**: In many cases, Kubernetes cluster operations require periodic maintenance of Nodes or system log aggregation for each Node. Usually, such workloads are not user-facing but are crucial to the proper functionality of each Node in the cluster. In Kubernetes, DaemonSet objects are used to manage node-local workloads. *Chapter 13, DaemonSet – Maintaining Pod Singletons on Nodes*, provides more details about these types of objects.

With this brief summary regarding the different types of workloads in Kubernetes, we can dive deeper into managing stateless workloads using Deployment objects. In short, they provide declarative and controlled updates for Pods and ReplicaSets. You can declaratively perform operations such as the following by using them:

- Perform a rollout of a new ReplicaSet.

- Change the Pod template and perform a controlled rollout. The old ReplicaSet will be gradually scaled down, whereas the new ReplicaSet will scale up at the same rate.

- Perform a rollback to an earlier version of the Deployment object.

- Scale ReplicaSet up or down, without needing to make any changes to the Pod definition.

- Pause and resume the rollout of the new ReplicaSet, if there is a need to introduce fixes.

In this way, Deployment objects provide an end-to-end pipeline for managing your stateless components running in Kubernetes clusters. Usually, you will combine them with Service objects, as presented in *Chapter 7, Exposing Your Pods with Services*, to achieve high fault tolerance, health monitoring, and intelligent load balancing for traffic coming into your application.

Now, let's have a closer look at the anatomy of the Deployment object specification and how to create a simple example deployment in our Kubernetes cluster.

Creating a Deployment object

First, let's take a look at the structure of an example Deployment YAML manifest file, `nginx-deployment.yaml`, that maintains three replicas of an `nginx` pod:

```yaml
apiVersion: apps/v1
kind: Deployment
metadata:
  name: nginx-deployment-example
spec:
  replicas: 3
  selector:
    matchLabels:
      app: nginx
      environment: test
  minReadySeconds: 10
  strategy:
    type: RollingUpdate
    rollingUpdate:
      maxUnavailable: 1
      maxSurge: 1
  template:
    metadata:
      labels:
        app: nginx
        environment: test
    spec:
      containers:
      - name: nginx
        image: nginx:1.17
        ports:
        - containerPort: 80
```

As you can see, the structure of the Deployment spec is almost identical to ReplicaSet, although it has a few extra parameters for configuring the strategy for rolling out new versions. The specification has four main components:

- `replicas`: Defines the number of Pod replicas that should run using the given `template` and matching label `selector`. Pods may be created or deleted to maintain the required number. This property is used by the underlying ReplicaSet.

- `selector`: A label selector, which defines how to identify Pods that the underlying ReplicaSet owns. This can include set-based and equality-based selectors. In the case of Deployments, the underlying ReplicaSet will also use a generated `pod-template-hash` label to ensure that there are no conflicts between different child ReplicaSets when you're rolling out a new version. Additionally, this generally prevents accidental acquisitions of bare Pods, which could easily happen with simple ReplicaSets. Nevertheless, Kubernetes does not prevent you from defining overlapping pod selectors between different Deployments or even other types of controllers. However, if this happens, they may conflict and behave unexpectedly.

- `template`: Defines the template for Pod creation. Labels used in `metadata` must match our `selector`.

- `strategy`: Defines the details of the strategy that will be used to replace existing Pods with new ones. You will learn more about such strategies in the following sections. In this example, we showed the default `RollingUpdate` strategy. In short, this strategy works by slowly replacing the Pods of the previous version, one by one, by using the Pods of the new version. This ensures zero downtime and, together with Service objects and readiness probes, provides traffic load balancing to pods that can serve the incoming traffic.

> **Important Note**
>
> The Deployment spec provides a high degree of reconfigurability to suit your needs. We recommend referring to the official documentation for all the details: `https://kubernetes.io/docs/reference/generated/kubernetes-api/v1.19/#deploymentspec-v1-apps`.

To better understand the relationship of Deployment, its underlying child ReplicaSet, and Pods, please look at the following diagram:

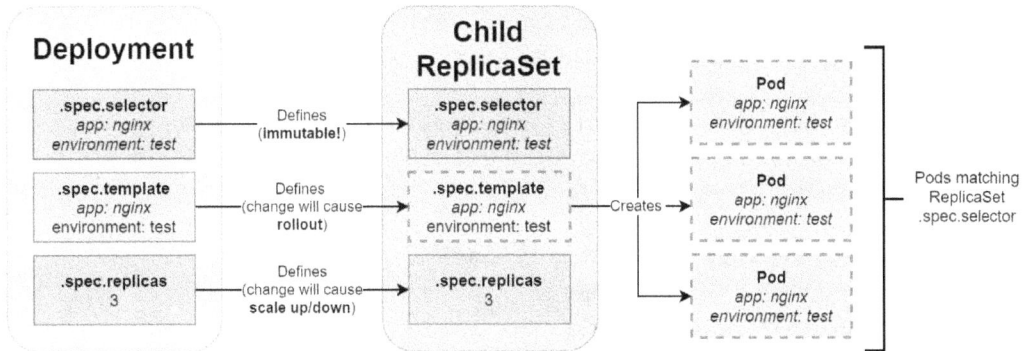

Figure 11.1 – Kubernetes Deployment

As you can see, once you have defined and created a Deployment, it is not possible to change its `selector`. This is desired because otherwise, you could easily end up with orphaned ReplicaSets. There are two important actions that you can perform on existing Deployment objects:

- **Modify template**: Usually, you would like to change the Pod definition to a new image version of your application. This will cause a rollout to begin, according to the rollout `strategy`.

- **Modify replicas number**: Just changing the number will cause ReplicaSet to gracefully scale up or down.

Now, let's declaratively apply our example Deployment YAML manifest file, `nginx-deployment.yaml`, to the cluster using the `kubectl apply` command:

```
$ kubectl apply -f ./nginx-deployment.yaml --record
```

Using the `--record` flag is useful for tracking the changes that are made to the objects, as well as to inspect which commands caused these changes. You will then see an additional automatic annotation, `kubernetes.io/change-cause`, which contains information about the command.

Immediately after the Deployment object has been created, use the `kubectl rollout` command to track the status of your Deployment in real time:

```
$ kubectl rollout status deployment nginx-deployment-example
Waiting for deployment "nginx-deployment-example" rollout to
finish: 0 of 3 updated replicas are available...
```

```
Waiting for deployment "nginx-deployment-example" rollout to
finish: 0 of 3 updated replicas are available...
Waiting for deployment "nginx-deployment-example" rollout to
finish: 0 of 3 updated replicas are available...
deployment "nginx-deployment-example" successfully rolled out
```

This is a useful command that can give us a lot of insight into what is happening with an ongoing Deployment rollout. You can also use the usual kubectl get or kubectl describe commands:

```
$ kubectl get deploy nginx-deployment-example
NAME                        READY    UP-TO-DATE    AVAILABLE    AGE
nginx-deployment-example    3/3      3             3
6m21s
```

As you can see, the Deployment has been successfully created and all three Pods are now in the ready state.

> **Tip**
>
> Instead of typing deployment, you can use the deploy abbreviation when using kubectl commands.

You may also be interested in seeing the underlying ReplicaSets:

```
$ kubectl get rs
NAME                                    DESIRED    CURRENT    READY
AGE
nginx-deployment-example-5549875c78     3          3          3
8m7s
```

Please take note of the additional generated hash, 5549875c78, in the name of our ReplicaSet, which is also the value of the pod-template-hash label, which we mentioned earlier.

Lastly, you can see the pods in the cluster that were created by the Deployment object using the following command:

```
$ kubectl get pods
NAME                                         READY    STATUS
RESTARTS    AGE
nginx-deployment-example-5549875c78-5srkn    1/1      Running    0
11m
```

```
nginx-deployment-example-5549875c78-h5n76    1/1       Running   0
11m
nginx-deployment-example-5549875c78-mn5zn    1/1       Running   0
11m
```

Congratulations – you have created and inspected your first Kubernetes Deployment! Next, we will take a look at how Service objects are used to expose your Deployment to external traffic coming into the cluster.

Exposing Deployment Pods using Service objects

Service objects were covered in detail in *Chapter 7, Exposing Your Pods with Services*, so in this section, we will provide a brief recap about the role of Services and how they are usually used with Deployments. Services are Kubernetes objects that allow you to expose your Pods, whether this is to other Pods in the cluster or end users. They are the crucial building blocks for highly available and fault-tolerant Kubernetes applications, since they provide a load balancing layer that actively routes incoming traffic to ready and healthy Pods.

The Deployment objects, on the other hand, provide Pod replication, automatic restarts when failures occur, easy scaling, controlled version rollouts, and rollbacks. But there is a catch: Pods that are created by ReplicaSets or Deployments have a finite life cycle. At some point, you can expect them to be terminated; then, new Pod replicas with new IP addresses will be created in their place. So, what if you have a Deployment running web server Pods that need to communicate with Pods that have been created as a part of another Deployment such as backend Pods? Web server Pods cannot assume anything about the IP addresses or the DNS names of backend Pods, as they may change over time. This issue can be resolved with Service objects, which provide reliable networking for a set of Pods.

In short, Services target a set of Pods, and this is determined by label selectors. These label selectors work on the same principle that you have learned about for ReplicaSets and Deployments. The most common scenario is exposing a Service for an existing Deployment by using the same label selector. The Service is responsible for providing a reliable DNS name and IP address, as well as for monitoring selector results and updating the associated Endpoint object with the current IP addresses of the matching Pods. For internal cluster communication, this is usually achieved using simple ClusterIP Services, whereas to expose them to external traffic, you can use the NodePort Service or, more commonly in cloud deployments, the LoadBalancer Service.

To visualize how Service objects interact with Deployment objects in Kubernetes, please look at the following diagram:

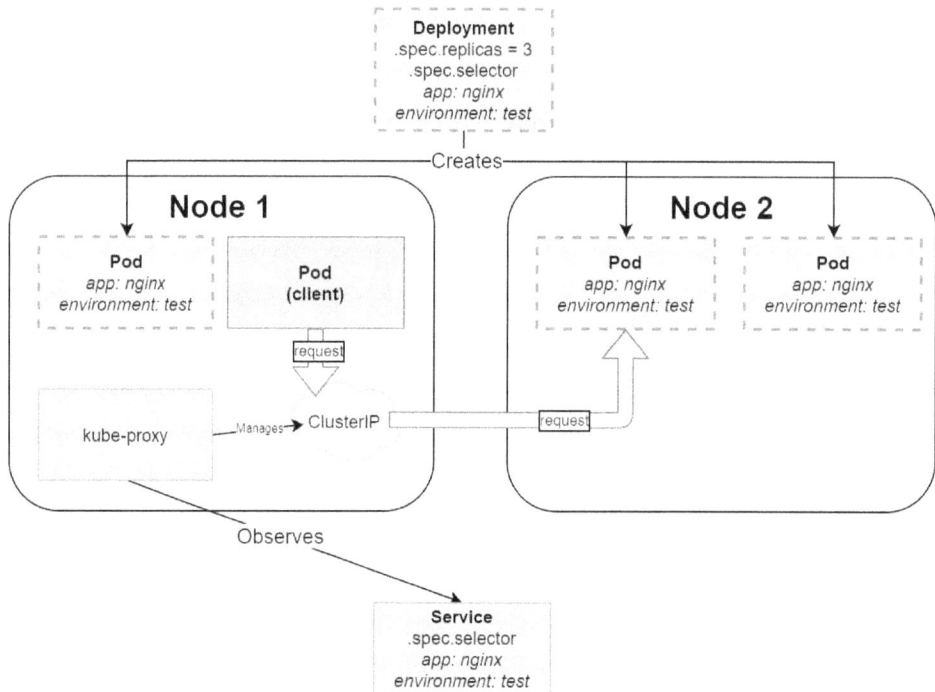

Figure 11.2 – Client Pod performing requests to the Kubernetes Deployment, exposed by the ClusterIP Service

This diagram visualizes how any client Pod in the cluster can transparently communicate with the `nginx` Pods that are created by our Deployment object and exposed using the **ClusterIP** Service. ClusterIPs are essentially virtual IP addresses that are managed by the `kube-proxy` service (process) that is running on each Node. `kube-proxy` is responsible for all the clever routing logic in the cluster and ensures that the routing is entirely transparent to the client Pods – they do not need to be aware if they are communicating with the same Node, a different Node, or even an external component. The role of the Service object is to define a set of ready Pods that should be *hidden* behind a stable ClusterIP. Usually, the internal clients will not be calling the Service pods using the ClusterIP, but they will use a DNS short name, which is the same as Service name; for example, `nginx-service-example`. This will be resolved to the ClusterIP by the cluster's internal DNS service. Alternatively, they may use a DNS **Fully Qualified Domain Name (FQDN)** in the form of `<serviceName>.<namespaceName>.svc.<clusterDomain>`; for example, `nginx-service-example.default.svc.cluster.local`.

> **Important Note**
>
> For LoadBalancer or NodePort Services that expose Pods to external traffic,
> the principle is similar as internally, they also provide a ClusterIP for internal
> communication. The difference is that they also configure more components so
> that external traffic can be routed to the cluster.

Now that you're equipped with the necessary knowledge about Service objects and their
interactions with Deployment objects, let's put what we've learned into practice!

Creating a Service declaratively

In this section, we are going to expose our `nginx-deployment-example` Deployment
using the `nginx-service-example` Service object, which is of the `LoadBalancer`
type, by performing the following steps:

1. Create an `nginx-service.yaml` manifest file with the following content:

```yaml
apiVersion: v1
kind: Service
metadata:
  name: nginx-service-example
spec:
  selector:
    app: nginx
    environment: test
  type: LoadBalancer
  ports:
  - port: 80
    protocol: TCP
    targetPort: 80
```

The label selector of the Service is the same as the one we used for our Deployment
object. The specification of the Service instructs us to expose our Deployment on
port 80 of the cloud load balancer, and then route the traffic from target port 80 to
the underlying Pods.

> **Important Note**
>
> Depending on how your Kubernetes cluster is deployed, you may not be able
> to use the `LoadBalancer` type. In that case, you may need to use the
> `NodePort` type for this exercise or stick to the simple `ClusterIP` type and
> skip the part about external access. For local development deployments such
> as `minikube`, you will need to use the `minikube service` command
> to access your Service. You can find more details in the documentation:
> `https://minikube.sigs.k8s.io/docs/commands/`
> `service/`.

2. Now, use the `kubectl get` or `kubectl describe` command to gather
    information about, and the status of, our new service and associated load balancer:

```
$ kubectl describe service nginx-service-example
...
LoadBalancer Ingress:       40.88.196.15
...
Events:
  Type      Reason                Age     From
  Message
  ----      ------                ----    ----
  -------
  Normal    EnsuringLoadBalancer  10m     service-controller
  Ensuring load balancer
  Normal    EnsuredLoadBalancer   11m     service-controller
  Ensured load balancer
```

Please note that creating the external cloud load balancer may take a bit of time, so
you may not see an external IP address immediately. In this case, the external IP is
`40.88.196.15`.

> **Tip**
>
> You can use the `svc` abbreviation in the `kubectl` commands instead of
> typing `service`.

3. Now, navigate to your favorite web browser and open the address for your Service; for example `http://40.88.196.15` (your address will be different, depending on the output of the `kubectl describe` command). With that, we have exposed port `80`, which is a simple HTTP endpoint, not HTTPS, which is a secure version of HTTP. For demonstration purposes, this is enough, but for production workloads, you need to ensure you have set up properly HTTPS for your load balancer and inside the cluster. You should see the default `nginx` welcome page, served by one of the three Pods in our Deployment object:

Welcome to nginx!

If you see this page, the nginx web server is successfully installed and working. Further configuration is required.

For online documentation and support please refer to nginx.org. Commercial support is available at nginx.com.

Thank you for using nginx.

Figure 11.3 – Web page served by nginx running in Deployment pods

This shows how Services are used to expose Deployment Pods to external traffic. Now, let's perform an experiment that demonstrates how internal traffic is handled by Services for other client Pods:

1. Create an interactive `busybox` Pod and start the Bourne shell process. The following command will create the Pod and immediately attach it to your terminal so that you can interact with it from within the Pod:

```
$ kubectl run -i --tty busybox --image=busybox --rm
--restart=Never -- sh
```

2. When the container shell prompt appears, download the default web page served by nginx Pods in the deployment. Use the `nginx-deployment-example` Service name as the short DNS name:

```
$ wget http://nginx-service-example && cat index.html
```

3. Alternatively, you can use DNS FQDN:

```
$ rm ./index.html
$ wget http://nginx-service-example.default.svc.cluster.
local && cat index.html
```

4. In both cases, you will see the same HTML source code for our default web page.

5. Use the `exit` command to exit the session and automatically remove the `busybox` Pod.

Now, we will quickly show you how to achieve a similar result using imperative commands to create a Service for our Deployment object.

Creating a Service imperatively

A similar effect can be achieved using the imperative `kubectl expose` command – a Service will be created for our Deployment object named `nginx-deployment-example`. Use the following command:

```
$ kubectl expose deployment --type=LoadBalancer nginx-
deployment-example
service/nginx-deployment-example exposed
```

This will create a Service with the same name as the Deployment object; that is, `nginx-deployment-example`. If you would like to use a different name, as shown in the declarative example, you can use the `--name=nginx-service-example` parameter. Additionally, port `80`, which will be used by the Service, will be the same as the one that was defined for the Pods. If you want to change this, you can use the `--port=<number>` and `--target-port=<number>` parameters.

Please note that this imperative command is recommended for use in development or debugging scenarios only. For production environments, you should leverage declarative *Infrastructure-as-Code* and *Configuration-as-Code* approaches as much as possible.

Role of readiness, liveness, and startup probes

In Kubernetes, there are three types of probes that you can configure for each container running in a Pod:

*   **Readiness probe**: This is used to determine whether a given container is ready to accept traffic. A Pod is considered ready only if all its containers are ready. Pods that are not ready will be removed from Service Endpoints until they become ready again. In other words, such Pods will not be considered while load balancing traffic via Service. This is the most important probe from the Service object's perspective – setting it up properly ensures that the end users will experience zero downtime during rollouts and rollbacks, and also during any failures of individual Pods.

- **Liveness probe**: This is used to detect whether a container needs to be restarted. This can help in situations where a container has been stuck in a deadlock or other issues, where the container process is alive but unable to operate properly. Restarting the container may increase the availability of Pods in that case. We briefly covered this probe in *Chapter 10, Running Production-Grade Kubernetes Workloads*, as it complements the role of ReplicaSet.

- **Startup probe**: This is an additional probe that's used for determining whether a container has been fully started – the readiness and liveness probes are disabled until this probe returns successfully. This is especially useful for containers that have a long startup time due to some initialization. In this way, you can avoid premature kills being made by the liveness probe.

All these probes are incredibly useful when you're configuring your Deployments – always try to predict possible life cycle scenarios for the processes running in your containers and configure the probes accordingly for your Deployments.

Probes can have different forms. For example, they can be running a command (`exec`) inside the container and verifying whether the exit code is successful. Alternatively, they can be HTTP GET requests (`httpGet`) to a specific endpoint of the container or attempting to open a TCP socket (`tcpSocket`) and checking if a connection could be established. Usually, `httpGet` probes are used on dedicated health endpoints (for liveness) or ready endpoints (for readiness) that are exposed by the process running in the container. These endpoints would encapsulate the logic of the actual health or readiness check.

Please note that, by default, no probes are configured on containers running in Pods. Kubernetes will serve traffic to Pod containers behind the Service, but only if the containers have successfully started, and restart them if they have crashed using the default always-restart policy. This means that it is your responsibility to figure out what type of probes and what settings you need for your particular case. You will also need to understand the possible consequences and caveats of incorrectly configured probes – for example, if your liveness probe is too restrictive and has timeouts that are too small, it may wrongfully restart your containers and decrease the availability of your application.

Now, let's demonstrate how you can configure a **readiness probe** on our Deployment and how it works in real time.

> **Important Note**
>
> If you are interested in the configuration details for other types of probes, please refer to the official documentation: `https://kubernetes.io/docs/tasks/configure-pod-container/configure-liveness-readiness-startup-probes/`. We have only covered the readiness probe in this section as it is the most important for interactions between Service objects and Deployment objects.

The `nginx` Deployment that we use is very simple and does not need any dedicated readiness probe. Instead, we will arrange the container's setup so that we can have the container's readiness probe fail or succeed on demand. The idea is to create an empty file called `/usr/share/nginx/html/ready` during container setup, which will be served on the `/ready` endpoint by `nginx` (just like any other file) and configure a readiness probe of the `httpGet` type to query the `/ready` endpoint for a successful HTTP status code. Now, by deleting or recreating the `ready` file using the `kubectl exec` command, we can easily simulate failures in our Pods that cause the readiness probe to fail or succeed.

Please follow these steps to configure and test the readiness probe:

1. Delete the existing Deployment using the following command:

    ```
    $ kubectl delete deployment nginx-deployment-example
    ```

2. Copy our previous YAML manifest for the Deployment:

    ```
    $ cp nginx-deployment.yaml nginx-deployment-readinessprobe.yaml
    ```

3. Open the file and modify the following parts of the manifest:

    ```
    apiVersion: apps/v1
    kind: Deployment
    metadata:
      name: nginx-deployment-example
    spec:
      ...
      template:
        ...
        spec:
          containers:
          - name: nginx
    ```

```
        image: nginx:1.17
        ports:
        - containerPort: 80
        command:
        - /bin/sh
        - -c
        - |
          touch /usr/share/nginx/html/ready
          echo "You have been served by Pod with IP
    address: $(hostname -i)" > /usr/share/nginx/html/index.
    html
          nginx -g "daemon off;"
        readinessProbe:
          httpGet:
            path: /ready
            port: 80
          initialDelaySeconds: 5
          periodSeconds: 2
          timeoutSeconds: 10
          successThreshold: 1
          failureThreshold: 2
```

There are multiple parts changing in the Deployment manifest, all of which have been highlighted. First, we overridden the default container entry point command using `command` and passed additional arguments. `command` is set to `/bin/sh` to execute a custom shell command. The additional arguments are constructed in the following way:

- `-c` is an argument for `/bin/sh` that instructs it that what follows is a command to be executed in the shell.

- `touch /usr/share/nginx/html/ready` is the first command that's used in the container shell. This will create an empty `ready` file that can be served by `nginx` on the `/ready` endpoint.

- `echo "You have been served by Pod with IP address:` `$(hostname -i)" > /usr/share/nginx/html/index.html` is the second command that sets the content of `index.html` to information about the internal cluster Pod's IP address. `hostname  -i` is the command that's used to get the container IP address. This value will be different for each Pod running in our Deployment.

- nginx -g "daemon off;". Finally, we execute the default entrypoint command for the nginx:1.17 image. This will start the nginx web server as the main process in the container.

> **Important Note**
> Usually, you would perform such customization using a new Docker image, which inherits from the nginx:1.17 image as a base. The method shown here is being used for demonstration purposes and shows how flexible the Kubernetes runtime is.

The second set of changes we made in the YAML manifest for the Deployment were for the definition of readinessProbe, which is configured as follows:

- The probe is of the httpGet type and executes an HTTP GET request to the /ready HTTP endpoint on port 80 of the container.

- initialDelaySeconds: This is set to 5 seconds and configures the probe to start querying after 5 seconds from container start.

- periodSeconds: This is set to 2 seconds and configures the probe to query in 2-second intervals.

- timeoutSeconds: This is set to 10 seconds and configures a number of seconds, after which the HTTP GET request times out.

- successThreshold: This is set to 1 and configures the minimum number of consecutive success queries of the probe before it is considered to be successful once it has failed.

- failureThreshold: This is set to 2 and configures the minimum number of consecutive failed queries of the probe before it is considered to have failed. Setting it to a value that's greater than 1 ensures that the probe is not providing false positives.

To create the deployment, follow these steps:

1. Apply the new YAML manifest file to the cluster using the following command:

```
$ kubectl apply -f ./nginx-deployment-readinessprobe.yaml
--record
```

2. Use the `kubectl describe` command to get the external load balancer IP
    address of the `nginx-deployment-example` Service, which we created at the
    beginning of this section and is still running (if you are recreating the Service, you
    may need to wait until it has been configured):

```
$ kubectl describe svc nginx-service-example
...
LoadBalancer Ingress:      52.188.43.251
...
Endpoints:                 10.244.0.43:80,10.244.1.50:80,1
0.244.1.51:80
```

We will use `52.188.43.251` as the IP address in our examples. You can also see
that the service has three Endpoints that map to our Deployment Pods, all of which
are ready to serve traffic.

3. Navigate to `http://52.188.43.251` in your favorite web browser. Please turn
    off web browser caching while making these requests to avoid any cached responses
    to your queries. In Chrome, you can simply do this for your current tab by opening
    Developer Tools by pressing *F12* and checking the **Disable Cache** checkbox. Press
    F5 to refresh. You will notice that the responses iterate over three different Pod IP
    addresses. This is, because our Deployment has been configured to have three Pod
    replicas. Each time you perform a request, you may hit a different Pod:

```
You have been served by Pod with IP address: 10.244.0.43
... (a few F5 hits later)
You have been served by Pod with IP address: 10.244.1.51
... (a few F5 hits later)
You have been served by Pod with IP address: 10.244.1.50
```

4. You can cross-check this with the `kubectl get pod` command to see the actual
    addresses of the Pods:

```
$ kubectl get pod -o wide
NAME                                          READY
STATUS       RESTARTS    AGE    IP ...
nginx-deployment-example-85cd4bb66f-94r4q    1/1
Running    0            11m    10.244.1.51 ...
nginx-deployment-example-85cd4bb66f-95bwd    1/1
Running    0            11m    10.244.1.50 ...
nginx-deployment-example-85cd4bb66f-ssccm    1/1
Running    0            11m    10.244.0.43 ...
```

5. Now, let's simulate a readiness failure for the first Pod. In our case, this is `nginx-deployment-example-85cd4bb66f-94r4q`, which has an IP address of `10.244.1.51`. To do this, we need to simply delete the `ready` file inside the container using the `kubectl exec` command:

```
$ kubectl exec -it nginx-deployment-example-85cd4bb66f-
94r4q -- rm /usr/share/nginx/html/ready
```

6. The readiness probe will now start to fail, but not immediately! We have set it up so that it needs to fail at least two times, and each check is performed in 2-second intervals. So, if you are quick, you can go to your web browser and try refreshing the external address of our Service a few times. However, you may still see that you were served by `10.244.1.51`. Later, you will notice that you are only served by two other Pods that are still ready.

7. Now, if you describe the `nginx-service-example` Service, you will see that it only has two Endpoints available, as expected:

```
$ kubectl describe svc nginx-service-example
...
Endpoints:                  10.244.0.43:80,10.244.1.50:80
```

8. In the events for the Pod, you can also see that it is considered not ready:

```
$ kubectl describe pod nginx-deployment-example-
85cd4bb66f-94r4q
...
Events:
  Type      Reason       Age                          From
Message
  ----      -------      ----                         ----
-------
  Warning   Unhealthy    2m21s (x151 over 7m21s)   kubelet
Readiness probe failed: HTTP probe failed with
statuscode: 404
```

9. We can push this even further. Delete the `ready` files in the other two Pods (`nginx-deployment-example-85cd4bb66f-95bwd` and `nginx-deployment-example-85cd4bb66f-ssccm`, in our case) to make the whole Service fail:

```
$ kubectl exec -it nginx-deployment-example-85cd4bb66f-
95bwd -- rm /usr/share/nginx/html/ready
```

```
$ kubectl exec -it nginx-deployment-example-85cd4bb66f-
ssccm -- rm /usr/share/nginx/html/ready
```

10. Now, when you refresh the web page, you will see that the request is pending and that eventually, it will fail with a timeout. We are now in a pretty bad state – we have a total readiness failure for all the Pod replicas in our Deployment!

11. Finally, let's make one of our Pods ready again by recreating the file. You can refresh the web page so that the request is pending and, at the same time, execute the necessary command to create the `ready` file:

```
$ kubectl exec -it nginx-deployment-example-85cd4bb66f-
ssccm -- touch /usr/share/nginx/html/ready
```

12. After about 2 seconds (this is the probe interval), the pending request in the web browser should succeed and you will be presented with a nice response from `nginx`:

```
You have been served by Pod with IP address: 10.244.0.43
```

Congratulations – you have successfully configured and tested the readiness probe for your Deployment Pods! This should give you a good insight into how the probes work and how you can use them with Services that expose your Deployments.

Next, we will take a brief look at how you can scale your Deployments.

Scaling a Deployment object

The beauty of Deployments is that you can almost instantly scale them up or down, depending on your needs. When the Deployment is exposed behind a Service, the new Pods will be automatically discovered as new Endpoints when you scale up, or automatically removed from the Endpoints list when you scale down. The steps for this are as follows:

1. First, let's scale up our Deployment declaratively. Open the `nginx-deployment-readinessprobe.yaml` YAML manifest file and modify the number of replicas:

```
apiVersion: apps/v1
kind: Deployment
metadata:
  name: nginx-deployment-example
spec:
  replicas: 10
...
```

2. Apply these changes to the cluster using the `kubectl apply` command:

```
$ kubectl apply -f ./nginx-deployment-readinessprobe.yaml
--record
deployment.apps/nginx-deployment-example configured
```

3. Now, if you check the Pods using the `kubectl get pods` command, you will immediately see that new Pods are being created. Similarly, if you check the output of the `kubectl describe` command for the Deployment, you will see the following in the events:

```
$ kubectl describe deploy nginx-deployment-example
...
Events:
  Type     Reason            Age      From
Message
  ----     ------            ----     ----
-------
  Normal   ScalingReplicaSet  21s     deployment-controller
Scaled up replica set nginx-deployment-example-85cd4bb66f
to 10
```

4. You can achieve the same result using the `imperative` command, which is only recommended for development scenarios:

```
$ kubectl scale deploy nginx-deployment-example
--replicas=10
deployment.apps/nginx-deployment-example scaled
```

5. To scale down our Deployment declaratively, simply modify the `nginx-deployment-readinessprobe.yaml` YAML manifest file and change the number of replicas:

```
apiVersion: apps/v1
kind: Deployment
metadata:
  name: nginx-deployment-example
spec:
  replicas: 2
...
```

6. Apply the changes to the cluster using the `kubectl apply` command:

```
$ kubectl apply -f ./nginx-deployment-readinessprobe.yaml
--record
```

7. You can achieve the same result using imperative commands. For example, you can execute the following command:

```
$ kubectl scale deploy nginx-deployment-example
--replicas=2
```

If you describe the Deployment, you will see that this scaling down is reflected in the events:

```
$ kubectl describe deploy nginx-deployment-example
...
Events:
  Type      Reason           Age     From
Message
  ----      ------           ----    ----
-------
  Normal   ScalingReplicaSet  5s      deployment-controller
Scaled down replica set nginx-deployment-example-
85cd4bb66f to 2
```

Deployment events are very useful if you want to know the exact timeline of scaling and the other operations that can be performed with the Deployment object.

> **Important Note**
> It is possible to autoscale your deployments using `HorizontalPodAutoscaler`. This will be covered in *Chapter 20, Autoscaling Kubernetes Pods and Nodes*.

Next, you will learn how to delete a Deployment from your cluster.

Deleting a Deployment object

To delete a Deployment object, you can do two things:

- Delete the Deployment object, along with the Pods that it owns. This can be done by first scaling down automatically.

- Delete the Deployment object and leave the other Pods unaffected.

To delete the Deployment object and its Pods, you can use the regular `kubectl delete` command:

```
$ kubectl delete deploy nginx-deployment-example
```

You will see that the Pods get terminated and that the Deployment object is then deleted.

Now, if you would like to delete just the Deployment object, you need to use the `--cascade=orphan` option for `kubectl delete`:

```
$ kubectl delete deploy nginx-deployment-example
--cascade=orphan
```

After executing this command, if you inspect what pods are in the cluster, you will still see all the Pods that were owned by the `nginx-deployment-example` Deployment.

How does a Deployment object manage revisions and version rollout?

So far, we have only covered making one possible modification to a living Deployment – we have scaled up and down by changing the `replicas` parameter in the specification. However, this is not all we can do! It is possible to modify the Deployment's Pod template (`.spec.template`) in the specification and, in this way, trigger a rollout. This rollout may be caused by a simple change, such as changing the labels of the Pods, but it may be also a more complex operation when the container images in the Pod definition are changed to a different version. This is the most common scenario as it enables you, as a Kubernetes cluster operator, to perform a controlled, predictable rollout of a new version of your image and effectively create a new revision of your Deployment.

Your Deployment uses a rollout strategy, which can be specified in a YAML manifest using `.spec.strategy.type`. Kubernetes supports two strategies out of the box:

- `RollingUpdate`: This is the default strategy and allows you to roll out a new version of your application in a controlled way. This type of strategy uses two ReplicaSets internally. When you perform a change in the Deployment spec that causes a rollout, Kubernetes will create a new ReplicaSet with a new Pod template scaled to zero Pods initially. The old, existing ReplicaSet will remain unchanged at this point. Next, the old ReplicaSet will be scaled down gradually, whereas the new ReplicaSet will be scaled up gradually at the same time. The number of Pods that may be unavailable (readiness probe failing) is controlled using the `.spec.strategy.rollilngUpdate.maxUnavailable` parameter. The maximum number of extra Pods that can be scheduled above the desired number of Pods in the Deployment is controlled by the `.spec.strategy.rollilngUpdate.maxSurge` parameter. Additionally, this type of strategy offers automatic revision history, which can be used for quick rollbacks in case of any failures.

- `Recreate`: This is a simple strategy that's useful for development scenarios where all the old Pods have been terminated and replaced with new ones. This instantly deletes any existing underlying ReplicaSet and replaces it with a new one. You should not use this strategy for production workloads unless you have a specific use case.

> **Tip**
> Consider the Deployment strategies as basic building blocks for more advanced Deployment scenarios. For example, if you are interested in Blue/Green Deployments, you can easily achieve this in Kubernetes by using a combination of Deployments and Services while manipulating label selectors. You can find out more about this in the official Kubernetes blog post: `https://kubernetes.io/blog/2018/04/30/zero-downtime-deployment-kubernetes-jenkins/`.

Now, we will perform a rollout using the `RollingUpdate` strategy. The `Recreate` strategy, which is much simpler, can be exercised similarly.

Updating a Deployment object

First, let's recreate the Deployment that we used previously for our readiness probe demonstration:

1. Make a copy of the previous YAML manifest file:

    ```
    $ cp nginx-deployment-readinessprobe.yaml nginx-
    deployment-rollingupdate.yaml
    ```

2. Ensure that you have a strategy of the RollingUpdate type, called readinessProbe, set up and an image version of nginx:1.17. This should already be set up in the nginx-deployment-readinessprobe.yaml manifest file, if you completed the previous sections:

    ```
    apiVersion: apps/v1
    kind: Deployment
    metadata:
      name: nginx-deployment-example
    spec:
      replicas: 3
    ...
      minReadySeconds: 10
      strategy:
        type: RollingUpdate
        rollingUpdate:
          maxUnavailable: 1
          maxSurge: 1
      template:
    ...
        spec:
          containers:
          - name: nginx
            image: nginx:1.17
    ...
            readinessProbe:
    ...
    ```

3. In this example, we are using `maxUnavailable` set to `1`, which means that we allow only one Pod out of three, which is the target number, to be unavailable (not ready). This means that, at any time, there must be at least two Pods ready to serve traffic. Similarly, `maxSurge` set to `1` means that we allow one extra Pod to be created above the target number of three Pods during the rollout. This effectively means that we can have up to four Pods (ready or not) present in the cluster during the rollout. Please note that it is also possible to set up these parameters as percentage values (such as `25%`), which is very useful in autoscaling scenarios. Additionally, `minReadySeconds` (which is set to `10`) provides an additional time span for which the Pod has to be ready before it can be *announced* as successful during the rollout.

4. Apply the manifest file to the cluster:

```
$ kubectl apply -f ./nginx-deployment-rollingupdate.yaml
--record
```

With the deployment ready in the cluster, we can start rolling out a new version of our application. We will change the image in the Pod template for our Deployment to a newer version and observe what happens during the rollout. To do this, follow these steps:

1. Modify the container image that was used in the Deployment to `nginx:1.18`:

```
apiVersion: apps/v1
kind: Deployment
metadata:
  name: nginx-deployment-example
spec:
...
  template:
...
    spec:
      containers:
      - name: nginx
        image: nginx:1.18
```

2. Apply the changes to the cluster using the following command:

```
$ kubectl apply -f ./nginx-deployment-rollingupdate.yaml
--record
deployment.apps/nginx-deployment-example configured
```

3. Immediately after that, use the `kubectl rollout status` command to see the
   progress in real time:

```
$ kubectl rollout status deployment nginx-deployment-
example
Waiting for deployment "nginx-deployment-example" rollout
to finish: 2 out of 3 new replicas have been updated...
Waiting for deployment "nginx-deployment-example" rollout
to finish: 2 of 3 updated replicas are available...
deployment "nginx-deployment-example" successfully rolled
out
```

4. The rollout will take a bit of time because we have configured `minReadySeconds`
   on the Deployment specification and `initialDelaySeconds` on the Pod
   container readiness probe.

5. Similarly, using the `kubectl describe` command, you can see the events for the
   Deployment that inform us of how the ReplicaSets were scaled up and down:

```
$ kubectl describe deploy nginx-deployment-example
...
Events:
  Type      Reason             Age     From
Message
  ----      ------             ----    ----
-------
  Normal    ScalingReplicaSet  3m56s   deployment-controller
Scaled up replica set nginx-deployment-example-85cd4bb66f
to 3
  Normal    ScalingReplicaSet  3m22s   deployment-controller
Scaled up replica set nginx-deployment-example-54769f6df8
to 1
  Normal    ScalingReplicaSet  3m22s   deployment-controller
Scaled down replica set nginx-deployment-example-
85cd4bb66f to 2
  Normal    ScalingReplicaSet  3m22s   deployment-controller
Scaled up replica set nginx-deployment-example-54769f6df8
to 2
  Normal    ScalingReplicaSet  3m5s    deployment-controller
Scaled down replica set nginx-deployment-example-
85cd4bb66f to 0
```

```
   Normal   ScalingReplicaSet   3m5s    deployment-controller
   Scaled up replica set nginx-deployment-example-54769f6df8
   to 3
```

6. Now, let's take a look at the ReplicaSets in the cluster:

```
$ kubectl get rs
NAME                                          DESIRED    CURRENT
READY     AGE
nginx-deployment-example-54769f6df8     3          3
3            6m44s
nginx-deployment-example-85cd4bb66f     0          0
0            7m18s
```

7. You will see something interesting here: the old ReplicaSet remains in the cluster
    but has been scaled down to zero pods! The reason for this is that we're keeping
    the Deployment revision history – each revision has a matching ReplicaSet that
    can be used if we need to roll back. The number of revisions that are kept for each
    Deployment is controlled by the `.spec.revisionHistoryLimit` parameter –
    by default, it is set to `10`. Revision history is important, especially if you are making
    imperative changes to your Deployments. If you are using the declarative model
    and always committing your changes to a source code repository, then the revision
    history may be less relevant.

8. Lastly, we can check if the Pods were indeed updated to a new image version. Use
    the following command and verify one of the Pods in the output:

```
$ kubectl describe pods
...
Containers:
  nginx:
    Container ID:   docker://
a3ff03abf8f76e0d128f5561b6b8fd0c7a355f0fb8a4d3d9ef45ed9
ee8adf23c
    Image:          nginx:1.18
```

This shows that we have indeed performed a rollout of the new `nginx` container image
version!

> **Tip**
>
> You can change the Deployment container image *imperatively* using the
> `kubectl set image deployment nginx-deployment-`
> `example nginx=nginx:1.18 --record` command. This approach
> is only recommended for non-production scenarios, and it works well with
> *imperative* rollbacks.

Next, you will learn how to roll back a deployment.

Rolling back a Deployment object

If you are using a declarative model to introduce changes to your Kubernetes cluster and are committing each change to your source code repository, performing a rollback is very simple and involves just reverting the commit and applying the configuration again. Usually, the process of applying changes is performed as part of the CI/CD pipeline for the source code repository, instead of the changes being manually applied by an operator. This is the easiest way to manage Deployments, and this is generally recommended in the Infrastructure-as-Code and Configuration-as-Code paradigms.

> **Tip**
>
> One very good example of using a declarative model in practice is Flux
> (`https://fluxcd.io/`), which is a project that's currently incubating
> at CNCF. Flux is the core of the approach known as **GitOps**, which is a way
> of implementing continuous deployment for cloud-native applications. It
> focuses on a developer-centric experience when operating the infrastructure by
> using tools developers are already familiar with, including Git and continuous
> deployment tools.

The Kubernetes CLI still provides an imperative way to roll back a deployment using revision history. Imperative rollbacks can also be performed on Deployments that have been updated declaratively. Now, we will demonstrate how to use `kubectl` for rollbacks. Follow these steps:

1. First, let's imperatively roll out another version of our Deployment. This time, we will update the `nginx` image to version `1.19`:

   ```
   $ kubectl set image deployment nginx-deployment-example
   nginx=nginx:1.19 --record
   deployment.apps/nginx-deployment-example image updated
   ```

2. Using `kubectl rollout status`, wait for the end of the deployment:

```
$ kubectl rollout status deployment nginx-deployment-
example

...

deployment "nginx-deployment-example" successfully rolled
out
```

3. Now, let's suppose that the new version of the application image, `1.19`, is causing problems and that your team decides to roll back to the previous version of the image, which was working fine.

4. Use the following `kubectl rollout history` command to see all the revisions that are available for the Deployment:

```
$ kubectl rollout history deploy nginx-deployment-example
deployment.apps/nginx-deployment-example
REVISION   CHANGE-CAUSE
1          kubectl apply --filename=./nginx-deployment-
rollingupdate.yaml --record=true
2          kubectl apply --filename=./nginx-deployment-
rollingupdate.yaml --record=true
3          kubectl set image deployment nginx-deployment-
example nginx=nginx:1.19 --record=true
```

5. As you can see, we have three revisions. The first revision is our initial creation of the Deployment. The second revision is the declarative update of the Deployment to the `nginx:1.18` image. Finally, the third revision is our last imperative update to the Deployment that caused the `nginx:1.19` image to be rolled out. CHANGE-CAUSE is the contents of the `kubernetes.io/change-cause` annotation, which is added when you use the `--record` flag for your `kubectl` commands.

6. The revisions that were created as a declarative change do not contain too much information in CHANGE-CAUSE. To find out more about the second revision, you can use the following command:

    ```
    $ kubectl rollout history deploy nginx-deployment-example
    --revision=2
    deployment.apps/nginx-deployment-example with revision #2
    Pod Template:
    ...
      Containers:
       nginx:
        Image:      nginx:1.18
    ```

7. Now, let's perform a rollback to this revision. Because it is the previous revision, you can simply execute the following command:

    ```
    $ kubectl rollout undo deploy nginx-deployment-example
    deployment.apps/nginx-deployment-example rolled back
    ```

8. This would be equivalent of executing a rollback to a specific revision number:

    ```
    $ kubectl rollout undo deploy nginx-deployment-example
    --to-revision=2
    ```

9. Again, as in the case of a normal rollout, you can use the following command to follow the rollback:

    ```
    $ kubectl rollout status deploy nginx-deployment-example
    ```

Please note that you can also perform rollbacks on currently ongoing rollouts. This can be done in both ways; that is, declaratively and imperatively.

> **Tip**
> If you need to pause and resume the ongoing rollout of a Deployment, you can use the `kubectl rollout pause deployment nginx-deployment-example` and `kubectl rollout resume deployment nginx-deployment-example` commands.

Congratulations – you have successfully rolled back your Deployment. In the next section, we will provide you with a set of best practices for managing Deployment objects in Kubernetes.

Deployment object best practices

This section will summarize known best practices when working with Deployment objects in Kubernetes. This list is by no means complete, but it is a good starting point for your journey with Kubernetes.

Use declarative object management for Deployments

In the DevOps world, it is a good practice to stick to declarative models when introducing updates to your infrastructure and applications. This is atthe core of the *Infrastructure-as-Code* and *Configuration-as-Code* paradigms. In Kubernetes, you can easily perform declarative updates using the `kubectl apply` command, which can be used on a single file or even a whole directory of YAML manifest files.

> **Tip**
>
> To delete objects, it is still better to use imperative commands. It is more predictable and less prone to errors. Declaratively deleting resources in your cluster is only useful in CI/CD scenarios, where the whole process is entirely automated.

The same principle also applies to Deployment objects. Performing a rollout or rollback when your YAML manifest files are versioned and kept in a source control repository is easy and predictable. Using the `kubectl rollout undo` and `kubectl set image deployment` commands is generally not recommended in production environments. Using these commands gets much more complicated when more than one person is working on operations in the cluster.

Do not use the Recreate strategy for production workloads

Using the `Recreate` strategy may be tempting as it provides instantaneous updates for your Deployments. However, at the same time, this will mean downtime for your end users. This is because all the existing Pods for the old revision of the Deployment will be terminated at once and replaced with the new Pods. There may be a significant delay before the new pods become ready, and this means downtime. This downtime can be easily avoided by using the `RollingUpdate` strategy in production scenarios.

Do not create Pods that match an existing Deployment label selector

It is possible to create Pods with labels that match the label selector of some existing Deployment. This can be done using bare Pods or another Deployment or ReplicaSet. This leads to conflicts, which Kubernetes does not prevent, and makes the existing deployment *believe* that it has created the other Pods. The results may be unpredictable and in general, you need to pay attention to how you label the resources in your cluster. We advise you to use semantic labeling here, which you can learn more about in the official documentation: `https://kubernetes.io/docs/concepts/configuration/overview/#using-labels`.

Carefully set up your container probes

The liveness, readiness, and startup probes of your Pod containers can provide a lot of benefits but at the same time, if they have been misconfigured, they can cause outages, including cascading failures. You should always be sure that you understand the consequences of each probe going into a failed state and how it affects other Kubernetes resources, such as Service objects.

There are a couple of established best practices for readiness probes that you should consider:

- Use this probe whenever your containers may not be ready to serve traffic as soon as the container is started.

- Ensure that you check things such as cache warm-ups or your database migration status during readiness probe evaluation. You may also consider starting the actual process of a warm-up if it has not been started yet, but use this approach with caution – a readiness probe will be executed constantly throughout the life cycle of a pod, which means you shouldn't perform any costly operations for every request. Alternatively, you may want to use a startup probe for this purpose.

- For microservice applications that expose HTTP endpoints, consider configuring the `httpGet` readiness probe. This will ensure that every basis is covered when a container is successfully running but the HTTP server has not been fully initialized.

- It is a good idea to use a separate, dedicated HTTP endpoint for readiness checks in your application. For example, a common convention is using `/health`.

- If you are checking the state of your dependencies (external database, logging services, and so on) in this type of probe, be careful with shared dependencies, such as databases. In this case, you should consider using a probe timeout, which is greater than the maximum allowed timeout for the external dependency – otherwise, you may get cascading failures and lower availability instead of occasionally increased latency.

Similar to readiness probes, there are a couple of guidelines on how and when you should use liveness probes:

- Liveness probes should be used with caution. Incorrectly configuring this probe can result in cascading failures in your services and container restart loops.

- Do not use liveness probes unless you have a good reason to do so. A good reason may be, for example, if there's a known issue with a deadlock in your application that has an unknown root cause.

- Execute simple and fast checks that determine the status of the process, not its dependencies. In other words, you do not want to check you status of your external dependencies in the liveness probe, since this can lead to cascading failures due to an avalanche of container restarts and overloading a small subset of Service Pods.

- If the process running in your container can crash or exit whenever it encounters an unrecoverable error, you probably do not need a liveness probe at all.

- Use conservative settings for `initialDelaySeconds` to avoid any premature container restarts and falling into a restart loop.

These are the most important points concerning probes for Pods. Now, let's discuss how you should tag your container images.

Use meaningful and semantic image tags

Managing deployment rollbacks and inspecting the history of rollouts requires that we use good tagging for the container images. If you rely on the `latest` tag, performing a rollback will not be possible because this tag points to a different version of the image as time goes on. It is a good practice to use semantic versioning for your container images. Additionally, you may consider tagging the images with a source code hash, such as a Git commit hash, to ensure that you can easily track what is running in your Kubernetes cluster.

Migrating from older versions of Kubernetes

If you are working on workloads that were developed on older versions of Kubernetes, you may notice that, starting with Kubernetes 1.16, you can't apply the Deployment to the cluster because of the following error:

```
error: unable to recognize "deployment": no matches for kind
"Deployment" in version "extensions/v1beta1"
```

The reason for this is that in version 1.16, the Deployment object was removed from the `extensions/v1beta1` API group, according to the API versioning policy. You should use the `apps/v1` API group instead, which Deployment has been part of since 1.9. It is becoming a stable feature.

This also shows an important rule to follow when you work with Kubernetes: always follow the API versioning policy and try to upgrade your resources to the latest API groups when you migrate to a new version of Kubernetes. This will save you unpleasant surprises when the resource is eventually deprecated in older API groups.

Summary

In this chapter, you learned how to work with **stateless** workloads and applications on Kubernetes using Deployment objects. First, you created an example Deployment and exposed its Pods using a Service object of the `LoadBalancer` type for external traffic. Next, you learned how to scale and manage Deployment objects in the cluster. The management operations we covered included rolling out a new revision of a Deployment and rolling back to an earlier revision in case of a failure. Lastly, we equipped you with a set of known best practices when working with Deployment objects.

The next chapter will extend this knowledge with details about managing **stateful** workloads and applications. While doing so, we will introduce a new Kubernetes object: StatefulSet.

Further reading

For more information regarding Deployments and Services, please refer to the following Packt books:

- *The Complete Kubernetes Guide*, by *Jonathan Baier, Gigi Sayfan, Jesse White* (`https://www.packtpub.com/virtualization-and-cloud/complete-kubernetes-guide`).

- *Getting Started with Kubernetes – Third Edition*, by *Jonathan Baier, Jesse White* (`https://www.packtpub.com/virtualization-and-cloud/getting-started-kubernetes-third-edition`).

- *Kubernetes for Developers*, by *Joseph Heck* (`https://www.packtpub.com/virtualization-and-cloud/kubernetes-developers`).

- *Hands-On Kubernetes on Windows*, by *Piotr Tylenda* (`https://www.packtpub.com/product/hands-on-kubernetes-on-windows/9781838821562`).

- You can also refer to the excellent official Kubernetes documentation (`https://kubernetes.io/docs/home/`), which is always the most up-to-date source of knowledge for Kubernetes in general.

12
StatefulSet – Deploying Stateful Applications

In the previous chapter, we explained how you can use a Kubernetes cluster to run *stateless* workloads and applications and how to use Deployment objects for this purpose. Running stateless workloads in the cloud is generally easier to handle, as any container replica can handle the request without taking any dependencies on the results of previous operations by the end user. In other words, every container replica would handle the request in an identical way; all you need to care about is proper load balancing.

However, stateless workloads are only the easier part of running cloud applications. The main complexity is in managing the *state* of applications. By *state,* we mean any stored *data* that the application or component needs to serve the requests and it can be modified by these requests. The most common example of a stateful component in applications is a database – for example, it can be a **relational MySQL database** or a **NoSQL MongoDB database**. In Kubernetes, you can use a dedicated object to run *stateful* workloads and applications: StatefulSet. When managing StatefulSet objects, you will usually need to work with PVs which have been covered in *Chapter 9, Persistent Storage in Kubernetes.* This chapter will provide you with knowledge about the role of StatefulSet in Kubernetes and how to create and manage StatefulSet objects to release new versions of your stateful applications.

In this chapter, we will cover the following topics:

- Introducing the StatefulSet object

- Managing StatefulSet

- Releasing a new version of an app deployed as a StatefulSet

- StatefulSet best practices

Technical requirements

For this chapter, you will need the following:

- Kubernetes cluster deployed. You can use either a local or a cloud-based cluster, but to fully understand the concepts, we recommend using a **multi-node**, cloud-based Kubernetes cluster. The cluster must support the creation of PersistentVolumeClaims – any cloud-based cluster will be sufficient, or locally, for example, `minikube` with a `k8s.io/minikube-hostpath` provisioner.

- A Kubernetes CLI (`kubectl`) installed on your local machine and configured to manage your Kubernetes cluster.

Kubernetes cluster deployment (local and cloud-based) and `kubectl` installation have been covered in *Chapter 3, Installing Your First Kubernetes Cluster.*

You can download the latest code samples for this chapter from the official GitHub repository at `https://github.com/PacktPublishing/The-Kubernetes-Bible/tree/master/Chapter12`.

Introducing the StatefulSet object

You may wonder why running stateful workloads in the distributed cloud is generally considered **harder** than stateless ones. In classic three-tier applications, all the states would be stored in a database (*data tier* or *persistence layer*) and there would be nothing special about it. For SQL servers, you would usually add a failover setup with data replication, and in case you require superior performance, you would scale *vertically* by simply purchasing better hardware for hosting. Then, at some point, you might think about clustered SQL solutions, introducing *data sharding* (horizontal data partitions). But still, from the perspective of a web server running your application, the database would be just a single connection string to read and write the data. The database would be responsible for persisting a *mutable state*.

> **Important note**
>
> Remember that every application *as a whole* is, in some way, stateful unless it
> serves static content only or just transforms user input. But this does not mean
> that *every* component in the application is stateful. A web server that runs the
> application logic will be a *stateless* component, but the database where this
> application stores user input and user sessions will be a *stateful* component.

We will first explain how you approach managing state in containers and what we actually
consider an application or system state.

Managing state in containers

Now, imagine how this could work if you deployed your SQL server (single instance) in a
container. The first thing you would notice is that after restarting the container, you would
lose the data stored in the database – each time it is restarted, you get a fresh instance of
the SQL server. Containers are *ephemeral*. This doesn't sound too useful for our use case.
Fortunately, containers come with the option to mount data **volumes**. A volume can be,
for example, a *host's directory*, which will be *mounted* to a specific path in the container's
filesystem. Whatever you store in this path will be kept in the host filesystem even after the
container is terminated or restarted. In a similar way, you can use NFS share or a cloud-
managed disk instance as a volume. Now, if you configure your SQL server to put its data
files in the path where the volume is mounted, you achieve data persistence even if the
container restarts. The container itself is still ephemeral, but the data (state) is *not*.

This is a high-level overview of how the state can be persisted for plain containers, without
involving Kubernetes. But before we move on to Kubernetes, we need to clarify what we
actually regard as a **state**.

If you think about it, when you have a web server that serves just simple *static* content
(which means it is always the same, as a simple HTML web page), there is still some data
that is persisted, for example, the HTML files. However, this is *not* state: user requests
cannot modify this data, so *previous* requests from the user will not influence what is the
result of the *current* request. In the same way, configuration files for your web server are
not their state or log files written on the disk (well, that is arguable, but from the end user's
perspective, it is not).

Now, if you have a web server that keeps user sessions and stores information about whether the user is logged in, then this is indeed the state. Depending on this information, the web server will return different web pages (responses) based on whether the user is logged in. Let's say that this web server runs in a container – there is a catch whether it is *the* stateful component in your application. If the web server process stores user sessions as a file in the container (warning: this is probably quite a bad design), then the web server container is a *stateful* component. But if it stores user sessions in a database or a Redis cache running in separate containers, then the web server is *stateless*, and the database or Redis container becomes the stateful component.

This is briefly how it looks from a single container perspective. We need now to zoom out a bit and take a look at state management in **Kubernetes Pods**.

Managing state in Kubernetes Pods

In Kubernetes, the concept of container volumes is extended by **PersistentVolumes (PVs)**, **PersistentVolumeClaims (PVCs)**, and **StorageClasses (SCs)**, which are dedicated, storage-related objects. PVC aims to *decouple* Pods from the actual storage. PVC is a Kubernetes object that models a request for the storage of a specific type, class, or size – think of saying *I would like 10 GB of read/write-once SSD storage*. To fulfill such a request, a PV object is required, which is a piece of real storage that has been provisioned by the cluster's automation process – think of this as a directory on the host system or cloud-managed disk. PV types are implemented as plugins, similarly to volumes in Docker. Now, the whole process of provisioning PV can be *dynamic* – it requires the creation of an SC object and for this to be used when defining PVCs. When creating a new SC, you provide a *provisioner* (or plugin) with specific parameters, and each PVC using the given SC will automatically create a PV using the selected provisioner. The provisioners may, for example, create cloud-managed disks to provide the backing storage. On top of that, containers of a given Pod can share data using the same PV and mount it to their filesystem.

This is just a brief overview of what Kubernetes provides for state storage. We have covered this in more detail in *Chapter 9, Persistent Storage in Kubernetes*.

On top of the management of state in a single Pod and its containers, there is the management of state in *multiple replicas* of a Pod. Let's think about what would happen if we used a Deployment object to run multiple Pods with MySQL Server. First, you would need to ensure that the state is persisted on a volume in a container – for this, you can use PVs in Kubernetes. But then you actually get multiple, separate MySQL servers, which is not very useful if you would like to have high availability and fault tolerance. If you expose such a deployment using a service, it will also be useless because each time, you may hit a different Pod and get different data. So, you arrive either at designing a *multi-node failover setup* with replication between the master and replicas or a complex *cluster with data sharding*. In any case, your individual MySQL Server Pod replicas need to have a *unique identity* and, preferably, *predictable network names* so that the Nodes and clients can communicate.

> **Tip**
> When designing your cloud-native application for the Kubernetes cluster, always analyze all the pros and cons of storing the state of the application as stateful components *running in Kubernetes*. Generally, it is easier to *outsource* the hard part of managing such components to dedicated, cloud-managed services (for example, Amazon S3, Azure CosmosDB, or Google Cloud SQL). Then, the Kubernetes cluster can be used just for running stateless components.

This is where StatefulSet comes in. Let's take a closer look at this Kubernetes object.

StatefulSet and how it differs from a Deployment object

Kubernetes StatefulSet is a similar concept to a Deployment object – it also provides a way of managing and scaling a set of Pods, but it provides guarantees about the *ordering and uniqueness* (unique identity) of the Pods. In the same way as Deployment, it uses a Pod template to define what each replica should look like. You can scale it up and down and perform rollouts of new versions. But now, in StatefulSet, the individual Pod replicas are *not interchangeable*. The unique, persistent identity for each Pod is maintained during any rescheduling or rollouts – this includes **Pod name** and its **cluster DNS names**. This unique, persistent identity can be used for clearly identifying PVs that are assigned to each Pod, even if replaced following a failure. For this, StatefulSet provides another type of template in its specification named `volumeClaimTemplates`. This template can be used for the dynamic creation of the PVCs of a given SC. By doing this, the whole process of storage provisioning is fully dynamic – you just create a StatefulSet and the underlying storage objects are managed by the StatefulSet controller.

> **Tip**
> Cluster DNS names of individual Pods in StatefulSet remain the same, but their cluster IP addresses are not guaranteed to stay the same. This means that if you need to connect to individual Pods in the StatefulSet, you should use cluster DNS names.

Basically, you can use StatefulSet for applications that require the following:

- Persistent storage managed by the Kubernetes cluster (this is the main use case, but not the only one)
- Stable and unique network identifiers (usually DNS names) for each Pod replica
- Ordered deployment and scaling
- Ordered, rolling updates

As you can see, StatefulSets can be seen as a more predictable version of a Deployment object, with the possibility to use persistent storage provided by PVCs. To summarize, the key differences between StatefulSet and Deployment are as follows:

- StatefulSet ensures a *deterministic* (sticky) name for Pods, which consists of `<statefulSetName>-<ordinal>`. For Deployments, you would have a *random* name consisting of `<deploymentName>-<podTemplateHash>-<randomHash>`.
- For StatefulSet objects, the Pods are started and terminated in a *specific* and *predictable* order while scaling the ReplicaSet.
- In terms of storage, Kubernetes creates PVCs based on `volumeClaimTemplates` of the StatefulSet specification for each Pod in the StatefulSet and always attaches this to the Pod with *the same* name. For Deployment, if you choose to use persistentVolumeClaim in the Pod template, Kubernetes will create a single PVC and attach the same to all the Pods in the deployment. This may be useful in certain scenarios, but is not a common use case.
- You need to create a **headless** Service object that is responsible for managing the *deterministic network identity* (cluster DNS names) for Pods. The headless Service allows us to return *all* Pods IP addresses behind the service as DNS `A records` instead of a single DNS `A record` with a `ClusterIP` Service. A headless Service is only required if you are not using a regular service. The specification of StatefulSet requires having the Service name provided in `.spec.serviceName`.

Now, let's take a look at a concrete example of StatefulSet that deploys `nginx` Pods with the backing of persistent storage.

Managing StatefulSet

To demonstrate how StatefulSet objects work, we will modify our `nginx` deployment and adapt it to be a StatefulSet. A significant part of the StatefulSet specification is the same as for Deployments. As we would like to demonstrate how automatic management of PVCs works in StatefulSet objects, we will use `volumeClaimTemplates` in the specification to create PVCs and PVs, which the Pods will consume. Each Pod will internally mount its assigned PV under the `/usr/share/nginx/html` path in the container filesystem, which is the default location of `nginx` files that are served over HTTP. In this way, we can demonstrate how the *state* is persisted, even if we forcefully restart Pods.

> **Important note**
>
> The example that we are going to use in this chapter is for demonstration purposes only and is meant to be as simple as possible. If you are interested in *complex* examples, such as deploying and managing distributed databases in StatefulSets, please take a look at the official Kubernetes blog post about deploying the Cassandra database: `https://kubernetes.io/docs/tutorials/stateful-application/cassandra/`. Usually, the main source of complexity in such cases is handling the joining and removal of Pod replicas when scaling the StatefulSet.

We will now go through all the YAML manifests required to create our StatefulSet and apply them to the cluster.

Creating a StatefulSet

First, let's take a look at the StatefulSet YAML manifest file named `nginx-statefulset.yaml` (the full version is available in the official GitHub repository for the book: `https://github.com/PacktPublishing/Kubernetes-for-Beginners/blob/master/Chapter12/01_statefulset-example/nginx-statefulset.yaml`):

```
apiVersion: apps/v1
kind: StatefulSet
metadata:
  name: nginx-statefulset-example
spec:
  replicas: 3
  serviceName: nginx-headless
  selector:
    matchLabels:
```

```
      app: nginx-stateful
      environment: test
  # (to be continued in the next paragraph)
```

The first part of the preceding file is very similar to the Deployment object specification, where you need to provide the number of `replicas` and a `selector` for Pods. There is one new parameter, `serviceName`, that we will explain shortly.

The next part of the file concerns specification of the Pod template that is used by the StatefulSet:

```
  # (continued)
    template:
      metadata:
        labels:
          app: nginx-stateful
          environment: test
      spec:
        containers:
        - name: nginx
          image: nginx:1.17
          ports:
          - containerPort: 80
          volumeMounts:
          - name: nginx-data
            mountPath: /usr/share/nginx/html
          command:
          - /bin/sh
          - -c
          - |
            echo "You have been served by Pod with IP address:
$(hostname -i)" > /usr/share/nginx/html/index.html
            nginx -g "daemon off;"
  # (to be continued in the next paragraph)
```

If you look closely, you can observe that the structure is the same as for Deployments. Also, the last part of the file contains `volumeClaimTemplates`, which is used to define templates for PVC used by the Pod:

```
# (continued)
  volumeClaimTemplates:
  - metadata:
      name: nginx-data
    spec:
      accessModes: [ "ReadWriteOnce" ]
      resources:
        requests:
          storage: 1Gi
```

As you can see, in general, the structure of the StatefulSet spec is similar to Deployment, although it has a few extra parameters for configuring PVCs and associated Service objects. The specification has five main components:

- `replicas`: Defines the number of Pod replicas that should run using the given `template` and the matching label `selector`. Pods may be created or deleted to maintain the required number.

- `serviceName`: The name of the service that governs StatefulSet and provides the network identity for the Pods. This Service must be created before StatefulSet is created. We will create the `nginx-headless` Service in the next step.

- `selector`: A **label selector**, which defines how to identify Pods that the StatefulSet owns. This can include *set-based* and *equality-based* selectors.

- `template`: Defines the template for Pod creation. Labels used in `metadata` must match the `selector`. Pod names are not random and follow the `<statefulSetName>-<ordinal>` convention.

- `volumeClaimTemplates`: Defines the template for PVC that will be created for each of the Pods. Each Pod in the StatefulSet object will get its own PVC that is assigned to a given Pod name persistently. In our case, it is a 1 GB volume with the `ReadWriteOnce` access mode. This access mode allows the volume to be mounted for reads and writes by a *single* Node only. We did not specify `storageClassName`, so the PVCs will be provisioned using the default SC in the cluster. PVC names are not random and follow the `<volumeClaimTemplateName>-<statefulSetName>-<ordinal>` convention.

> **Tip**
>
> The default SC in your cluster is marked with the `storageclass.`
> `kubernetes.io/is-default-class` annotation. Whether you
> have a default SC, and how it is defined, depends on your cluster deployment.
> For example, in the Azure Kubernetes Service cluster, it will be an SC named
> `default` that uses the `kubernetes.io/azure-disk` provisioner.
> In `minikube`, it will be an SC named `standard` that uses the `k8s.io/`
> `minikube-hostpath` provisioner.

The specification also contains other fields that are related to rolling out new revisions of StatefulSet – we will explain these in more detail in the next section.

Apart from that, in our Pod template, we have used a similar *override* of the command for the `nginx` container, as in the case of Deployment in the previous chapter. The command creates an `index.html` file in `/usr/share/nginx/html/` with information about what the IP address of the Pod is that serves the request. After that, it starts the `nginx` web server with the standard entry point command for the image.

Next, let's have a look at our *headless* Service named `nginx-headless`. Create a `nginx-headless-service.yaml` file with the following content:

```
apiVersion: v1
kind: Service
metadata:
  name: nginx-headless
spec:
  selector:
    app: nginx-stateful
    environment: test
  clusterIP: None
  ports:
  - port: 80
    protocol: TCP
    targetPort: 80
```

The specification is very similar to the normal Service that we created previously for the Deployment, the only difference is that it has the value None for the clusterIP field. This will result in the creation of a headless Service, nginx-headless. A headless Service allows us to return *all* Pods' IP addresses behind the Service as DNS A records instead of a single DNS A record with a clusterIP Service. We will demonstrate what this means in practice in the next steps.

Lastly, let's create a LoadBalancer Service YAML manifest file, nginx-client-service.yaml. This Service will allow us to reach the StatefulSet from an external network using a web browser:

```yaml
apiVersion: v1
kind: Service
metadata:
  name: nginx-client
spec:
  selector:
    app: nginx-stateful
    environment: test
  type: LoadBalancer
  ports:
  - port: 80
    protocol: TCP
    targetPort: 80
```

The preceding specification is the same as we previously used for the Deployment demonstration. The only difference is the name, which is now nginx-client.

With all the YAML manifest files, we can start deploying our example StatefulSet! Perform the following steps:

1. Create a headless Service, nginx-headless, using the following command:

   ```
   $ kubectl apply -f ./nginx-headless-service.yaml
   ```

2. Create a LoadBalancer Service, nginx-client, using the following command:

   ```
   $ kubectl apply -f ./nginx-client-service.yaml
   ```

3. Create a StatefulSet object, `nginx-stateful-example`, using the following command:

    ```
    $ kubectl apply -f ./nginx-statefulset.yaml
    ```

4. Now, you can use the `kubectl describe` command to observe the creation of the StatefulSet object:

    ```
    $ kubectl describe statefulset nginx-statefulset-example
    ```

5. Alternatively, you can use `sts` as an abbreviation for StatefulSet when using the `kubectl` commands.

6. If you use the `kubectl get pods` command, you can see that the three desired Pod replicas have been created. Note that this can take a bit of time as the Pods have to get the PVs provisioned based on their PVCs:

    ```
    $ kubectl get pods
    NAME                          READY   STATUS     RESTARTS
    AGE
    nginx-statefulset-example-0   1/1     Running    0
    7m
    nginx-statefulset-example-1   1/1     Running    0
    5m
    nginx-statefulset-example-2   1/1     Running    0
    4m
    ```

    Please note the ordered, deterministic Pod naming – this is the key for providing a unique identity to the Pods in the StatefulSet object.

7. If you describe one of the Pods, you will see more details about the associated PV and PVC:

    ```
    $ kubectl describe pod nginx-statefulset-example-1
    ...
    Volumes:
      nginx-data:
        Type:       PersistentVolumeClaim (a reference to a
    PersistentVolumeClaim in the same namespace)
        ClaimName:  nginx-data-nginx-statefulset-example-1
    ...
    Events:
      Type    Reason                Age    From
    ```

```
Message
  ----      ------                     ----   ----
-------
   Normal   Scheduled                  7m      default-scheduler
Successfully assigned default/nginx-statefulset-example-1
to aks-nodepool1-77120516-vmss000001
   Normal   SuccessfulAttachVolume 7m      attachdetach-
controller  AttachVolume.Attach succeeded for volume
"pvc-6b2b1ad8-5b08-4d3e-bac9-6f7ec7de7d40"
   Normal   Pulled                     6m      kubelet
Container image "nginx:1.17" already present on machine
   Normal   Created                    6m      kubelet
Created container nginx
   Normal   Started                    6m      kubelet
Started container nginx
```

As you can see, the PVC used by this Pod is named `nginx-data-nginx-statefulset-example-1`. Additionally, right after the Pod was scheduled on its target Node, the PV, `pvc-6b2b1ad8-5b08-4d3e-bac9-6f7ec7de7d40`, has been provisioned based on the PVC and attached to the Pod. After that, the actual container, which internally mounts this PV, has been created.

8. Using the `kubectl get` command, we can reveal more details about the PVC:

```
$ kubectl get pvc nginx-data-nginx-statefulset-example-1
NAME                                         STATUS    VOLUME
CAPACITY    ACCESS MODES    STORAGECLASS    AGE
nginx-data-nginx-statefulset-example-1    Bound
pvc-6b2b1ad8-5b08-4d3e-bac9-6f7ec7de7d40    1Gi          RWO
default         43m
```

9. And finally, let's take a look at the PV that was provisioned:

```
$ kubectl describe pv pvc-6b2b1ad8-5b08-4d3e-bac9-
6f7ec7de7d40
...
Source:
    Type:        AzureDisk (an Azure Data Disk mount on
the host and bind mount to the pod)
    DiskName:    kubernetes-dynamic-pvc-6b2b1ad8-5b08-
4d3e-bac9-6f7ec7de7d40
    DiskURI:     /subscriptions/cc9a8166-829e-401e-
```

```
a004-76d1e3733b8e/resourceGroups/mc_k8sforbeginners-rg_
k8sforbeginners-aks_eastus/providers/Microsoft.Compute/
disks/kubernetes-dynamic-pvc-6b2b1ad8-5b08-4d3e-bac9-
6f7ec7de7d40
...
```

In our example, as we are demonstrating this using Azure Kubernetes Service, the PV was provisioned using `AzureDisk`, and you can also see the actual resource ID that you can find in Azure Portal.

We have successfully created the StatefulSet object and now it is time to verify whether it works as expected in a basic scenario. Please follow these steps:

1. Get the external IP address of the `nginx-client` Service so that we can use it to access via a web browser:

```
$ kubectl describe svc nginx-client
...
LoadBalancer Ingress:     104.45.176.241
...
```

2. In our example, it is `104.45.176.241`, and we will use it in subsequent steps. Navigate to `http://104.45.176.241` in your favorite web browser:

You have been served by Pod with IP address: 10.244.1.103

Figure 12.1 – Successful request to the nginx web server running in StatefulSet

If you refresh the page a few times (with the browser cache disabled), you will notice that you are served by three different Pods. This is as expected – we are currently running three Pod replicas in our StatefulSet object.

We will now take a quick look at how the *headless* Service behaves.

Using the headless Service and stable network identities

Let's do an experiment that demonstrates how the *headless* Service is used to provide stable and predictable network identities for our Pods:

1. Create an *interactive* busybox Pod and start the Bourne shell process. The following command will create the Pod and immediately attach your terminal so that you can interact from *within* the Pod:

```
$ kubectl run -i --tty busybox --image=busybox:1.28 --rm
--restart=Never -- sh
```

2. First, we need to check how our normal service, nginx-client, with ClusterIP assigned, is resolved by the cluster DNS:

```
$ nslookup nginx-client
Server:      10.0.0.10
Address 1: 10.0.0.10 kube-dns.kube-system.svc.cluster.
local

Name:        nginx-client
Address 1: 10.0.147.198 nginx-client.default.svc.cluster.
local
```

As you would expect, the response from DNS is that we have A record that points to the service ClusterIP 10.0.147.198.

3. Perform a similar check for the headless Service, nginx-headless:

```
$ nslookup nginx-headless
Server:      10.0.0.10
Address 1: 10.0.0.10 kube-dns.kube-system.svc.cluster.
local

Name:        nginx-headless
Address 1: 10.244.1.103 nginx-statefulset-example-0.
nginx-headless.default.svc.cluster.local
Address 2: 10.244.0.77 nginx-statefulset-example-1.nginx-
headless.default.svc.cluster.local
Address 3: 10.244.1.104 nginx-statefulset-example-2.
nginx-headless.default.svc.cluster.local
```

In this case, we have received three `A records` that point directly to Pod IP addresses. Additionally, they have `CNAME records` in the form of `<podName>-<serviceName>.<namespace>.svc.cluster.local`. So, the difference here is that a Service that has `ClusterIP` will get load balancing to a *virtual IP* level (which, on Linux, is usually handled at a kernel level by `iptables` rules configured by `kube-proxy`) whereas, in the case of the headless Service, the responsibility for load balancing or choosing the target Pod is on the *client* making the request.

4. Having *predictable* FQDNs for Pods in the StatefulSet gives us the option to send the requests directly to individual Pods, without guessing their IP addresses or names. Let's try getting the contents served by the `nginx-statefulset-example-0` Pod using its short DNS name provided by the headless Service:

```
$ wget http://nginx-statefulset-example-0.nginx-headless
&& cat index.html
Connecting to nginx-statefulset-example-0.nginx-headless
(10.244.1.103:80)
...
You have been served by Pod with IP address: 10.244.1.103
```

As expected, you have connected directly to the Pod IP address and have been served by the proper Pod.

5. Now, we will show that this DNS name remains unchanged even if a Pod is restarted. The IP of the Pod will change, but the DNS name will not. What is more, the PV that is mounted will also stay the same, but we will investigate this in the next paragraphs. In another shell window, outside of the container, execute the following command to force a restart of the `nginx-statefulset-example-0` Pod:

```
$ kubectl delete pod nginx-statefulset-example-0
```

6. In the `busybox` shell, execute the same command for getting contents served by the `nginx-statefulset-example-0` Pod:

```
$ rm index.html && wget http://nginx-statefulset-
example-0.nginx-headless && cat index.html
Connecting to nginx-statefulset-example-0.nginx-headless
(10.244.1.113:80)
...
You have been served by Pod with IP address: 10.244.1.113
```

You can see that we used the same DNS name to call the Pod. We have been served by a Pod with the *same* name, but now with a *different* IP address.

This explains how the headless Services can be leveraged to get a stable and predictable network identity that will not change when a Pod is restarted. You may wonder what the actual use of this is and why it is important for StatefulSet objects. There are a couple of possible use cases:

- Deploying clustered databases, such as etcd or *MongoDB*, requires specifying network addresses of other Nodes in the database cluster. This is especially necessary if there are no *automatic discovery* capabilities provided by the database. In such cases, stable DNS names provided by headless Services help to run such clusters on Kubernetes as StatefulSets. There is still the problem of changing the configuration when Pod replicas are added or removed from the StatefulSet during scaling. In some cases, this is solved by the *sidecar container pattern*, which monitors the Kubernetes API to dynamically change the database configuration.

- If you decide to implement your own storage solution running as StatefulSet with advanced data sharding, you will most likely need mappings of logical shards to physical Pod replicas in the cluster. Then, the stable DNS names can be used as part of this mapping. They will guarantee that queries for each logical shard are performed against a proper Pod, irrespective of whether it was rescheduled to another Node or restarted.

Finally, let's take a look at the state persistence for Pods running in StatefulSet.

State persistence

To demonstrate how persisting state in StatefulSets works, we will use the kubectl exec commands to introduce changes to the mounted PV. This is the easiest way in which we can show that the files are persisted during Pod restarts or reschedules – in real-world use cases, changes to the state would be done by the actual application, for example, a database container writing files to the container filesystem. Perform the following steps:

1. Use the kubectl exec command to create a file named state.html in the /usr/share/nginx/html directory in each of three nginx Pod containers. This path is where the PV is mounted based on volumeClaimTemplates of the StatefulSet object:

```
$ kubectl exec -it nginx-statefulset-example-0 -- /bin/
sh -c "echo State of Pod 0 > /usr/share/nginx/html/state.
html"
$ kubectl exec -it nginx-statefulset-example-1 -- /bin/
```

```
sh -c "echo State of Pod 1 > /usr/share/nginx/html/state.
html"
$ kubectl exec -it nginx-statefulset-example-2 -- /bin/
sh -c "echo State of Pod 2 > /usr/share/nginx/html/state.
html"
```

2. Navigate in your web browser to the external IP of the Service and the state.
html file. In our example, it is http://104.45.176.241/state.html. You
will see that you have been served one of the files that we created, depending on
which Pod you have hit. If you refresh the page, with the cache disabled, you will see
that the contents change depending on the Pod:

104.45.176.241/state.html   ×   +

←   C   ⌂   ⚠ Not secure | 104.45.176.241/state.html

State of Pod 2

Figure 12.2 – Accessing the state.html file persisted in StatefulSet

3. Use the kubectl get pods command to see the current IP addresses of the
Pods in the StatefulSet object:

```
$ kubectl get pods -o wide
NAME                             ...    IP              ...
nginx-statefulset-example-0     ...    10.244.1.113 ...
nginx-statefulset-example-1     ...    10.244.0.77  ...
nginx-statefulset-example-2     ...    10.244.1.104 ...
```

4. Now, let's simulate failure to our Pods and use the kubectl delete command in
order to delete all three of them. This will cause them to be recreated by StatefulSet:

```
$ kubectl delete pod nginx-statefulset-example-0 nginx-
statefulset-example-1 nginx-statefulset-example-2
pod "nginx-statefulset-example-0" deleted
pod "nginx-statefulset-example-1" deleted
pod "nginx-statefulset-example-2" deleted
```

5. StatefulSet object will recreate the Pods in an *ordered* fashion, waiting for each of them to be *ready*. You can use the `kubectl get` command with an additional `-w` flag to follow the process in real time and eventually, you will be presented with all Pods being ready:

```
$ kubectl get pods -o wide -w
NAME                                ...   AGE      IP
...
nginx-statefulset-example-0    ...   3m58s    10.244.1.115
...
nginx-statefulset-example-1    ...   3m49s    10.244.0.79
...
nginx-statefulset-example-2    ...   92s      10.244.1.116
...
```

6. As you can see, the Pods have received new IPs. The containers that they are running are freshly created, but the existing PVC was used to mount PV to the containers. Once PVC is created by StatefulSet, it will not be deleted when you delete or scale the StatefulSet object. This ensures that data is not lost, unless you explicitly delete the PVC and PV yourself. If you navigate to the external IP of the Service for the StatefulSet object to get `state.html`, you will see that the file is still being served, even though it is not present in the original container image:

State of Pod 1

Figure 12.3 – Accessing the state.html file persisted in StatefulSet after the Pods restart

7. Lastly, let's show that when accessing the `state.html` file via the headless Service, we are getting the same result:

```
$ kubectl run -i --tty busybox --image=busybox:1.28 --rm
--restart=Never -- sh
If you don't see a command prompt, try pressing enter.
/ # wget http://nginx-statefulset-example-0.nginx-
headless/state.html && cat state.html
Connecting to nginx-statefulset-example-0.nginx-headless
(10.244.1.115:80)
```

```
...
State of Pod 0
```

This demonstration shows how StatefulSet can provide state and data persistence for your containerized applications using stable network identities and PVCs. Next, we will take a look at scaling the StatefulSet object.

Scaling StatefulSet

In the case of StatefulSets, you can do similar *scaling* operations as for Deployment objects by changing the number of `replicas` in the specification or using the `kubectl scale` imperative command. The new Pods will be automatically discovered as new Endpoints for the Service when you scale up, or automatically removed from the Endpoints list when you scale down.

However, there are a few differences when compared to Deployment objects:

- When you deploy a StatefulSet object of N replicas, the Pods during deployment are created sequentially, in order from 0 to N-1. In our example, during the creation of a StatefulSet object of three replicas, the first `nginx-statefulset-example-0` Pod is created, followed by `nginx-statefulset-example-1`, and finally `nginx-statefulset-example-2`.

- When you scale *up* the StatefulSet, the new Pods are also created sequentially and in an ordered fashion.

- When you scale *down* the StatefulSet, the Pods are terminated sequentially, in *reverse order*, from N-1 to 0. In our example, while scaling down the StatefulSet object to zero replicas, the first `nginx-statefulset-example-2` Pod is terminated, followed by `nginx-statefulset-example-1`, and finally `nginx-statefulset-example-0`.

- During scaling up of the StatefulSet object, before the next Pod is created in the sequence, all its predecessors must be *running* and *ready*.

- During scaling *down* of the StatefulSet object, before the next Pod is terminated in the reverse sequence, all its predecessors must be completely *terminated* and *deleted*.

- Also, in general, before *any* scaling operation is applied to a Pod in a StatefulSet object, all its predecessors must be running and ready. This means that if, during scaling down from four replicas to one replica, the `nginx-statefulset-example-0` Pod were to suddenly fail, then no further scaling operation would be performed on `nginx-statefulset-example-1`, `nginx-statefulset-example-2`, and `nginx-statefulset-example-3` Pods. Scaling would resume when the `nginx-statefulset-example-0` Pod becomes ready again.

> **Tip**
>
> This sequential behavior of scaling operations can be relaxed by changing the `.spec.podManagementPolicy` field in the specification. The default value is `OrderedReady`. If you change it to `Parallel`, the scaling operations will be performed on Pods in parallel, similar to what you know from Deployment objects. Note that this affects only scaling operations. The way of updating StatefulSet object with `updateStrategy` of the `RollingUpdate` type does not change.

Equipped with this knowledge, let's *scale up* our StatefulSet *declaratively*:

1. Open the `nginx-statefulset.yaml` manifest file and modify the number of replicas:

    ```
    apiVersion: apps/v1
    kind: StatefulSet
    metadata:
      name: nginx-statefulset-example
    spec:
      replicas: 5
    ...
    ```

2. Apply the changes to the cluster using the `kubectl apply` command:

    ```
    $ kubectl apply -f ./nginx-statefulset.yaml –record
    statefulset.apps/nginx-statefulset-example configured
    ```

3. If you now check the Pods using the `kubectl get pods -w` command, you will see the sequential, ordered creation of new Pods. The `nginx-statefulset-example-4` Pod will not start creating until `nginx-statefulset-example-3` has been created and becomes ready.

Similarly, if you check the output of the `kubectl describe` command for the StatefulSet object, you will see the following in the events:

```
$ kubectl describe sts nginx-statefulset-example
...
Events:
  Type      Reason            Age                  From
Message
  ----      ------            ----                 ----
-------
  Normal    SuccessfulCreate  2m31s
statefulset-controller   create Pod nginx-statefulset-
example-3 in StatefulSet nginx-statefulset-example
successful
  Normal    SuccessfulCreate  117s
statefulset-controller   create Claim nginx-data-nginx-
statefulset-example-4 Pod nginx-statefulset-example-4 in
StatefulSet nginx-statefulset-example success
  Normal    SuccessfulCreate  117s
statefulset-controller   create Pod nginx-statefulset-
example-4 in StatefulSet nginx-statefulset-example
successful
```

You can achieve the same result using the *imperative* command, which is recommended only for development scenarios:

```
$ kubectl scale sts nginx-statefulset-example
--replicas=5
statefulset.apps/nginx-statefulset-example scaled
```

4. To perform the scaling down of our StatefulSet object *declaratively*, simply modify the `nginx-statefulset.yaml` manifest file and change the number of replicas:

```
apiVersion: apps/v1
kind: StatefulSet
metadata:
  name: nginx-statefulset-example
spec:
  replicas: 2
...
```

5. Apply the changes to the cluster using the `kubectl apply` command:

```
$ kubectl apply -f ./nginx-statefulset.yaml --record
```

You can achieve the same result using imperative commands by executing the following command:

```
$ kubectl scale sts nginx-statefulset-example --replicas=2
```

If you describe the StatefulSet object, you will see in the events that the scaling down is reflected:

```
$ kubectl describe sts nginx-statefulset-example
...
Events:
   Type      Reason              Age                   From
Message
   ----      -------             ----                  ----
-------
   Normal   SuccessfulDelete   61s (x2 over 32m)     statefulset-
controller   delete Pod nginx-statefulset-example-3 in
StatefulSet nginx-statefulset-example successful
   Normal   SuccessfulDelete   30s                   statefulset-
controller   delete Pod nginx-statefulset-example-2 in
StatefulSet nginx-statefulset-example successful
```

Of course, if you scale the StatefulSet object back to three replicas, and attempt to get `state.html` for the `nginx-statefulset-example-2` Pod replica, you will get the expected, persisted file contents. PVs and PVCs are not deleted during any Pod operations in the StatefulSet object.

Next, we will demonstrate how you can delete a StatefulSet object.

Deleting a StatefulSet

To delete a StatefulSet object, there are two possibilities:

- Delete the StatefulSet together with Pods that it owns.
- Delete the StatefulSet and leave the Pods unaffected.

In both cases, the PVCs and PVs that were created for the Pods using `volumeClaimTemplates` will *not* be deleted. This ensures that state data is not lost accidentally unless you explicitly clean up the PVCs and PVs.

> **Important note**
>
> Currently, there is a **Kubernetes Enhancement Proposal (KEP)** 1847 that proposes a way of requesting the automatic deletion of PVCs for StatefulSet object. This KEP is likely to appear in future releases. You can find more details at `https://github.com/kubernetes/enhancements/tree/33e16e4d192153d8b41d1e5d91659612d6d633f4/keps/sig-apps/1847-autoremove-statefulset-pvcs.`

To delete the StatefulSet object together with Pods, you can use the regular `kubectl delete` command:

```
$ kubectl delete sts nginx-statefulset-example
```

You will see that the Pods will be terminated first, followed by the StatefulSet object. Please note that this operation is different from *scaling down* the StatefulSet object to zero replicas and then deleting it. If you delete StatefulSet object with existing Pods, there are no guarantees regarding the order of termination of the individual Pods. In most cases, they will be terminated at once.

Optionally, if you would like to delete just the StatefulSet object, you need to use the `--cascade=orphan` option for `kubectl delete`:

```
$ kubectl delete sts nginx-statefulset-example --cascade=orphan
```

After this command, if you inspect what Pods are in the cluster, you will still see all the Pods that were owned by the `nginx-statefulset-example` StatefulSet.

Lastly, if you would like to clean up PVCs and PVs after deleting the StatefulSet object, you need to perform this step manually.

> **Important note**
>
> Please note that if you want to perform verifications of state persistence after exercising the new version rollout in the next section, you should not yet delete the PVCs. Otherwise, you will lose the `state.html` files stored in the PVs.

Use the following command:

```
$ kubectl delete pvc nginx-data-nginx-statefulset-example-0
nginx-data-nginx-statefulset-example-1 nginx-data-nginx-
statefulset-example-2
```

This command will delete PVCs and associated PVs.

Next, let's take a look at releasing new versions of apps deployed as StatefulSets and how StatefulSet revisions are managed.

Releasing a new version of an app deployed as a StatefulSet

We have just covered the *scaling* of StatefulSets in the previous section by making changes to the `.spec.replicas` number in the specification. Everything you have learned about sequential and ordered changes to the Pods plays an important role in rolling out a new revision of a StatefulSet object when using the `RollingUpdate` strategy. There are many similarities between StatefulSets and Deployment objects – we have covered the details of Deployment updates in *Chapter 11, Deployment – Deploying Stateless Applications*. Making changes to the StatefulSet Pod *template* (`.spec.template`) in the specification will also cause the rollout of a new revision for StatefulSet. Usually, you will change the image used by the Pod container to a new version – this is how you perform the release of a new version of an app deployed as StatefulSet.

StatefulSets support two types of *update strategies* that you define using the `.spec.updateStrategy.type` field in the specification:

- `RollingUpdate`: The default strategy, which allows you to roll out a new version of your application in a controlled way. This is slightly different to the `RollingUpdate` strategy known from Deployment objects. For StatefulSet, this strategy will terminate and recreate Pods in a sequential and ordered fashion and make sure that the Pod is recreated and in a ready state before proceeding to the next one.

- `OnDelete`: This strategy implements the legacy behavior of StatefulSet updates prior to Kubernetes 1.7. However, it is still useful! In this type of strategy, the StatefulSet will *not* automatically update the Pod replicas by recreating them. You need to manually delete a Pod replica to get the new Pod template applied. This is useful in scenarios when you need to perform additional manual actions or verifications before proceeding to the next Pod replica. For example, if you are running a *Cassandra cluster* or *etcd cluster* in a StatefulSet, you may want to verify whether the new Pod has correctly joined the existing cluster following removal of the previous version of the Pod. Of course, it is possible to perform similar checks using the Pod template life cycle `postStart` and `preStop` **hooks** while using the `RollingUpdate` strategy, but this requires more sophisticated error handling in the hooks.

Let's now take a closer look at the `RollingUpdate` strategy, as it is the most important and the most commonly used update strategy for StatefulSets. The key thing about this is that the strategy respects all the StatefulSet guarantees that we explained in the previous section regarding scaling. The rollout is done in reverse order, for example, the first Pod, `nginx-statefulset-example-2`, is recreated with the new Pod template, followed by `nginx-statefulset-example-1`, and finally `nginx-statefulset-example-0`.

If the process of rollout fails (not necessarily the Pod that was currently recreated), the StatefulSet controller is going to restore any failed Pod to its *current version*. This means that Pods that have already received a *successful* update to the current version will remain at the current version, whereas the Pods that have not yet received the update will remain at the previous version. In this way, the StatefulSet attempts to always keep applications healthy and consistent. However, this can also lead to *broken* rollouts of StatefulSets. If one of the Pod replicas *never* becomes running and ready, then the StatefulSet will stop the rollout and wait for *manual* intervention. Applying the template again to the previous revision of StatefulSet is not enough – this operation will not proceed as StatefulSet will wait for the failed Pod to become ready. The only resolution is manual deletion of the failed Pod and then the StatefulSet can apply the previous revision of the Pod template.

Lastly, the `RollingUpdate` strategy also provides the option to execute *staged* rollouts using the `.spec.updateStrategy.rollingUpdate.partition` field. This field defines a number for which all the Pod replicas that have a *lesser ordinal* number will not be updated, and, even if they are deleted, they will be recreated at the previous version. So, in our example, if `partition` were to be set to `1`, this means that during the rollout, only `nginx-statefulset-example-1` and `nginx-statefulset-example-2` would be updated, whereas `nginx-statefulset-example-0` would remain unchanged and running on the previous version. By controlling the `partition` field, you can easily roll out a single *canary* replica and perform *phased* rollouts. Please note that the default value is `0`, which means that all Pod replicas will be updated.

Now, we will release a new version of our `nginx` web server using the `RollingUpdate` strategy.

Updating StatefulSet

We will now demonstrate how to do a rollout of a new image version for a Pod container using the StatefulSet YAML manifest file that we created previously:

1. Make a copy of the previous YAML manifest file:

    ```
    $ cp nginx-statefulset.yaml nginx-statefulset-
    rollingupdate.yaml
    ```

2. Ensure that you have `RollingUpdate` strategy type and partition set to `0`. Please note that if you have attempted to create the StatefulSet object with a different strategy first, you will not be able to modify it without deleting the StatefulSet beforehand:

    ```
    apiVersion: apps/v1
    kind: StatefulSet
    metadata:
      name: nginx-statefulset-example
    spec:
      replicas: 3
      serviceName: nginx-headless
      podManagementPolicy: OrderedReady
      updateStrategy:
        type: RollingUpdate
        rollingUpdate:
          partition: 0
    ...
      template:
    ...
        spec:
          containers:
          - name: nginx
            image: nginx:1.17
    ```

    These values are the default ones, but it is worth specifying them explicitly to understand what is really happening.

3. Apply the manifest file to the cluster:

    ```
    $ kubectl apply -f ./nginx-statefulset-rollingupdate.yaml
    --record
    ```

When the StatefulSet is ready in the cluster, we can now roll out a new version of the `nginx` container image for our StatefulSet object. To do that, please perform the following steps:

1. Modify the container image used in the StatefulSet Pod template to `nginx:1.18`:

   ```
   apiVersion: apps/v1
   kind: StatefulSet
   metadata:
     name: nginx-statefulset-example
   spec:
   ...
     template:
   ...
       spec:
         containers:
         - name: nginx
           image: nginx:1.18
   ```

2. Apply the changes to the cluster using the following command:

   ```
   $ kubectl apply -f ./nginx-statefulset-rollingupdate.yaml
   --record
   statefulset.apps/nginx-statefulset-example configured
   ```

3. Immediately after that, use the `kubectl rollout status` command to see the progress in real time. This process will be a bit longer than in the case of Deployment objects because the rollout is performed in a sequential and ordered fashion:

   ```
   $ kubectl rollout status sts nginx-statefulset-example
   Waiting for partitioned roll out to finish: 0 out of 3
   new pods have been updated...
   Waiting for 1 pods to be ready...
   Waiting for 1 pods to be ready...
   Waiting for partitioned roll out to finish: 1 out of 3
   new pods have been updated...
   Waiting for 1 pods to be ready...
   Waiting for 1 pods to be ready...
   Waiting for partitioned roll out to finish: 2 out of 3
   new pods have been updated...
   ```

```
Waiting for 1 pods to be ready...
Waiting for 1 pods to be ready...
partitioned roll out complete: 3 new pods have been
updated...
```

4. Similarly, using the `kubectl describe` command, you can see events for the StatefulSet that demonstrate precisely what the order of Pod replica recreation was:

```
$ kubectl describe sts nginx-statefulset-example
...
Events:
  Type     Reason            Age                       From
  Message
  ----     ------            ----                      ----
  -------
  Normal   SuccessfulDelete  3m12s
statefulset-controller  delete Pod nginx-statefulset-
example-2 in StatefulSet nginx-statefulset-example
successful
  Normal   SuccessfulCreate  2m34s (x2 over 12m)
statefulset-controller  create Pod nginx-statefulset-
example-2 in StatefulSet nginx-statefulset-example
successful
  Normal   SuccessfulDelete  2m25s
statefulset-controller  delete Pod nginx-statefulset-
example-1 in StatefulSet nginx-statefulset-example
successful
  Normal   SuccessfulCreate  105s (x2 over 13m)
statefulset-controller  create Pod nginx-statefulset-
example-1 in StatefulSet nginx-statefulset-example
successful
  Normal   SuccessfulDelete  99s
statefulset-controller  delete Pod nginx-statefulset-
example-0 in StatefulSet nginx-statefulset-example
successful
  Normal   SuccessfulCreate  64s (x2 over 13m)
statefulset-controller  create Pod nginx-statefulset-
example-0 in StatefulSet nginx-statefulset-example
successful
```

As expected, the rollout was done in *reverse* order. The first Pod to recreate was `nginx-statefulset-example-2`, followed by `nginx-statefulset-example-1`, and finally `nginx-statefulset-example-0`. Also, because we have used the default `partition` value of `0`, all the Pods were updated. This is because all ordinal numbers of Pod replicas are greater than or equal to `0`.

5. Now we can verify that the Pods were recreated with the new image. Execute the following command to verify the first Pod replica in the StatefulSet object:

```
$ kubectl describe pod nginx-statefulset-example-0
...
Containers:
  nginx:
    Container ID:   docker://031627cbea4c60194f7a2774ef965
ad52f6460f07bc44f7b426ff74a4ccad479
    Image:          nginx:1.18
```

6. And finally, we can verify that the *state* was persisted because the existing PVCs were used for the new Pods. Please note that this will only work properly if you haven't deleted the PVCs for the StatefulSet manually in the previous section:

```
$ kubectl run -i --tty busybox --image=busybox:1.28 --rm
--restart=Never -- sh
If you don't see a command prompt, try pressing enter.
/ # wget http://nginx-statefulset-example-0.nginx-
headless/state.html && cat state.html
Connecting to nginx-statefulset-example-0.nginx-headless
(10.244.1.135:80)
...
State of Pod 0
```

As you can see, the rollout of a new version of `nginx` was completed successfully and the state has been persisted even though the Pods were recreated.

Tip

You can change the StatefulSet container image *imperatively* using the `kubectl set image sts nginx-statefulset-example nginx=nginx:1.18 --record` command. This approach is only recommended for non-production scenarios. In general, StatefulSets are much easier to manage declaratively than imperatively.

Now, we will show how you can use the `partition` field to do a *phased* rollout with a *canary*. Assume we would like to update the `nginx` image version to `1.19`. You would like to make sure that the change is working properly in your environment, using a canary deployment, which is a single Pod replica updated to the new image version. Please refer to the following steps:

1. Modify the `nginx-statefulset-rollingupdate.yaml` manifest file so that the Pod container image version is `nginx:1.19` and the `partition` number is equal to current `replicas`, in our case 3:

    ```yaml
    apiVersion: apps/v1
    kind: StatefulSet
    metadata:
      name: nginx-statefulset-example
    spec:
      replicas: 3
      serviceName: nginx-headless
      podManagementPolicy: OrderedReady
      updateStrategy:
        type: RollingUpdate
        rollingUpdate:
          partition: 3
    ...
      template:
    ...
        spec:
          containers:
          - name: nginx
            image: nginx:1.19
    ```

    When the `partition` number is the same as the number of `replicas`, we can apply the YAML manifest to the cluster and no changes to the Pods will be introduced yet. This is called **staging a rollout**.

2. Apply the manifest file to the cluster:

    ```
    $ kubectl apply -f ./nginx-statefulset-rollingupdate.yaml
    --record
    statefulset.apps/nginx-statefulset-example configured
    ```

3. Now, let's create a *canary* for our new version. Decrease the `partition` number by one to `2` in the manifest file. This means that all Pod replicas with an ordinal number of less than 2 will not be updated – in our case, the `nginx-statefulset-example-2` Pod only, and all others will remain unchanged:

```
apiVersion: apps/v1
kind: StatefulSet
metadata:
  name: nginx-statefulset-example
spec:
  replicas: 3
  serviceName: nginx-headless
  podManagementPolicy: OrderedReady
  updateStrategy:
    type: RollingUpdate
    rollingUpdate:
      partition: 2
...
```

4. Apply the manifest file to the cluster again:

```
$ kubectl apply -f ./nginx-statefulset-rollingupdate.yaml
--record
statefulset.apps/nginx-statefulset-example configured
```

5. Use the `kubectl rollout status` command to follow the process. As expected, only one Pod will be recreated:

```
$ kubectl rollout status sts nginx-statefulset-example
Waiting for 1 pods to be ready...
partitioned roll out complete: 1 new pods have been
updated...
```

6. If you describe the `nginx-statefulset-example-1` and `nginx-statefulset-example-2` Pods, you can see that the first one is using the old version of the image, whereas the second is using the new one:

```
$ kubectl describe pod nginx-statefulset-example-1
...
Containers:
  nginx:
```

```
     Container ID:   docker://031627cbea4c60194f7a2774ef965
ad52f6460f07bc44f7b426ff74a4ccad479
     Image:          nginx:1.18
 . . .


$ kubectl describe pod nginx-statefulset-example-2
 . . .
Containers:
  nginx:
     Container ID:   docker://1c3b1e4dc7bb048d407f8aef2da91
8bf8ef2a8e8c2258b53b847c06bff2efbc5
     Image:          nginx:1.19
 . . .
```

7. At this point, you would like to perform verifications and smoke tests on your canary. For that, we will try getting the state.html file directly via the *headless* Service:

```
$ kubectl run -i --tty busybox --image=busybox:1.28 --rm
--restart=Never -- sh
If you don't see a command prompt, try pressing enter.
/ # wget http://nginx-statefulset-example-2.nginx-
headless/state.html && cat state.html
Connecting to nginx-statefulset-example-2.nginx-headless
(10.244.1.139:80)
 . . .
State of Pod 2
```

8. Canary looks good, so we can continue with a *phased* rollout of our new version. For a phased rollout, you may use any *lower* partition number in the manifest. You can do a few small, phased rollouts or just proceed with a full rollout. Let's do a full rollout by decreasing partition to 0:

```
apiVersion: apps/v1
kind: StatefulSet
metadata:
  name: nginx-statefulset-example
spec:
  replicas: 3
  serviceName: nginx-headless
```

```
    podManagementPolicy: OrderedReady
    updateStrategy:
      type: RollingUpdate
      rollingUpdate:
        partition: 0
...
```

9. Apply the manifest file to the cluster again:

```
$ kubectl apply -f ./nginx-statefulset-rollingupdate.yaml
--record
statefulset.apps/nginx-statefulset-example configured
```

10. Observe the next phase of the rollout using the kubectl rollout status command:

```
$ kubectl rollout status sts nginx-statefulset-example
Waiting for partitioned roll out to finish: 1 out of 3
new pods have been updated...
Waiting for 1 pods to be ready...
Waiting for 1 pods to be ready...
Waiting for partitioned roll out to finish: 2 out of 3
new pods have been updated...
Waiting for 1 pods to be ready...
Waiting for 1 pods to be ready...
partitioned roll out complete: 3 new pods have been
updated...
```

As you can see, the phased rollout to the nginx:1.19 image version was completed successfully.

> **Tip**
>
> It is possible to do phased rollouts *imperatively*. To do that, you need to control the partition number using the kubectl patch command, for example, kubectl patch sts nginx-statefulset-example -p '{"spec":{"updateStrategy":{"type":"RollingUpdate","rollingUpdate":{"partition":3}}}}'. However, this is much less readable and error prone than *declarative* changes.

We will now take a look at how you can do rollbacks of StatefulSets.

Rolling back StatefulSet

In the previous *Chapter 11*, *Deployment – Deploying Stateless Applications*, we have described how you can do *imperative* rollbacks to Deployments. For StatefulSets, you can do exactly the same operations. To do that, you need to use the `kubectl rollout undo` commands. However, especially for StatefulSets, we recommend using a *declarative* model for introducing changes to your Kubernetes cluster. In this model, you usually commit each change to the source code repository, and performing rollback is very simple and involves just reverting the commit and applying the configuration again. Usually, the process of applying changes is performed as part of the CI/CD pipeline for the source code repository, instead of manually applying the changes by an operator. This is the easiest way to manage StatefulSets, and generally recommended in Infrastructure-as-Code and Configuration-as-Code paradigms.

> **Important note**
>
> When performing rollbacks to StatefulSets, you must be fully aware of the consequences of operations such as *downgrading* to an earlier version of the container image while persisting the state. For example, if your rollout to a new version has introduced *data schema changes* to the state, then you will not be able to safely roll back to an earlier version unless you ensure that the *downward migration* of state data is implemented!

In our example, if you would like to roll back to the `nginx:1.18` image version for our StatefulSet, you would either modify the YAML manifest file manually or revert the commit in your source code repository if you use one. Then, all you would need to do is execute the `kubectl apply` command to the cluster.

In the last section, we will provide you with a set of best practices for managing StatefulSets in Kubernetes.

StatefulSet best practices

This section summarizes known best practices when working with StatefulSet objects in Kubernetes. The list is by no means complete, but is a good starting point for your journey with Kubernetes.

Use declarative object management for StatefulSets

It is a good practice in DevOps world to stick to declarative models for introducing updates to your infrastructure and applications. Using the declarative way of updates is the core concept for paradigms such as Infrastructure-as-Code and Configuration-as-Code. In Kubernetes, you can easily perform declarative updates using the `kubectl apply` command, which can be used on a single file or even a whole directory of YAML manifest files.

> **Tip**
> To delete objects, it is still better to use imperative commands. It is more predictable and less prone to errors. The declarative deletion of resources in the cluster is useful only in CI/CD scenarios, where the whole process is entirely automated.

The same principle applies also to StatefulSets. Performing a rollout or rollback when your YAML manifest files are versioned and kept in a source control repository is easy and predictable. Using the `kubectl rollout undo` and `kubectl set image deployment` commands is generally not recommended in production environments. Using these commands gets much more complicated when more than one person is working on operations in the cluster.

Do not use the TerminationGracePeriodSeconds Pod with a 0 value for StatefulSets

The specification of Pod allows you to set `TerminationGracePeriodSeconds`, which informs `kubelet` how much time it should allow for a Pod to gracefully terminate when it attempts to terminate it. If you set `TerminationGracePeriodSeconds` to `0`, this will effectively make Pods terminate *immediately*, which is strongly discouraged for StatefulSets. StatefulSets often need graceful cleanup or `preStop` life cycle hooks to run before the container is removed, otherwise there is a risk that the state of StatefulSet will become inconsistent.

Scale down StatefulSets before deleting

When you delete a StatefulSet and you intend to reuse the PVCs later, you need to ensure that the StatefulSet terminates gracefully, in an ordered manner, so that any subsequent redeployment will not fail because of an inconsistent state in PVCs. If you perform the `kubectl delete` operation on your StatefulSet, all the Pods will be terminated *at once*. This is often not desired and you should first scale down the StatefulSet gracefully to zero replicas and then delete the StatefulSet itself.

Ensure state compatibility during StatefulSet rollbacks

If you ever intend to use StatefulSet rollbacks, you need to be aware of the consequences of operations such as downgrading to an earlier version of the container image while persisting the state. For example, if your rollout to a new version has introduced data schema changes in the state, then you will not be able to safely roll back to an earlier version unless you ensure that the downward migration of state data is implemented. Otherwise, your rollback will just recreate Pods with the older versions of the container image and they will fail to start properly because of incompatible state data.

Do not create Pods that match an existing StatefulSet label selector

It is possible to create Pods with labels that match the label selector of some existing StatefulSet. This can be done using bare Pods or another Deployment or ReplicaSet. This leads to conflicts, which Kubernetes does not prevent, and makes the existing StatefulSet *believe* that it has created the other Pods. The results may be unpredictable and, in general, you need to pay attention to how you organize your labeling of resources in the cluster. It is advised to use semantic labeling – you can learn more about this approach in the official documentation: `https://kubernetes.io/docs/concepts/configuration/overview/#using-labels`.

Summary

This chapter has demonstrated how to work with *stateful* workloads and applications on Kubernetes using StatefulSets. You first learned what the approaches to persisting state in containers and in Kubernetes Pods are and, based on that, we have described how a StatefulSet object can be used to persist the state. Next, we created an example StatefulSet, together with a *headless* Service. Based on that, you learned how PVCs and PVs are used in StatefulSets to ensure that the state is persisted between Pod restarts. Next, you learned how you can scale the StatefulSet and how to introduce updates using *canary* and *phased* rollouts. And finally, we provided you with a set of known best practices when working with StatefulSets.

In the next chapter, you will learn more about managing special workloads where you need to maintain exactly one Pod per each Node in Kubernetes – we will introduce a new Kubernetes object: DaemonSet.

Further reading

For more information regarding StatefulSets and persistent storage management in Kubernetes, please refer to the following Packt books:

- *The Complete Kubernetes Guide*, by *Jonathan Baier, Gigi Sayfan, Jesse White* (`https://www.packtpub.com/virtualization-and-cloud/complete-kubernetes-guide`)

- *Getting Started with Kubernetes – Third Edition*, by *Jonathan Baier, Jesse White* (`https://www.packtpub.com/virtualization-and-cloud/getting-started-kubernetes-third-edition`)

- *Kubernetes for Developers*, by *Joseph Heck* (`https://www.packtpub.com/virtualization-and-cloud/kubernetes-developers`)

- *Hands-On Kubernetes on Windows*, by *Piotr Tylenda* (`https://www.packtpub.com/product/hands-on-kubernetes-on-windows/9781838821562`)

You can also refer to the excellent official Kubernetes documentation (`https://kubernetes.io/docs/home/`), which is always the most up-to-date source of knowledge regarding Kubernetes in general.

13
DaemonSet – Maintaining Pod Singletons on Nodes

The previous chapters have explained and demonstrated how to use the most common Kubernetes controllers for managing Pods, such as ReplicaSet, Deployment, and StatefulSet. Generally, when running cloud application components that contain the actual *business logic* you will need either Deployments or StatefulSets for controlling your Pods. In some cases, when you need to run batch workloads as part of your application, you will use Jobs and CronJobs.

However, in some cases, you will need to run components that have a supporting function and, for example, execute maintenance tasks or aggregate logs and metrics. More specifically, if you have any tasks that need to be executed for each Node in the cluster, they can be performed using a DaemonSet. This is the last type of Pod management controller that we are going to introduce in this part of the book. The purpose of a DaemonSet is to ensure that *each* Node (unless specified otherwise) runs a *single* replica of a Pod. If you add a new Node to the cluster, it will automatically get a Pod replica scheduled. Similarly, if you remove a Node from the cluster, the Pod replica will be terminated – the DaemonSet will execute all required actions.

In this chapter, we will cover the following topics:

- Introducing the DaemonSet object
- Creating and managing DaemonSets
- Common use cases for DaemonSets
- Alternatives to DaemonSets

Technical requirements

For this chapter, you will need the following:

- A Kubernetes cluster deployed. You can use either a local or cloud-based cluster, but in order to fully understand the concepts we recommend using a *multi-node*, cloud-based Kubernetes cluster.
- The Kubernetes CLI (`kubectl`) installed on your local machine and configured to manage your Kubernetes cluster.

Kubernetes cluster deployment (local and cloud-based) and `kubectl` installation were covered in *Chapter 3, Installing Your First Kubernetes Cluster*.

You can download the latest code samples for this chapter from the official GitHub repository: `https://github.com/PacktPublishing/The-Kubernetes-Bible/tree/master/Chapter13`.

Introducing the DaemonSet object

The term **daemon** in operating systems has a long history and, in short, is used to describe a program that runs as a background process, without interactive control from the user. In many cases, daemons are responsible for handling maintenance tasks, serving network requests, or monitoring hardware activities. These are often processes that you want to run reliably, all the time, in the background, from the time you boot the operating system to when you shut it down.

> Tip
> Daemons are associated in most cases with Unix-like operating systems. In Windows, you will more commonly encounter the term *Windows service*.

In Kubernetes, you may need a similar functionality where your Pods behave like classic operating system daemons on each of the Nodes in the cluster. For this, Kubernetes offers a dedicated Pod management controller named **DaemonSet**. The role of a DaemonSet is straightforward: run a *single* Pod replica on *each* of the Nodes in the cluster and manage them automatically. There are variety of use cases that require such Node-local facilities and usually they serve important and fundamental roles for the whole cluster – we will discuss some common use cases in the sections coming up, but generally you need them for the following:

- Node monitoring in the cluster.

- Logs and telemetry gathering from individual Nodes and sometimes Pods running on a Node.

- Managing cluster storage – this is especially important for handling requests from provisioners for `PersistentVolumeClaims` and `PersistentVolumes`.

All you have learned in the previous chapters about ReplicaSets, Deployments, and StatefulSets applies more or less to the DaemonSet. Its specification requires you to provide a Pod template, Pod label selectors, and optionally Node selectors if you want to schedule the Pods only on a subset of Nodes.

Depending on the case, you may not need to *communicate* with the DaemonSet from other Pods or from an external network. For example, if the job of your DaemonSet is just to perform a periodic cleanup of the filesystem on the Node, it is unlikely you would like to communicate with such Pods. If your use case requires any ingress or egress communication with the DaemonSet Pods, then you have the following common patterns:

- **Map container ports to host ports**: Since the DaemonSet Pods are guaranteed to be singletons on cluster Nodes, it is possible to use mapped host ports. The clients must know the Node IP addresses.

- **Pushing data to a different service**: In some cases, it may be enough that the DaemonSet is responsible for sending updates to other services without needing to allow ingress traffic.

- **Headless service matching DaemonSet Pod label selectors**: This is a similar pattern to the case of StatefulSets, where you can use the cluster DNS to retrieve multiple `A record`s for Pods using the DNS name of the headless service.

- **Normal service matching DaemonSet Pod label selectors**: Less commonly, you may need to reach *any* Pod in the DaemonSet. Using a normal Service object, for example of the `ClusterIP` type, will allow you to communicate with a random Pod in the DaemonSet.

We will now show how you can create and manage an example DaemonSet in your cluster.

Creating and managing DaemonSets

In order to demonstrate how DaemonSets work, we will use `nginx` running in a Pod container that returns simple information about the Node IP address where it is currently scheduled. The IP address will be provided to the container using an *environment variable* and based on that, a modified version of `index.html` in `/usr/share/nginx/html` will be created. To access the DaemonSet endpoints, we will use a *headless* service, similar to what we did for `StatefulSet` in *Chapter 12*, *StatefulSet – Deploy Stateful Applications*. Most of the real use cases of DaemonSets are rather complex and involve mounting various system resources to the Pods. We will keep our DaemonSet example as simple as possible to show the principles.

> **Important note**
>
> If you would like to work on a real example of a DaemonSet, we have provided a working version of Prometheus `node-exporter` deployed as a DaemonSet behind a headless Service: `https://github.com/PacktPublishing/Kubernetes-for-Beginners/blob/master/Chapter13/02_daemonset-prometheus-nodeexporter/node-exporter.yaml`. When following the guide in this section, the only difference is that you need to use `node-exporter` as the Service name, use port `9100` and append the `/metrics` path for requests sent using `wget`. This DaemonSet exposes Node metrics in *Prometheus data model* format on port `9100` under the `/metrics` path.

We will now go through all the YAML manifests required to create our DaemonSet and apply them to the cluster.

Creating a DaemonSet

First, let's take a look at the StatefulSet YAML manifest file named `nginx-daemonset.yaml` (full version available in the official GitHub repository for the book: `https://github.com/PacktPublishing/Kubernetes-for-Beginners/blob/master/Chapter13/01_daemonset-example/nginx-daemonset.yaml`):

```
apiVersion: apps/v1
kind: DaemonSet
metadata:
  name: nginx-daemonset-example
spec:
  selector:
    matchLabels:
        app: nginx-daemon
        environment: test
# (to be continued in the next paragraph)
```

The first part of the preceding file contains the `metadata` and Pod label `selector` for the DaemonSet, quite similar to what you have seen in Deployments and StatefulSets. In the second part of the file, we present the Pod template that will be used by the DaemonSet:

```
# (continued)
  template:
    metadata:
      labels:
          app: nginx-daemon
          environment: test
    spec:
      containers:
      - name: nginx
        image: nginx:1.17
        ports:
        - containerPort: 80
        env:
        - name: NODE_IP
          valueFrom:
            fieldRef:
```

```
        fieldPath: status.hostIP
command:
- /bin/sh
- -c
- |
    echo "You have been served by Pod running on Node
with IP address: $(NODE_IP)" > /usr/share/nginx/html/index.html
    nginx -g "daemon off;"
```

As you can see, the structure of DaemonSet spec is similar to what you know from Deployments and StatefulSets. The general idea is the same, you need to configure the Pod template and use a proper label selector to match the Pod labels. Note that you do *not* see the `replicas` field here, as the number of Pods running in the cluster will be dependent on the number of Nodes in the cluster. The specification has two main components:

- `selector`: A **label selector**, which defines how to identify Pods that the DaemonSet owns. This can include *set-based* and *equality-based* selectors.

- `template`: This defines the template for Pod creation. Labels used in `metadata` must match the `selector`.

It is also common to specify `.spec.template.spec.nodeSelector` or `.spec.template.spec.tolerations` in order to control the Nodes where the DaemonSet Pods are deployed. We cover Pod scheduling in detail in *Chapter 19, Advanced Techniques for Scheduling Pods*. Additionally, you can specify `.spec.updateStrategy`, `.spec.revisionHistoryLimit`, and `.spec.minReadySeconds`, which are similar to what you have learned about Deployment objects.

> **Tip**
>
> If you run hybrid Linux-Windows Kubernetes clusters, one of the common use cases for Node selectors or Node affinity for DaemonSets is ensuring that the Pods are scheduled only on Linux Nodes or only on Windows Nodes. This makes sense as the container runtime and operating system are very different between such Nodes.

Apart from that, in our Pod template we have used a similar *override* of command for the nginx container as we did in the cases of Deployments and StatefulSets in the previous chapters. The command creates index.html in /usr/share/nginx/html/ with information about the IP address of the Node that runs the Pod serving the request. After that, it starts the nginx web server with the standard entrypoint command for the image. To provide the information about the Node IP address, we use an additional NODE_IP *environment variable* populated by status.hostIP of the Pod object (at runtime).

Next, let's take a quick look at the *headless* Service named nginx-daemon-headless. Create an nginx-daemon-headless-service.yaml file with the following YAML manifest:

```yaml
apiVersion: v1
kind: Service
metadata:
  name: nginx-daemon-headless
spec:
  selector:
    app: nginx-daemon
    environment: test
  clusterIP: None
  ports:
  - port: 80
    protocol: TCP
    targetPort: 80
```

As we explained in the case of the StatefulSet example, the specification is very similar to a normal Service, the only difference is that it has the None value for the clusterIP field. This will result in the creation of an nginx-daemon-headless headless Service. A headless Service allows us to return *all* Pods' IP addresses behind the Service as DNS A records instead of a single DNS A record with a Service clusterIP.

We have all the required YAML manifest files for our demonstration and we can proceed with applying the manifests to the cluster. Please follow these steps:

1. Create the `nginx-daemon-headless` headless Service using the following command:

   ```
   $ kubectl apply -f ./nginx-daemon-headless-service.yaml
   ```

2. Create an `nginx-daemonset-example` DaemonSet using the following command:

   ```
   $ kubectl apply -f ./nginx-daemonset.yaml
   ```

3. Now, you can use the `kubectl describe` command to observe the creation of the DaemonSet:

   ```
   $ kubectl describe daemonset nginx-daemonset-example
   ```

4. Alternatively, you can use `ds` as an abbreviation for `daemonset` when using the `kubectl` commands.

5. If you use the `kubectl get pods` command, you can see that there will be one Pod scheduled for each of the Nodes in the cluster:

   ```
   $ $ kubectl get pods -o wide
   NAME                           ... IP            NODE
   ...
   nginx-daemonset-example-5w8jx ... 10.244.1.144
   aks-nodepool1-77120516-vmss000000 ...
   nginx-daemonset-example-tzqmc ... 10.244.0.90
   aks-nodepool1-77120516-vmss000001 ...
   ```

In our case, we have two Nodes in the cluster, so exactly two Pods have been created.

We have successfully deployed the DaemonSet and we can now verify that it works as expected. To do that, follow the given steps:

1. First, we need to know the IP addresses of the individual Nodes in order to cross-check the output of further commands:

   ```
   $ kubectl get node -o wide
   NAME                              ...   INTERNAL-IP   ...
   aks-nodepool1-77120516-vmss000000  ...   10.240.0.4    ...
   aks-nodepool1-77120516-vmss000001  ...   10.240.0.5    ...
   ```

2. Create an *interactive* busybox Pod and start the Bourne shell process. The following command will create the Pod and immediately attach your terminal so that you can interact from *within* the Pod:

```
$ kubectl run -i --tty busybox --image=busybox:1.28 --rm
--restart=Never -- sh
```

3. We need to check how our nginx-daemon-headless headless Service is resolved by the cluster DNS:

```
$ nslookup nginx-daemon-headless
Server:       10.0.0.10
Address 1: 10.0.0.10 kube-dns.kube-system.svc.cluster.
local

Name:         nginx-daemon-headless
Address 1: 10.244.1.144 10-244-1-144.nginx-daemon-
headless.default.svc.cluster.local
Address 2: 10.244.0.90 10-244-0-90.nginx-daemon-headless.
default.svc.cluster.local
```

We have been provided with two DNS names for the individual Pods in the DaemonSet: 10-244-1-144.nginx-daemon-headless.default. svc.cluster.local, which belongs to the aks-nodepool1-77120516- vmss000000 Node, and 10-244-0-90.nginx-daemon-headless. default.svc.cluster.local, which belongs to the aks-nodepool1- 77120516-vmss000001 Node.

4. We can now use the DNS names (also in short-name form) to communicate with the Pods of the DaemonSet. First, we will try to connect to the 10-244-1-144. nginx-daemon-headless Pod:

```
$ wget http://10-244-1-144.nginx-daemon-headless && cat
index.html
Connecting to 10-244-1-144.nginx-daemon-headless
(10.244.1.144:80)
...
You have been served by Pod running on Node with IP
address: 10.240.0.4
```

As expected, the Pod is scheduled on the aks-nodepool1-77120516- vmss000000 Node, which has an IP address of 10.240.0.4 (you can cross-check this with earlier commands' output).

5. Let's do a similar check for the other Pod in the cluster, `10-244-0-90.nginx-daemon-headless`:

    ```
    $ rm index.html && wget http://10-244-0-90.nginx-daemon-
    headless && cat index.html
    Connecting to 10-244-0-90.nginx-daemon-headless
    (10.244.0.90:80)
    ...
    You have been served by Pod running on Node with IP
    address: 10.240.0.5
    ```

    And again, as expected, we have been served by a Pod running on the `aks-nodepool1-77120516-vmss000001` Node.

This demonstrates the most important principles underlying how DaemonSet Pods are scheduled and how you can interact with them using headless Services. We will now show how you can modify the DaemonSet to roll out a new version of a container image for the Pods.

Modifying a DaemonSet

Updating a DaemonSet can be done in a similar way as for Deployments. If you modify the *Pod template* of the DaemonSet, this will trigger a *rollout* of a new revision of DaemonSet according to its `updateStrategy`. There are two strategies available:

*   `RollingUpdate`: The default strategy, which allows you to roll out a new version of your daemon in a controlled way. It is similar to rolling updates in Deployments in that you can define `.spec.updateStrategy.rollingUpdate.maxUnavailable` to control how many Pods in the clusters are unavailable at most during the rollout (defaults to `1`) and `.spec.minReadySeconds` (defaults to `0`). It is guaranteed that *at most one* Pod of DaemonSet will be in running state on each node in the cluster during the rollout process.

*   `OnDelete`: This strategy implements the legacy behavior of StatefulSet updates prior to Kubernetes 1.6. In this type of strategy, the DaemonSet will *not* automatically update the Pod by recreating them. You need to manually delete a Pod on a Node in order to get the new Pod template applied. This is useful in scenarios when you need to do additional manual actions or verifications before proceeding to the next Node.

The rollout of a new DaemonSet revision can be controlled in similar ways as for a Deployment object. You can use the `kubectl rollout status` command and perform *imperative* rollbacks using the `kubectl rollout undo` command. Let's demonstrate how you can *declaratively* update the container image in a DaemonSet Pod to a newer version:

1. Modify the `nginx-daemonset.yaml` YAML manifest file so that it uses `nginx:1.18` container image in the template:

```
apiVersion: apps/v1
kind: DaemonSet
metadata:
  name: nginx-daemonset-example
spec:
...
  template:
...
    spec:
      containers:
      - name: nginx
        image: nginx:1.18
```

2. Apply the manifest file to the cluster:

```
$ kubectl apply -f ./nginx-daemonset.yaml --record
```

3. Immediately after that, use the `kubectl rollout status` command to see the progress in real time:

```
$ kubectl rollout status ds nginx-daemonset-example
Waiting for daemon set "nginx-daemonset-example" rollout
to finish: 0 out of 2 new pods have been updated...
Waiting for daemon set "nginx-daemonset-example" rollout
to finish: 1 out of 2 new pods have been updated...
daemon set "nginx-daemonset-example" successfully rolled
out
```

4. Similarly, using the `kubectl describe` command, you can see events for the DaemonSet that exactly show what the order was of the Pod recreation:

```
$ kubectl describe ds nginx-daemonset-example
...
Events:
  Type      Reason           Age    From
Message
  ----      ------           ----   ----
-------
  Normal    SuccessfulDelete  113s   daemonset-controller
Deleted pod: nginx-daemonset-example-5w8jx
  Normal    SuccessfulCreate  74s    daemonset-controller
Created pod: nginx-daemonset-example-jsh7x
  Normal    SuccessfulDelete  73s    daemonset-controller
Deleted pod: nginx-daemonset-example-tzqmc
  Normal    SuccessfulCreate  41s    daemonset-controller
Created pod: nginx-daemonset-example-kgqbj
```

You can see that the Pods were replaced one by one. This is because we had the default value of `.spec.updateStrategy.rollingUpdate.maxUnavailable`, which is 1.

> **Tip**
> You can change the DaemonSet container image *imperatively* using the `kubectl set image ds nginx-daemonset-example nginx=nginx:1.18 --record` command. This approach is recommended only for non-production scenarios.

Additionally, DaemonSet will automatically create Pods if a new Node joins the cluster (providing that it matches the selector and affinity parameters). If a Node is removed from the cluster, the Pod will be terminated also. The same will happen if you modify the labels or taints on a Node so that it matches the DaemonSet – a new Pod will be created for that Node. If you modify the labels or taints for a Node in a way that it no longer matches the DaemonSet, the existing Pod will be terminated.

Next, we will show how you can delete a DaemonSet.

Deleting a DaemonSet

In order to delete a DaemonSet object, there are two possibilities:

- Delete the DaemonSet together with Pods that it owns.
- Delete the DaemonSet and leave the Pods unaffected.

To delete the DaemonSet together with Pods, you can use the regular the `kubectl delete` command:

```
$ kubectl delete ds nginx-daemonset-example
```

You will see that the Pods will first get terminated and then the DaemonSet will be deleted.

Now, if you would like to delete just the DaemonSet, you need to use the `--cascade=orphan` option with `kubectl delete`:

```
$ kubectl delete ds nginx-daemonset-example --cascade=orphan
```

After this command, if you inspect what Pods are in the cluster, you will still see all the Pods that were owned by the `nginx-daemonset-example` DaemonSet.

> **Important note**
>
> If you are draining a node using the `kubectl drain` command and this node is running Pods owned by a DaemonSet, you need to pass the `--ignore-daemonsets` flag to drain the node completely.

Let's now take a look at the most common use cases for DaemonSets in Kubernetes.

Common use cases for DaemonSets

At this point, you may wonder what is the actual use of the DaemonSet and what are the real-life use cases for this Kubernetes object? In general, DaemonSets are used either for very fundamental functions of the cluster, without which it is not useable, or for helper workloads performing maintenance or data collection. We have summarized the common and interesting use cases for DaemonSets in the following points:

- Depending on your cluster deployment, the `kube-proxy` core service may be deployed as a DaemonSet instead of a regular operating system service. For example, in the case of **Azure Kubernetes Service** (**AKS**), you can see the definition of this DaemonSet using the `kubectl describe ds -n kube-system kube-proxy` command. This is a perfect example of a backbone service that needs to run as a singleton on each Node in the cluster. You can also see an example YAML manifest for `kube-proxy` here: `https://github.com/kubernetes/kubernetes/blob/master/cluster/addons/kube-proxy/kube-proxy-ds.yaml`.

- Another example of fundamental services running as DaemonSets is running an installation of **Container Network Interface** (**CNI**) plugins and agents for maintaining the network in a Kubernetes cluster. A good example of such a DaemonSet is the Flannel agent (`https://github.com/flannel-io/flannel/blob/master/Documentation/kube-flannel.yml`), which runs on each Node and is responsible for allocating a subnet lease to each host out of a larger, preconfigured address space. This of course depends on what type of networking is installed on the cluster.

- Cluster storage daemons will be often deployed as DaemonSets. A good example of a commonly used daemon is **Object Storage Daemon** (**OSD**) for Ceph, which is a distributed object, block, and file storage platform. OSD is responsible for storing objects on the local filesystem of each Node and providing access to them over the network. You can find an example manifest file here (as part of a Helm Chart template): `https://github.com/ceph/ceph-container/blob/master/examples/helm/ceph/templates/osd/daemonset.yaml`.

- Ingress controllers in Kubernetes are sometimes deployed as DaemonSets. We will take a closer look at Ingress in *Chapter 21*, *Advanced Traffic Routing with Ingress*. For example, when you deploy `nginx` as an Ingress controller in your cluster, you have an option to deploy it as a DaemonSet: `https://github.com/nginxinc/kubernetes-ingress/blob/master/deployments/daemon-set/nginx-ingress.yaml`. Deploying an Ingress controller as a DaemonSet is especially common if you do Kubernetes cluster deployments on bare-metal servers.

- Log gathering and aggregation agents are often deployed as DaemonSets. For example, `fluentd` can be deployed as a DaemonSet in a cluster. You can find multiple YAML manifest files with examples in the official repository: `https://github.com/fluent/fluentd-kubernetes-daemonset`.

- Agents for collecting Node metrics make a perfect use case for deployment as DaemonSets. A well-known example of such an agent is Prometheus node-exporter: `https://github.com/prometheus-operator/kube-prometheus/blob/main/manifests/node-exporter-daemonset.yaml`.

And the list goes on – as you can see, DaemonSet is another building block provided for engineers designing the workloads running on Kubernetes clusters. In many cases, DaemonSets are the hidden backbone of a cluster that makes it fully operational.

Next, let's discuss what possible alternatives there are to using DaemonSets.

Alternatives to DaemonSets

The reason for using DaemonSets is quite simple – you would like to have exactly one Pod with a particular function on each Node in the cluster. However, sometimes you should consider different approaches that may fit your needs better:

- In log-gathering scenarios, you need to evaluate if you want to design your log pipeline architecture based on DaemonSets or the *sidecar* container pattern. Both have their advantages and disadvantages, but in general, running sidecar containers may be easier to implement and be more robust, even though it may require more system resources.

- If you just want to run periodic tasks, and you do not need to do it on each Node in the cluster, a better solution can be using Kubernetes CronJobs. Again, it is important to know what the actual use case is and whether running a separate Pod on each Node is a must-have requirement.

- Operating system daemons (for example, provided by `systemd` in Ubuntu) can be used to do similar tasks as DaemonSets. The drawback of this approach is that you cannot manage these native daemons using the same tools as you manage Kubernetes clusters with, for example `kubectl`. But at the same time, you do not have the dependency on any Kubernetes service, which may be a good thing in some cases.

- Static Pods (`https://kubernetes.io/docs/tasks/configure-pod-container/static-pod/`) can be used to achieve a similar result. This type of Pod is created based on a specific directory watched by `kubelet` for static manifest files. Static Pods cannot be managed using `kubectl` and they are most useful for cluster bootstraping functions.

Finally, we can now summarize our knowledge about DaemonSets.

Summary

In this chapter, you have learned how to work with DaemonSets in Kubernetes, and how they are used to manage special types of workloads or processes that must run as a singleton on each Node in the cluster. You first created an example DaemonSet and learned what the most important parts of its specification are. Next, you practiced how to roll out a new revision of a DaemonSet to the cluster and saw how you can monitor the deployment. Additionally, we discussed what the most common use cases are for this special type of Kubernetes object and what alternatives there are that you could consider.

This was the last type of Pod management controller that we discuss in this part of the book. In the next part, you will learn all the details required to effectively deploy Kubernetes clusters in different cloud environments. We will first take a look at working with clusters deployed on Google Kubernetes Engine.

Further reading

For more information regarding DaemonSets and their use cases in Kubernetes, please refer to the following Packt books:

- *The Complete Kubernetes Guide*, by *Jonathan Baier, Gigi Sayfan, Jesse White* (`https://www.packtpub.com/virtualization-and-cloud/complete-kubernetes-guide`).

- *Getting Started with Kubernetes – Third Edition*, by *Jonathan Baier, Jesse White* (`https://www.packtpub.com/virtualization-and-cloud/getting-started-kubernetes-third-edition`).

- *Kubernetes for Developers*, by *Joseph Heck* (`https://www.packtpub.com/virtualization-and-cloud/kubernetes-developers`).

- *Hands-On Kubernetes on Windows*, by *Piotr Tylenda* (`https://www.packtpub.com/product/hands-on-kubernetes-on-windows/9781838821562`).

- You can also refer to the excellent official Kubernetes documentation (`https://kubernetes.io/docs/home/`), which is always the most up-to-date source of knowledge about Kubernetes in general.

Section 4:
Deploying
Kubernetes
on the Cloud

The easiest way to run Kubernetes in production is to use one of the top major cloud providers. Google Cloud Platform, Amazon Web Services, and Microsoft Azure offer top services that can automate the management of a Kubernetes cluster for you at scale.

This part of the book comprises the following chapters:

14
Kubernetes Clusters on Google Kubernetes Engine

In this chapter, we are going to look at launching our very own Kubernetes cluster in the first of the three public cloud providers that we will be covering in this title: **Google Cloud Platform** (**GCP**).

By the end of the chapter, we will have signed up to GCP and launched a Kubernetes workload using **Google Kubernetes Engine** (**GKE**), as well as having discussed some of the features that GKE has to offer.

We will be covering the following topics:

- What are GCP and GKE?
- Preparing your local environment
- Launching your first GKE cluster
- Deploying a workload and interacting with your cluster
- More about cluster nodes

Technical requirements

To follow along with this chapter, you will need a GCP account with a valid payment method attached to it.

> **Important note**
>
> Following the instructions in this chapter will incur a financial cost. It is therefore important that you terminate any resources you launch once you have finished using them.

All prices quoted in this chapter are correct at the time of writing this book, and we recommend that you review the current costs before you launch any resources.

What are GCP and GKE?

Before we roll up our sleeves and look at signing up for a GCP account and installing the tools, we will need to launch our GKE-powered cluster. We should also discuss GCP and how it came to be.

Google Cloud Platform

Of the *big three* public cloud providers, GCP is the newest. We will look at **Amazon Web Services** (**AWS**) and Microsoft Azure over the next two chapters.

Google's foray into public cloud technology started very differently from the other two providers. In April of 2008, Google launched the public preview of its Google App Engine, which was the first component of its cloud offering. Google App Engine, as a service, is still available to this day. The service allows developers to deploy their applications into Google-managed runtimes – these include PHP, Java, Ruby, Python, Node.js, and .NET, along with Google's own programming language, Go.

The next service under the GCP banner didn't arrive until May 2010, and this was Google Cloud Storage, followed by Google BigQuery and a preview version of its Prediction API. A year later, October 2011 saw the launch of Google Cloud SQL. Then, in June 2012, the Google Compute Engine preview was launched.

As you can see, 4 years had passed, and we then had what most would consider the core services that go into making a public cloud service. However, the majority of the services were still in preview – in fact, it wouldn't be until 2013 that a lot of these core services would move out of preview and into **generally available** (**GA**), which meant that it was possible to safely run production workloads at scale.

All of this was just a year before Google would launch Kubernetes. Towards the end of 2014, Google would bring out the first alpha of GKE.

Google Kubernetes Engine

Given that Kubernetes was developed at Google, and also given Google's vast experience of running container workloads at scale with the Borg project (as we discussed in *Chapter 1, Kubernetes Fundamentals*), it came as no surprise that Google was one of the first of the public cloud providers to offer its own Kubernetes offering in GKE.

In fact, after Kubernetes v1 came out and was then handed over to the **Cloud Native Computing Foundation** (**CNCF**) to maintain in July 2015, it was only a month later that the GKE service went GA.

The GKE service allows you to launch and manage a CNCF certified Kubernetes cluster powered by the compute, storage, and network services of GCP, and also allows for deep integration with the monitoring, identity, and access management functions of the platform.

Now that we know a little bit of the history behind the service, we can sign up and install some of the management tools we will be using to launch our cluster.

Preparing your environment

The first thing we need to do is get you access to GCP. To do this, you will either need to sign up for an account or log in to your existing one. Let's learn how.

Signing up for a GCP account

To sign up for a GCP account, you will need to visit `https://cloud.google.com/`. Here you should be greeted by a page that looks like the following:

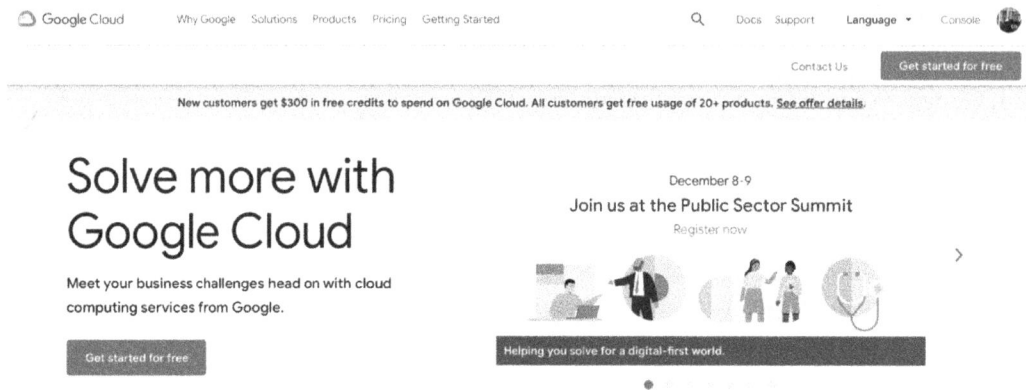

Figure 14.1 – The GCP welcome page

If you are already using a Google service such as Gmail, YouTube, or have an Android phone, then you will possess a Google account. You can use this account to enroll for GCP. As you can see in *Figure 14.1*, I am already logged into my own Google account, as indicated by the avatar on the top right-hand side of the screenshot.

At the time of writing, Google is offering $300 of credits for you to use over 90 days.

> **Important note**
>
> If you choose to take advantage of the free credits, you will still need to enter a valid payment method. Google does this to ensure that it is not an automated bot signing up for the account to abuse the credits they are offering. Once the credits have been used or expired, you will be given the option to upgrade your account to a paid account.

If you want to take advantage of this, then click on the **Get started for free** button and follow the onscreen prompts, making sure you read the terms and conditions. Once you have enrolled, you will be taken to the GCP console.

If you already have a GCP account, then log in to the GCP console directly at the following URL: `https://console.cloud.google.com/`.

Now that we have an account in place, let's learn how to create a project using the platform.

Creating a project

GCP has a concept whereby resources are launched into projects. if you have just signed up for an account, then a project called `My First Project` will have been created for you as part of the enrollment process.

If you are using an existing account, then I would recommend creating a new project to launch your GKE cluster in. To do this, click on the **Select** menu, which can be found in the top bar next to the GCP logo.

This will display all of your projects, as well as giving you the option to create a new project. To do so, follow these steps:

1. Click on **NEW PROJECT**.
2. You will be asked to give your new project a name, select the billing account you would like to attach the project to, and finally select the organization or folder you would like to place the project in. Fill in the required details.
3. Once these details have been entered, then you simply need to click on the **CREATE** button.

> **Important note**
>
> If you are using the automatically provided `My First Project`, you will need to make sure that the project is attached to a billing account before you proceed. To do this, click on the burger menu icon in the top left of the **Google Cloud Console** page, and select the **Billing** option. From here, follow the onscreen instructions to link the project to the billing account.

Now that we have a place to launch our resources, we can look at installing the GCP command-line tool.

Installing the GCP command-line interface

Here we are going to cover the GCP **Command-Line Interface** (**CLI**) on your local machine. Don't worry if you do not want to or are unable to install it, as there is a way that you can run the CLI in the cloud. But let's start with my operating system of choice, macOS.

Installing on macOS

If you are like me and you do a lot of work on macOS using Terminal, there is a high likelihood that you have installed and used Homebrew at some point.

> **Important note**
> Homebrew is a package manager for macOS that simplifies the installation of software on your machine. It works exclusively on the command line via the `brew` command.

If you don't have Homebrew installed, then you can install it by opening a Terminal session and running the following command:

```
$ /bin/bash -c "$(curl -fsSL https://raw.githubusercontent.com/
Homebrew/install/master/install.sh)zz
```

Once installed, you will be able to install the GCP CLI using the following command:

```
$ brew tap homebrew/cask
$ brew cask install google-cloud-sdk
```

You can test the installation by running the following command:

```
$ gcloud --version
```

If everything goes as planned, you should see something like the following output:

Figure 14.2 – Checking the version number on macOS

If you are having problems, you can check the documentation by running this code:

```
$ brew cask info google-cloud-sdk
```

Once Homebrew is installed and working, you can move onto the *Initialization* section of this chapter.

Installing on Windows

There are a few options to install the GCP CLI on Windows. The first is to open a PowerShell session and run the following command in order to install it using Chocolatey.

> **Important note**
>
> Like Homebrew on macOS, Chocolatey is a package manager that lets you easily and consistently install a wide variety of packages on Windows via PowerShell, using the same command syntax rather than having to worry about the numerous installation methods that exist on Windows.

If you don't have Chocolatey installed, then you can run the following command in a PowerShell session that has been launched with administrative privileges:

```
$ Set-ExecutionPolicy Bypass -Scope Process -Force; [System.
Net.ServicePointManager]::SecurityProtocol = [System.Net.
ServicePointManager]::SecurityProtocol -bor 3072; iex
((New-Object System.Net.WebClient).DownloadString('https://
chocolatey.org/install.ps1'))
```

Once Chocolatey is installed, or if you already have it installed, simply run the following:

```
$ choco install --ignore-checksum gcloudsdk
```

The other way you can install it is to download the installer from the following URL: https://dl.google.com/dl/cloudsdk/channels/rapid/GoogleCloudSDKInstaller.exe.

Once downloaded, run the executable by double-clicking on it and following the onscreen prompts.

Once installed, open a new PowerShell window and run the following command:

```
$ gcloud --version
```

If everything goes as planned, you should see something like the following output:

```
PS C:\Users\russ> gcloud --version
Google Cloud SDK 319.0.0
bq 2.0.62
core 2020.11.13
gsutil 4.55
PS C:\Users\russ>
```

Figure 14.3 – Checking the version number in PowerShell

Once installed and working, you can move onto the *Initialization* section of this chapter.

Installing on Linux

While the Google Cloud CLI packages are available for most distributions, we don't have the space to cover all of the various package managers here. Instead, we will use the install script provided by Google. To run this, you simply need to use the following commands:

```
$ curl https://sdk.cloud.google.com | bash
$ exec -l $SHELL
```

The script will ask you several questions during the installation. For most people, answering Yes will be fine. Once installed as per the macOS and Windows installations, you can run the following command:

```
$ gcloud -version
```

You should then see the following output:

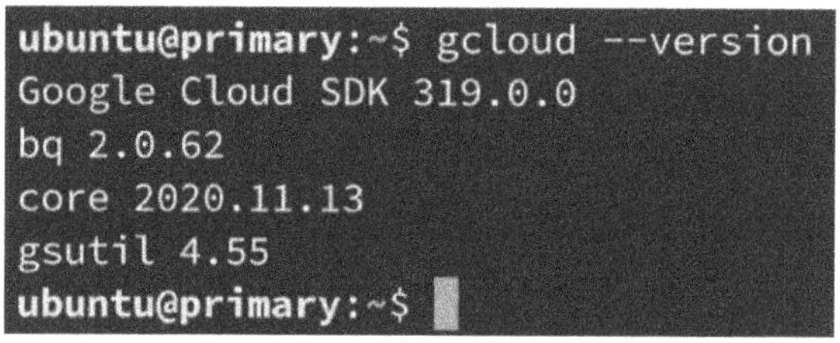

```
ubuntu@primary:~$ gcloud --version
Google Cloud SDK 319.0.0
bq 2.0.62
core 2020.11.13
gsutil 4.55
ubuntu@primary:~$
```

Figure 14.4 – Checking the version number on Linux

The one thing you might have noticed from the three installations is that, while the install method differs, once the package has been installed, we are using the same gcloud command, and also getting the same results back. From here, it shouldn't matter which operating system you are running, as the commands will apply to all three.

Cloud Shell

Before we started installing the Google Cloud CLI, I did mention that there was a fourth option. That option is Google Cloud Shell, which is built into Google Cloud Console. To access this, click on the Shell icon, which can be found on the right of the top menu.

Once configured, you should see what looks to be a web-based terminal from which you can run the following:

```
$ gcloud -version
```

The output differs slightly here as Google has provided the full suite of supporting tools. However, you will notice from the following screen that the versions do match the ones we installed locally:

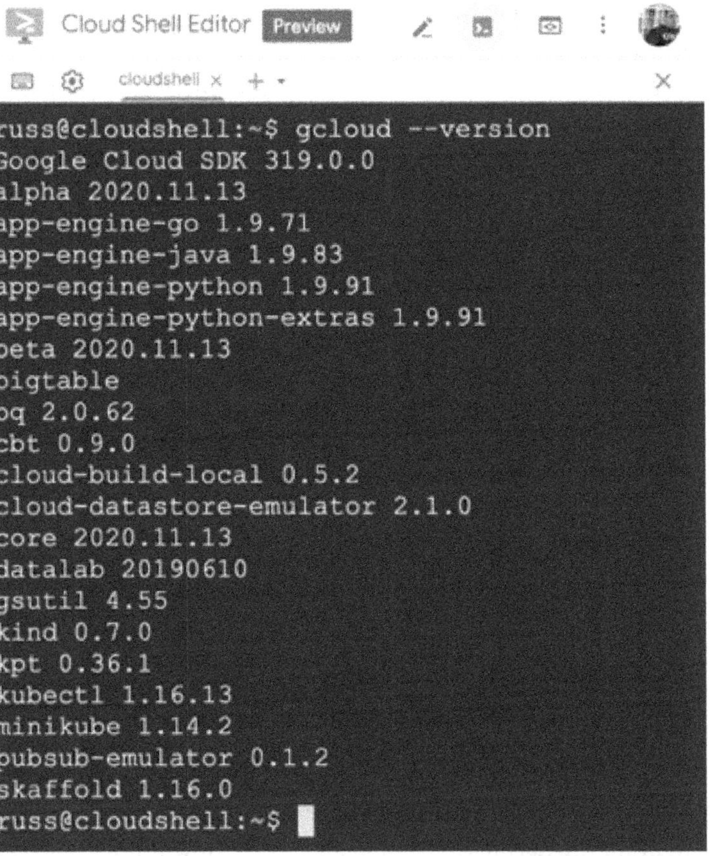

Figure 14.5 – Checking the version number in Google Cloud Shell

If you are using Google Cloud Shell, then you can skip the initialization step, as this has already been done for you.

Initialization

If you have chosen to install the client locally, then we will need to do one final step to link it to your GCP account. To do this, run the following command:

```
$ gcloud init
```

This will immediately run a quick network diagnostic to ensure that the client has the connectivity it needs to run. Once the diagnostic has passed, you will be prompted with the following question:

```
You must log in to continue. Would you like to log in (Y/n)?
```

Answering Y will open a browser window. If it doesn't, then copy and paste the provided URL into your browser where, once you have selected the account you wish to use, you will be presented with an overview of the permissions the client is requesting, as seen in the following figure:

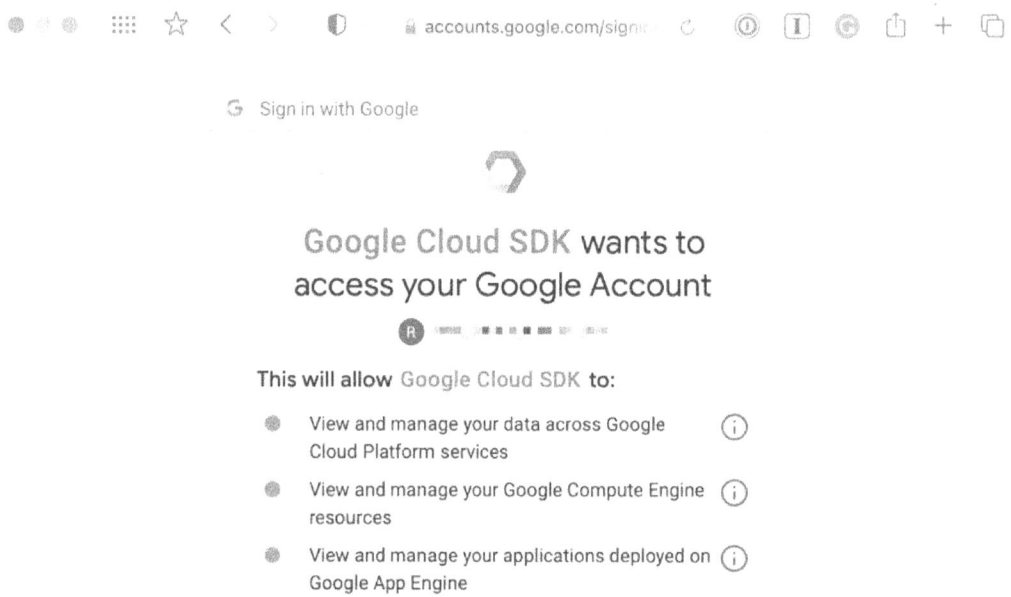

Figure 14.6 – Reviewing the permissions

If you are happy to grant the permissions, then click the **Allow** button. Back on your terminal, you will get confirmation of the user you are logged in as. Then you will be asked to pick a cloud project to use. The list will only contain the unique ID of the project, and not the friendly name that you saw or set up in Google Cloud Console earlier. If you have more than one project, please make sure you pick the correct project.

Should you need to update the project the client is using at any point, you can run the following command:

```
$ gcloud config set project PROJECT_ID
```

Make sure you replace PROJECT_ID with the unique ID of the project you wish to switch to.

Now that you have installed the Google Cloud CLI and have configured your account, we are ready to launch your GKE cluster.

Launching your first GKE cluster

As it is going to take a few minutes to launch the cluster, let's run the command to initiate the process and then talk through in a little more detail what is happening while it launches.

Before we launch our cluster, we need to make sure that the `container.googleapis.com` service is enabled. To do this, run the following command:

```
$ gcloud services enable container.googleapis.com
```

Once the service has been enabled, the command to launch a *two-node* cluster called `k8sforbeginners`, which will be hosted in a single zone in the *Central US* region, is as follows:

```
$ gcloud container clusters create k8sforbeginners
--num-nodes=2 --zone=us-central1-a
```

After about 5 minutes, you should see something that looks like the following output:

```
  gcloud container clusters create k8sforbeginners --num-nodes=2 --zone=us-central1-a
WARNING: Warning: basic authentication is deprecated, and will be removed in GKE control plane versi
ons 1.19 and newer. For a list of recommended authentication methods, see: https://cloud.google.com/
kubernetes-engine/docs/how-to/api-server-authentication
WARNING: Currently VPC-native is not the default mode during cluster creation. In the future, this w
ill become the default mode and can be disabled using `--no-enable-ip-alias` flag. Use `--[no-]enabl
e-ip-alias` flag to suppress this warning.
WARNING: Newly created clusters and node-pools will have node auto-upgrade enabled by default. This
can be disabled using the `--no-enable-autoupgrade` flag.
WARNING: Starting with version 1.18, clusters will have shielded GKE nodes by default.
WARNING: Your Pod address range (`--cluster-ipv4-cidr`) can accommodate at most 1008 node(s).
WARNING: Starting with version 1.19, newly created clusters and node-pools will have COS_CONTAINERD
as the default node image when no image type is specified.
Creating cluster k8sforbeginners in us-central1-a... Cluster is being health-checked (master is hea
lthy)...done.
Created [https://container.googleapis.com/v1/projects/wide-gamma-296319/zones/us-central1-a/clusters
/k8sforbeginners].
To inspect the contents of your cluster, go to: https://console.cloud.google.com/kubernetes/workload
_/gcloud/us-central1-a/k8sforbeginners?project=wide-gamma-296319
kubeconfig entry generated for k8sforbeginners.
NAME            LOCATION       MASTER_VERSION   MASTER_IP      MACHINE_TYPE   NODE_VERSION   NUM
_NODES  STATUS
k8sforbeginners  us-central1-a  1.16.13-gke.401  34.123.175.234  n1-standard-1  1.16.13-gke.401  2
         RUNNING
```

Figure 14.7 – Launching the cluster

Once the cluster has launched, you should be able to follow the URL in the output and view it in Google Cloud Console, as seen in the following figure:

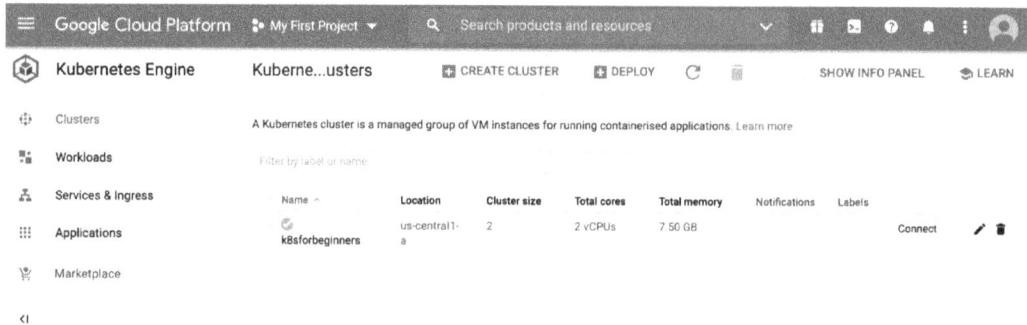

Figure 14.8 – Viewing the cluster in Google Cloud Console

Now that we have our cluster up and running, we can look at deploying a workload, then take a look at Google Cloud Console in a little more detail.

Deploying a workload and interacting with your cluster

One of the things to note from the feedback when we launched our cluster is the following output:

```
kubeconfig entry generated for k8sforbeginners.
```

As you may have already guessed, this has downloaded and configured all of the necessary information to connect the kubectl instance that you used to launch the cluster. You can confirm this by running the following command:

```
$ kubectl get nodes
```

The output you get from the command should show two nodes with a prefix of gke, so should appear something like the following Terminal output:

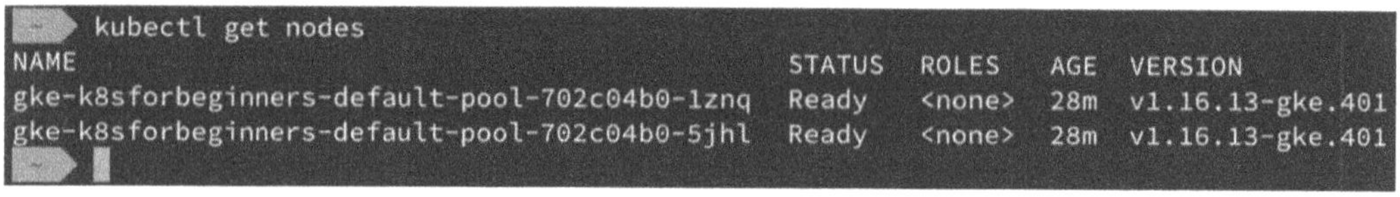

Figure 14.9 – Using kubectl to list the nodes

If you saw the preceding output and you are happy to proceed with the `kubectl` instance you are using, then you can skip the next section of the chapter and move straight onto *Launching an example workload*.

You can also find a link to the official GKE documentation in the *Further reading* section at the end of this chapter.

Configuring your local client

Should you need to configure another `kubectl` instance to connect to your cluster, then the GCP CLI has a command to do just that.

> **Important note**
> Running the command that follows assumes you have the GCP CLI installed and configured. If you don't have this, then please follow the instructions from the *Installing the Google Cloud Platform CLI* section of this chapter.

The command you need to run to download the credentials and configure `kubectl` is as follows:

```
$ gcloud container clusters get-credentials k8sforbeginners
--zone=us-central1-a
```

Should you need to switch to or from another configuration (or context as it is known), you can run the following commands. The first command lists the current context:

```
$ kubectl config current-context
```

The next command lists the names of all of the contexts that you have configured:

```
$ kubectl config get-contexts -o name
```

Once you know the name of the context that you need to use, you can run the following command, making sure to replace `context_name` with the name of the context that you change it to, as in the following:

```
$ kubectl config use-context context_name
```

So now that we have your `kubectl` control configured to talk to and interact with your GKE cluster, we can launch an example workload.

Launching an example workload

The example workload we are going to launch is the PHP / Redis Guestbook example, which is used throughout the official Kubernetes documentation:

1. The first step in launching the workload is to create the Redis Leader deployment and service. To do this, we use the following commands:

   ```
   $ kubectl apply -f https://raw.githubusercontent.com/
   GoogleCloudPlatform/kubernetes-engine-samples/master/
   guestbook/redis-leader-deployment.yaml
   $ kubectl apply -f https://raw.githubusercontent.com/
   GoogleCloudPlatform/kubernetes-engine-samples/master/
   guestbook/redis-leader-service.yaml
   ```

2. Next up, we need to repeat the process, but this time to launch the Redis Follower deployment and service, as shown here:

   ```
   $ kubectl apply -f https://raw.githubusercontent.com/
   GoogleCloudPlatform/kubernetes-engine-samples/master/
   guestbook/redis-follower-deployment.yaml
   $ kubectl apply -f https://raw.githubusercontent.com/
   GoogleCloudPlatform/kubernetes-engine-samples/master/
   guestbook/redis-follower-service.yaml
   ```

3. Now that we have Redis up and running, it is time to launch the frontend deployment and service – this is the application itself:

   ```
   $ kubectl apply -f https://raw.githubusercontent.com/
   GoogleCloudPlatform/kubernetes-engine-samples/master/
   guestbook/frontend-deployment.yaml
   $ kubectl apply -f https://raw.githubusercontent.com/
   GoogleCloudPlatform/kubernetes-engine-samples/master/
   guestbook/frontend-service.yaml
   ```

4. After a few minutes, you should be able to run the following command to get information on the service you have just launched:

   ```
   $ kubectl get service frontend
   ```

The output of the command should give you an external IP address that looks like the following Terminal output:

```
~ kubectl get service frontend
NAME       TYPE           CLUSTER-IP     EXTERNAL-IP     PORT(S)        AGE
frontend   LoadBalancer   10.3.247.54    34.72.194.254   80:31750/TCP   2m12s
~
```

Figure 14.10 – Getting information on the frontend service

Now that we have launched our application, copy the **EXTERNAL-IP** value and put the IP address into your browser. Here you should be presented with the **Guestbook** application. Try submitting a few messages, as shown in the following figure:

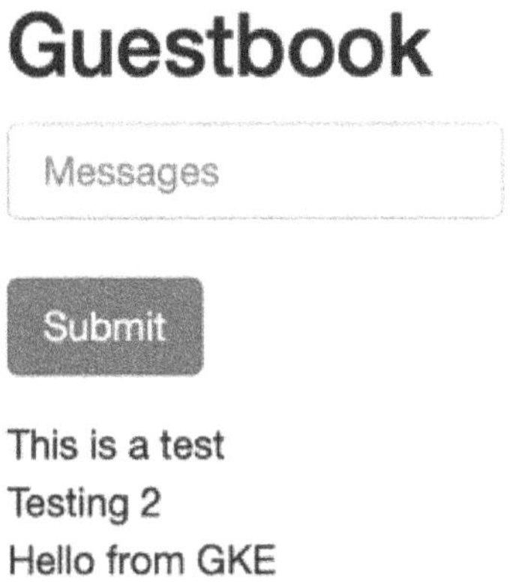

Guestbook

Messages

Submit

This is a test
Testing 2
Hello from GKE

Figure 14.11 – The Guestbook application with a few test messages

So, now that we have our workload launched, let's return to and explore Google Cloud Console.

Exploring Google Cloud Console

We have already seen our cluster in Google Cloud Console, so next, click on the **Workloads** link, which can be found in the left-hand-side menu of the **Kubernetes Engine** section in Google Cloud Console.

Workloads

Once the page loads, you should see something that resembles the following screen:

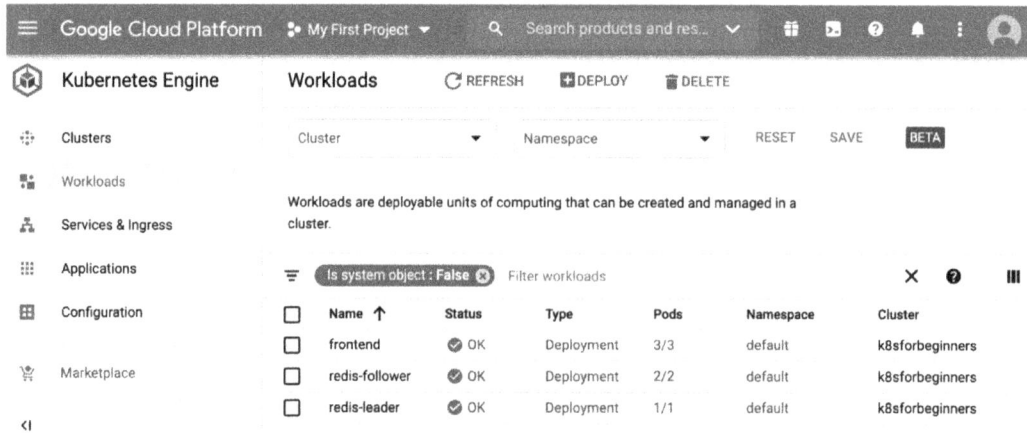

Figure 14.12 – Viewing the workload in the console

As you can see from the preceding screen, the three deployments are listed along with confirmation of the namespace they are in, as well as the cluster that the workload belongs to.

If we had more than one cluster with multiple namespaces and deployments, we would be able to use the filter to drill down into our GKE workloads.

Clicking on the `frontend` deployment will let you view more information:

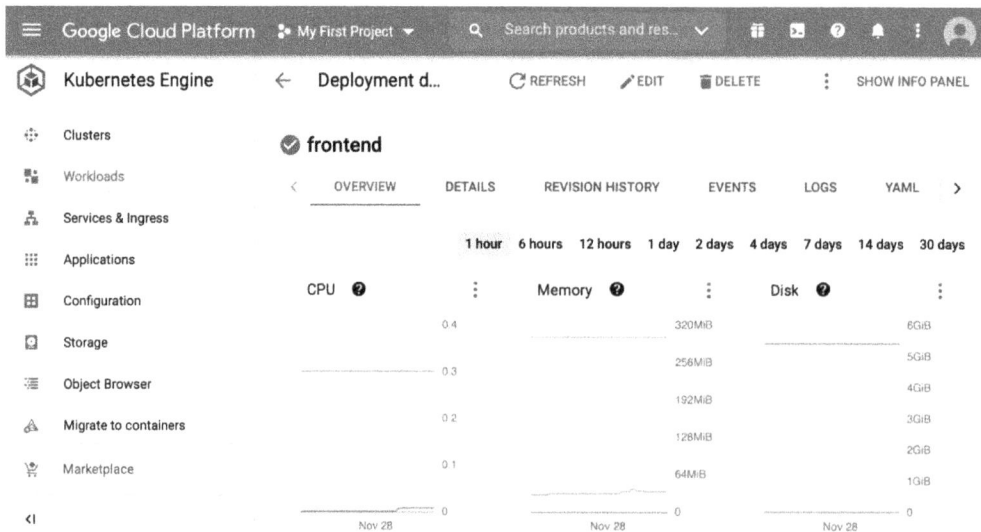

Figure 14.13 – Getting insights into a deployment

On this page, you will be able to drill further down into your deployment using the tabs below the deployment name, as follows:

- **OVERVIEW**: This view gives you, well, an overview of your deployment – as you can see from the previous screenshot, you can see the CPU, memory, and disk utilization, along with other information.

- **DETAILS**: This lists more details about the environment and deployment itself. Here, you can find out when it was created, annotations, labels (as well as details about the replicas), the update strategy, and pod information.

- **REVISION HISTORY**: Here, you will find a list of all of the revisions to the deployment. This is useful if your deployment is updated frequently and you need to keep track of when the deployment was updated.

- **EVENTS**: If you have any problems with your deployment, then this is the place you should look. All events – such as scaling, pod availability, and other errors – will be logged here.

- **LOGS**: This is a searchable list of the logs from all containers running the pod.

- **YAML**: This is an exportable YAML file containing the full deployment configuration.

This information is available for all deployments across all GKE clusters you have launched within the project.

Services & Ingress

In this section, we are going to look at the **Services & Ingress** section. As you may have already guessed, this lists all the services that are launched across all of your GKE clusters.

After clicking on **Services & Ingress**, you will be presented with a screen that looks similar to the following:

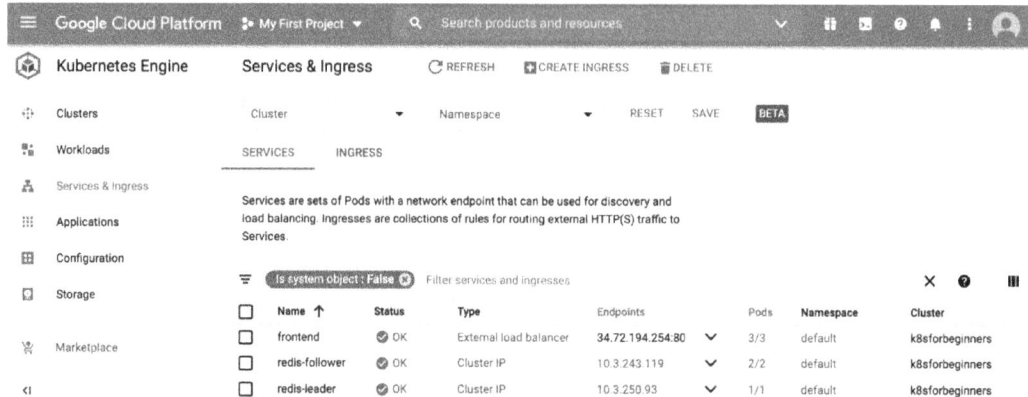

Figure 14.14 – Viewing the services in the console

As you can see, we have the three services we launched listed. However, the `frontend` service has a type of `External load balancer` and a public IP address listed, as opposed to the `redis-leader` and `redis-follower` services, which only have a *Cluster IP*. This is because, of the three services, we only want the `frontend` service to be accessible publicly as only our `frontend` service uses the two Redis ones.

When the external load balancer was launched (and as we are running our cluster in a public cloud provider), the Kubernetes scheduler knew to contact the Google Cloud API and launch a load balancer for use, and then configure it to talk back to our cluster nodes, exposing the deployment. This deployment is running on port `80` internally.

As before, clicking on one of the three running services will give you several bits of information:

- **OVERVIEW**: This view gives you a summary of the service configuration and utilization.
- **DETAILS**: Here you can find more detail on the service along with a link to view the load balancer resource that has been launched within our Google Cloud project.
- **EVENTS**: As before, here you can find any events that have affected the service. This is useful for troubleshooting.
- **LOGS**: This is a repeat of the logs shown in the **Workloads** section.
- **YAML**: Again, this is a way for you to export the full configuration for the service as a YAML file.

Although we are not going to use the other menu items, we should quickly discuss them.

Other menu options

There are five more options for us to discuss. So far, we haven't really launched a workload that would utilize them.

Applications

With no applications launched, this links you to a marketplace where you can choose and deploy pre-configured applications into your environment – at the time of writing, there are over 100 applications, which range from free and open source to commercial offerings:

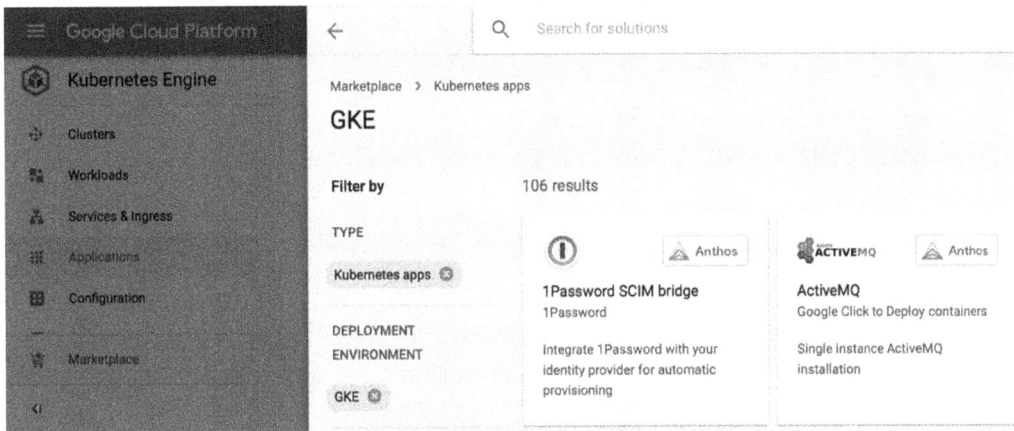

Figure 14.15 – Just a couple of the hundreds of applications available

The applications are categorized. They cover everything from Blog and CMS, databases, and analytics, all the way through to development tools and stacks. The applications are based on the Application Custom Resource Definition, which is being worked on by the Kubernetes Application Special Interest Group.

Configuration

Here you can manage ConfigMaps, such as environment variables, as well as `Secrets`, and share them amongst your GKE clusters.

Storage

From this section, you can manage, monitor, and review your persistent volume claims as well as storage classes associated with your GKE clusters.

Object browser

This allows you to browse all of the objects running in your clusters using a graphical representation of the Kubernetes API. From here you can generate YAML for any part of your GKE cluster.

Migrate to containers

This is an interesting tool; it inspects your virtual machine workload (which could be running on either Linux or Windows) and then figures out the critical parts of the application. It then attempts to containerize it. In the end, you will be left with a Dockerfile, along with all of the necessary data files and configurations.

Deleting your cluster

Once you have finished with your cluster, you can remove it by running the following commands. The first removes the service we created and the second removes the cluster itself:

```
$ kubectl delete service frontend
$ gcloud container clusters delete k8sforbeginners --zone=us-central1-a
```

It will take a few minutes to remove the cluster.

> **Important note**
>
> This should also delete any services that were launched by your workloads, such as load balancers. However, I do recommend checking Google Cloud Console for any orphaned services, to ensure that you do not get any unexpected costs.

So far throughout this chapter, we have been using the `--zone=us-central1-a` zone. This has been launching our cluster in a single availability zone in the US Central region. Let's discuss what other cluster options are available.

More about cluster nodes

At the end of the previous section, I mentioned availability zones and regions. Before we discuss some of the cluster deployment options, we should get a bit of a better understanding of what we mean by availability zones and regions:

- **Region**: A region is made up of zones. Zones have great low-latency network connectivity to other zones within the same region. This gives you a way of deploying highly available always-on fault-tolerant workloads.

- **Availablity zone**: Think of availability zones as separate data centers within a region. The zones have diverse networks and power, meaning that, should a single zone have an issue and you are running your workload across multiple zones, then your workload shouldn't be impacted.

  The one thing to note with zones is that you might find that not all machine types are available across all zones within a region. Therefore, please check before attempting to deploy your workload.

Google best practice recommends that, for optimum performance and availability, you should deploy your workload across the maximum number of zones within a single region. However, it is possible to split your workloads across multiple regions – all you have to do is take into account and allow for the increased latency between regions.

With that in mind, let's take a look at the command we used to launch the test cluster at the start of the chapter:

```
$ gcloud container clusters create k8sforbeginners
--num-nodes=2 --zone=us-central1-a
```

As we know, it will launch two nodes, but only in a single zone, as we have only passed one using the `--zone` flag, which in our case was the `us-central1` region and zone a.

Run the following command, but use the `--region` flag rather than the `--zone` one:

```
$ gcloud container clusters create k8sforbeginners
--num-nodes=2 --region=us-central1
```

Once launched, run the following:

```
$ kubectl get nodes
```

This will produce something that looks like the following output:

```
    kubectl get nodes
NAME                                                STATUS  ROLES    AGE    VERSION
gke-k8sforbeginners-default-pool-8aabe110-7slq      Ready   <none>   3m17s  v1.16.13-gke.401
gke-k8sforbeginners-default-pool-8aabe110-ss7h      Ready   <none>   3m25s  v1.16.13-gke.401
gke-k8sforbeginners-default-pool-c105f077-d8tp      Ready   <none>   3m24s  v1.16.13-gke.401
gke-k8sforbeginners-default-pool-c105f077-gvq2      Ready   <none>   3m10s  v1.16.13-gke.401
gke-k8sforbeginners-default-pool-ccc06659-3vj3      Ready   <none>   3m23s  v1.16.13-gke.401
gke-k8sforbeginners-default-pool-ccc06659-70k5      Ready   <none>   3m22s  v1.16.13-gke.401
```

Figure 14.16 – Viewing the nodes running in a region

As you can see, we have two nodes in each zone, giving us a total cluster size of six nodes. The reason for this is that when you define a region by default, your cluster is spread across three zones. You can override this behavior by using the `--node-locations` flag.

This makes our command look like the following:

```
$ gcloud container clusters create k8sforbeginners
--num-nodes=2 --region=us-central1 --node-locations
us-central1-a,us-central1-b,us-central1-c,us-central1-f
```

We are still using the `us-central1` region, but deploying to the a,b,c, and f zones. Running the `kubectl get nodes` command now shows the following:

```
    kubectl get nodes
NAME                                              STATUS  ROLES    AGE  VERSION
gke-k8sforbeginners-default-pool-25e6bcc5-9snm   Ready   <none>   19s  v1.16.13-gke.401
gke-k8sforbeginners-default-pool-25e6bcc5-ndw3   Ready   <none>   21s  v1.16.13-gke.401
gke-k8sforbeginners-default-pool-6c753ce5-sq4s   Ready   <none>   19s  v1.16.13-gke.401
gke-k8sforbeginners-default-pool-6c753ce5-vmv4   Ready   <none>   20s  v1.16.13-gke.401
gke-k8sforbeginners-default-pool-711b9aca-mvl7   Ready   <none>   18s  v1.16.13-gke.401
gke-k8sforbeginners-default-pool-711b9aca-xtd6   Ready   <none>   20s  v1.16.13-gke.401
gke-k8sforbeginners-default-pool-858f6808-phkm   Ready   <none>   19s  v1.16.13-gke.401
gke-k8sforbeginners-default-pool-858f6808-pvvv   Ready   <none>   19s  v1.16.13-gke.401
```

Figure 14.17 – Our cluster across four zones

As you can see, Google has made it straightforward to deploy your clusters across multiple zones. This means that you can deploy your workload across a fully redundant cluster with very little effort required.

To remove clusters deployed using the `--region` flag, you should use the following command:

```
$ gcloud container clusters delete k8sforbeginners --region=us-
central1
```

At the time of writing, the simple two-node cluster we launched at the start of the chapter has a cost of around $49 per month, with the price increasing to around $270 per month for the eight-node cluster we have just launched. For more information on cost, see the link to the Google Cloud Pricing Calculator in the *Further reading* section.

That concludes our look at GKE. Before we move onto the next public cloud provider, let's summarize what we have covered.

Summary

In this chapter, we discussed the origins of GCP and the GKE service, before walking through how to sign up for an account and how to install and configure the Google Cloud command-line tool.

We then launched a simple two-node cluster using a single command, then deployed and interacted with a workload using both the `kubectl` command and Google Cloud Console.

Finally, and again only using a single command, we redeployed our cluster to take advantage of multiple availability zones, quickly scaling to a fully redundant and highly available eight-node cluster running across four availability zones.

I am sure you will agree that Google has done an excellent job in making deploying and maintaining what is a complex infrastructure configuration a relatively trivial and quick task.

Also, once your workloads are deployed, managing them is exactly the same as you would if your cluster was deployed elsewhere – we really haven't had to make any allowances for our cluster being run on GCP.

In the next chapter, we are going to be looking at deploying a Kubernetes cluster in Amazon Web Services using Amazon Elastic Kubernetes Service. This is Amazon's fully managed Kubernetes offering.

Further reading

Here are links to more information on some of the topics and tools that we have covered in this chapter:

- Google Kubernetes Engine: `https://cloud.google.com/kubernetes-engine/`
- Google Kubernetes Engine Documentation: `https://cloud.google.com/kubernetes-engine/docs`
- Google Cloud Pricing Calculator: `https://cloud.google.com/products/calculator`
- The Guestbook Sample Application: `https://github.com/GoogleCloudPlatform/kubernetes-engine-samples/tree/master/guestbook`
- Migrate your VM workloads to Kubernetes: `https://cloud.google.com/migrate/anthos/docs/getting-started`

- The Kubernetes Application SIG: `https://github.com/kubernetes-sigs/application`
- The Google Cloud Kubernetes comic: `https://cloud.google.com/kubernetes-engine/kubernetes-comic`

15
Launching a Kubernetes Cluster on Amazon Web Services with Amazon Elastic Kubernetes Service

In the previous chapter, we took our first steps with launching a Kubernetes cluster in a public cloud. Now we know what the **Google Cloud Platform** (**GCP**) Kubernetes offering looks like, we are going to move on to the Amazon **Elastic Kubernetes Service** (**EKS**) by **Amazon Web Services** (**AWS**).

In this chapter, you will learn what is needed to do to set up an AWS account, installing the supporting toolsets on macOS, Windows, and also Linux before finally launching and interacting with an Amazon EKS cluster.

We will be covering the following topics:

- What are AWS and Amazon EKS?

- Preparing your local environment

- Launching your Amazon EKS cluster

- Deploying a workload and interacting with your cluster

- Deleting your Amazon EKS cluster

Technical requirements

To following along with this chapter, you will need an AWS account with a valid payment attached to it.

> **Important note**
>
> Following the instructions in this chapter will incur a cost and it is important that you terminate any resources you launch once you have finished with them.

All prices quoted in this chapter are correct at the time of writing this book, and we recommend that you review the current costs before you launch any resources.

What are AWS and Amazon EKS?

There is a very good chance you have already heard of AWS, as not only is it one of the first public cloud providers, but also, at the time of writing, it has the largest market share.

AWS

As you may have already guessed, AWS is owned and operated by Amazon. Amazon, the retailer, first started to dabble with cloud services way back in 2000 when they started to develop and deploy **application programming interfaces** (**APIs**) for their retail partners to consume.

Off the back of this work, Amazon realized that they would need to build a better and more standardized infrastructure platform to not only host the services they had been developing but to also ensure that they could quickly scale, as more of the Amazon retail outlet was consuming more of the software services and was growing at an expediential rate.

Chris Pinkham and Benjamin Black wrote a white paper in 2003 that was approved by Jeff Bezos in 2004, which described an infrastructure platform where compute and storage resources can all be deployed programmatically.

The first public acknowledgment of AWS's existence was made in late 2004; however, the term was used to describe a collection of tools and APIs that would allow third parties to interact with Amazon's retail product catalog rather than what we know today.

It wasn't until 2006 that a rebranded AWS was launched, starting in March with the **Simple Storage Service**, or **S3** for short, which was a service that allowed developers to write and serve individual files using a web API rather having to write and read from a traditional local filesystem.

The next service to launch, Amazon **Simple Queue Service (SQS)**, had formed part of the original AWS collection of tools—this was a distributed message system that again could be controlled and consumed by developers using an API.

The final service launched in 2006 was a beta of the Amazon **Elastic Compute Cloud** service, or Amazon **EC2** for short, which was limited to existing AWS customers—again, you could use the APIs developed by Amazon to launch resources.

This was the final piece of the puzzle for Amazon, and they now had the foundations of a public cloud platform they could use not only their own retail platform on but also sell space to other companies and the public, such as you and me.

If we fast forward from 2006, when there were three services, to the time of writing, which is late 2020, there are now over 170 services available. All of these 170+ services stick to the core principles that were laid out in the white paper written in 2003 each service is software-defined, meaning that all a developer has to do is to make a simple API request to launch, configure, and in some cases consume before finally being able to make a request terminate the service.

> **Tip**
> Services that are prefixed with *Amazon* are services that are standalone, unlike ones that are prefixed with *AWS*, which are services that are designed to be used alongside other AWS services.

Long gone are the days of having to order a service, have someone build and deploy, then hand it over to you—this takes deployment times down to seconds from what sometimes could take weeks.

Rather than discuss all 170+ services, which would be a collection of books all by itself, we should discuss the service we are going to be looking at in this chapter.

Amazon EKS

While AWS was the first of the major public cloud providers, it was one of the last to launch a standalone Kubernetes service. Amazon EKS was first announced in late 2017 and became generally available in the **United States** (**US**) East (N. Virginia) and US West (Oregon) regions in June 2018.

The service is built to work with and take advantage of other AWS services and features, such as the following:

- **AWS Identity and Access Management** (**IAM**): This service allows you to take control of and manage both end-user and programmatic access to other AWS services.

- **Amazon Route 53**: This is Amazon's **Domain Name System** (**DNS**) service. EKS can use it as a source of DNS for clusters, which means that service discovery and routing can easily be managed within your cluster.

- **Amazon Elastic Block Storage** (**EBS**): If you need persistent block storage for the containers running within your Amazon EKS instance, then it is provided by the same service used to provide block storage for the rest of your EC2 compute resources.

- **EC2 Auto Scaling**: Should your cluster need to scale, then the same technology to scale your EC2 instances is employed.

- **Multi Availability Zones** (**AZs**): The Amazon EKS management layer, as well as cluster nodes, can be configured to be spread across multiple AZs within a given region to bring **high availability** (**HA**) and resilience to your deployment.

Before we launch our Amazon EKS cluster, we are going to have to download, install, and configure a few tools.

Preparing your local environment

There are two sets of command-line tools we are going to install, but before we do, we should quickly discuss the steps to sign up for a new AWS account. If you already have an AWS account, then you can skip this task and move straight on to the *Installing the AWS command-line interface* section.

Signing up for an AWS account

Signing up for an AWS account is a straightforward process, as detailed here:

1. Head over to https://aws.amazon.com/ and then click on the **Create an AWS account** button, which can be found in the top right of the page.

    > **Important note**
    >
    > While Amazon offers a free tier for new users, it is limited to certain services, instance sizing, and also for 12 months. For information on the AWS Free Tier, see https://aws.amazon.com/free/.

2. Fill out the initial form that asks for an email address, your preferred password, and the AWS account name.

3. Once done, click on **Continue** and follow the onscreen instructions. These will involve you confirming both your payment details and also confirming your identity via an automated phone call.

Once you have your account created and enabled, you will be able to start using AWS services. In our case, we now need to install the command-line tools we will be using to launch our Amazon EKS cluster.

Installing the AWS command-line interface

Next on our list of tasks is to install the AWS **command-line interface** (**CLI**). As with the previous chapter, *Chapter 14, Kubernetes Clusters on Google Kubernetes Engine*, we will be targeting Windows, Linux, and also macOS, which we will be looking at first.

Installing on macOS

Installing the AWS CLI on macOS using Homebrew is as simple as running the following command:

```
$ brew install awscli
```

Once it's installed, running the following command:

```
$ aws --version
```

This will output the version of the AWS CLI, along with some of the support services it needs, as illustrated in the following screenshot:

```
aws --version
aws-cli/2.1.5 Python/3.9.0 Darwin/20.1.0 source/x86_64 prompt/off
```

Figure 15.1 – Checking the AWS CLI version number on macOS

Once it's installed, we can move on to the *AWS CLI configuration* section.

Installing on Windows

As with macOS, you can use a package manager to install the AWS CLI. As in *Chapter 14*, *Kubernetes Clusters on Google Kubernetes Engine*, we will be using Chocolatey. The command you need to run is shown here:

```
$ choco install awscli
```

Once using Chocolatey, running aws --version will give you similar output to what we saw on macOS, as illustrated in the following screenshot:

```
PS C:\Users\russ> aws --version
aws-cli/2.1.7 Python/3.7.9 Windows/10 exe/AMD64 prompt/off
PS C:\Users\russ>
```

Figure 15.2 – Checking the AWS CLI version number on Windows

Once it's installed, we can move on to the *AWS CLI configuration* section.

Installing on Linux

While there are packages available for each distribution, the easiest way of installing the AWS CLI on Linux is to download and run the installer.

> **Important note**
>
> These instructions assume that you have the curl and unzip packages installed, if you don't, please install them using your distribution's package manager—for example, on Ubuntu, you would need to run sudo apt install unzip curl to install both packages.

To download and install the AWS CLI, run the following commands:

```
$ curl "https://awscli.amazonaws.com/awscli-exe-linux-x86_64.
zip" -o "awscliv2.zip"
$ unzip awscliv2.zip
$ sudo ./aws/install
```

Once it's installed, you should be able to run `aws --version`, and you will get something like this:

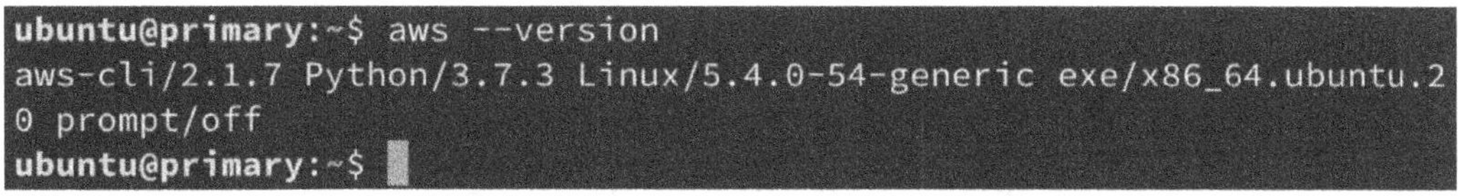

```
ubuntu@primary:~$ aws --version
aws-cli/2.1.7 Python/3.7.3 Linux/5.4.0-54-generic exe/x86_64.ubuntu.2
0 prompt/off
ubuntu@primary:~$
```

Figure 15.3 – Checking the AWS CLI version number on Linux

Once installed, we can move on to the *AWS CLI configuration* section.

AWS CLI configuration

Once you have the AWS CLI installed and you have checked that it is running okay by issuing the `aws --version` command, you will need to link the CLI to your AWS account.

To do this, you will need to log in to the AWS console—this can be accessed at `http://console.aws.amazon.com/`. Once logged in, type IAM into the **Find Services** search box and click on the link to be taken to the **Identity and Access Management (IAM)** page. We need to create a user with programmatic access; to do this, follow these steps:

1. Once the page has loaded, click on **Users**, which can be found under the **Access Management** section of the left-hand side menu.

2. Enter the username—I will be using `ekscluster`—then select both the **Programmatic access** and **AWS Management Console access** types. If you select the latter option, you will be given the option of setting an autogenerated or custom password and can decide whether the password should be reset after the user logs in.

    Personally, I set a custom password and unticked the option of resetting the password after the first login. To proceed to the next step, once you have set the details, click on the **Next: Permissions** button.

3. Rather than create a group, we are going to simply grant our user an existing policy. To do this, select **Attach existing policies directly** and select the **AdministratorAccess** policy, then click on **Next: Tags** to proceed to the next step.

4. As we are going to be removing this user at the end of the chapter, I am not going to enter any tags; however, you can, if you like. Click on the **Next: Review** button to move on to the final step.

5. Once you have reviewed the information presented to you, click on the **Create user** button.

6. Once the user has been created, click on the **Download .csv** button—this is the only time that the secret will be displayed, and once you close the page, you will have to generate a new one.

   Once you have the secret, click on the **Close** button. We now have the credentials we can use to authenticate our AWS CLI against our AWS account.

Return to your terminal and then run the following command to create a default profile:

```
$ aws configure
```

This will ask for a few bits of information, as follows:

- **AWS access key identifier (ID)**: This is the access key ID from the **comma-separated values (CSV)** file we downloaded.

- **AWS secret access key**: This is the key from the CSV file.

- **Default region name**: I entered us-east-1.

- **Default output format**: I left this blank.

To test that everything is working, you can run the following command:

```
$ aws ec2 describe-regions
```

This will list the AWS regions that are available, and the output should look like something like this:

```
REGIONS  ec2.eu-north-1.amazonaws.com    opt-in-not-required       eu-north-1
REGIONS  ec2.ap-south-1.amazonaws.com    opt-in-not-required       ap-south-1
REGIONS  ec2.eu-west-3.amazonaws.com     opt-in-not-required       eu-west-3
REGIONS  ec2.eu-west-2.amazonaws.com     opt-in-not-required       eu-west-2
REGIONS  ec2.eu-west-1.amazonaws.com     opt-in-not-required       eu-west-1
REGIONS  ec2.ap-northeast-2.amazonaws.com       opt-in-not-required       ap-northeast-2
REGIONS  ec2.ap-northeast-1.amazonaws.com       opt-in-not-required       ap-northeast-1
REGIONS  ec2.sa-east-1.amazonaws.com     opt-in-not-required       sa-east-1
REGIONS  ec2.ca-central-1.amazonaws.com  opt-in-not-required       ca-central-1
REGIONS  ec2.ap-southeast-1.amazonaws.com       opt-in-not-required       ap-southeast-1
REGIONS  ec2.ap-southeast-2.amazonaws.com       opt-in-not-required       ap-southeast-2
REGIONS  ec2.eu-central-1.amazonaws.com  opt-in-not-required       eu-central-1
REGIONS  ec2.us-east-1.amazonaws.com     opt-in-not-required       us-east-1
REGIONS  ec2.us-east-2.amazonaws.com     opt-in-not-required       us-east-2
:
```

Figure 15.4 – Testing the AWS CLI

Now we have the AWS CLI installed and configured to talk to our account, we need to install the next command-line tool.

Installing eksctl, the official CLI for Amazon EKS

While it is possible to launch an Amazon EKS cluster using the AWS CLI, it is complicated and there are a lot of steps. To get around this, Weaveworks have created a simple command-line tool that generates an AWS CloudFormation template and then launches your cluster.

> **Tip**
> AWS CloudFormation is Amazon's **Infrastructure-as-Code** (IaC) definition language that lets you define your AWS resources in such a way that they be can be deployed across multiple accounts or repeatedly in the same one. This is useful if you have to keep spinning up an environment—for example, as part of a **continuous integration** (CI) build.

Installation couldn't be easier—macOS users can run the following commands to install using Homebrew:

```
$ brew tap weaveworks/tap
$ brew install weaveworks/tap/eksctl
```

Windows users can use the following command:

```
$ choco install eksctl
```

All Linux users have to do is download the precompiled binaries and copy them into place, as follows:

```
$ curl --silent --location "https://github.com/weaveworks/
eksctl/releases/latest/download/eksctl_$(uname -s)_amd64.tar.
gz" | tar xz -C /tmp
$ sudo mv /tmp/eksctl /usr/local/bin
```

Once installed, you should be able to run `eksctl version` to get the version number. We are now ready to launch our cluster.

Launching your Amazon EKS cluster

For our test, we are going to use the defaults built into the `eksctl` command. These will launch an Amazon EKS cluster with the following attributes:

- In the `us-west-1` region.
- With two worker nodes, using the `m5.large` instance type.
- Uses the official AWS EKS **Amazon Machine Image** (**AMI**).
- In its own **virtual private cloud** (**VPC**), which is Amazon's networking service.
- With an automatically generated random name.

So, without further ado, let's launch our cluster by running the following command:

```
$ eksctl create cluster
```

You might want to go and make a drink or catch up on emails, as this process can take around 30 minutes to complete.

If you are not following along, here is the output I got when running the command. First of all, some basic information is displayed about the version of `eksctl` and which region will be used:

```
[i]   eksctl version 0.33.0
[i]   using region us-east-1
```

Now, by default, eksctl uses the us-west-2 region; however, as we set us-east-1 as the default when we configured the AWS CLI, it has used that setting. Next up, it will give some information on the networking and AZs it will be deploying resources into, as illustrated in the following code snippet:

```
[i]   setting availability zones to [us-east-1e us-east-1a]
[i]   subnets for us-east-1e - public:192.168.0.0/19
private:192.168.64.0/19
[i]   subnets for us-east-1a - public:192.168.32.0/19
private:192.168.96.0/19
```

It will now give details of which version of the AMI it is going to use, along with the Kubernetes version that image supports, as follows:

```
[i]   nodegroup "ng-6cd00965" will use "ami-0f4cae6ae56be18ee"
[AmazonLinux2/1.18]
[i]   using Kubernetes version 1.18
```

Now it knows all of the elements, it is going to create a cluster and with it make a start on the deployment, as follows:

```
[i]   creating EKS cluster "attractive-sheepdog-1607259336" in
"us-east-1" region with un-managed nodes
[i]   will create 2 separate CloudFormation stacks for cluster
itself and the initial nodegroup
[i]   if you encounter any issues, check CloudFormation console
or try 'eksctl utils describe-stacks --region=us-east-1
--cluster=attractive-sheepdog-1607259336'
```

As you can see, it has called my cluster attractive-sheepdog-1607259336; this will be referenced throughout the build. By default, logging is not enabled, as we can see here:

```
[i]   CloudWatch logging will not be enabled for cluster
"attractive-sheepdog-1607259336" in "us-east-1"
[i]   you can enable it with 'eksctl utils update-cluster-
logging --enable-types={SPECIFY-YOUR-LOG-TYPES-HERE (e.g. all)}
--region=us-east-1 --cluster=attractive-sheepdog-1607259336'
```

Now is the point where we wait, and the following messages are displayed:

```
[i]   Kubernetes API endpoint access will use default of
{publicAccess=true, privateAccess=false} for cluster
"attractive-sheepdog-1607259336" in "us-east-1"
[i]   2 sequential tasks: { create cluster control plane
"attractive-sheepdog-1607259336", 3 sequential sub-tasks: { no
tasks, create addons, create nodegroup "ng-6cd00965" } }
[i]   building cluster stack "eksctl-attractive-sheepdog-
1607259336-cluster"
[i]   deploying stack "eksctl-attractive-sheepdog-1607259336-
cluster"
```

Once deployed, it will download the cluster credentials and configure `kubectl`, as follows:

```
[✓]   saved kubeconfig as "/Users/russ.mckendrick/.kube/config"
[i]   no tasks
[✓]   all EKS cluster resources for "attractive-
sheepdog-1607259336" have been created
[i]   adding identity "arn:aws:iam::687011238589:role/eksctl-
attractive-sheepdog-160725-NodeInstanceRole-1FZFP968TXZG9" to
auth ConfigMap
```

The final step is to wait for the nodes to become available, as is happening here:

```
[i]   nodegroup "ng-6cd00965" has 0 node(s)
[i]   waiting for at least 2 node(s) to become ready in "ng-
6cd00965"
[i]   nodegroup "ng-6cd00965" has 2 node(s)
[i]   node "ip-192-168-23-50.ec2.internal" is ready
[i]   node "ip-192-168-50-107.ec2.internal" is ready
```

Now we have both nodes online and ready, it is time to display a message confirming that everything is ready, as follows:

```
[i]   kubectl command should work with "/Users/russ.mckendrick/.
kube/config", try 'kubectl get nodes'
[✓]   EKS cluster "attractive-sheepdog-1607259336" in "us-
east-1" region is ready0
```

Now that the cluster is ready, let's do as the output suggests and run `kubectl get nodes`. As expected, this gives us details on the two nodes that make up our cluster, as illustrated in the following screenshot:

```
~      kubectl get nodes
NAME                            STATUS   ROLES    AGE     VERSION
ip-192-168-23-50.ec2.internal   Ready    <none>   8m38s   v1.18.9-eks-d1db3c
ip-192-168-50-107.ec2.internal  Ready    <none>   8m41s   v1.18.9-eks-d1db3c
~
```

Figure 15.5 – Viewing the two cluster nodes

Now we have a cluster up and running, let's launch the same workload we launched when we launched our **Google Kubernetes Engine** (**GKE**) cluster.

Deploying a workload and interacting with your cluster

If you recall from the last chapter, we used the Guestbook example from the GCP GKE examples' GitHub repository.

First, we are going to deploy the workload before we then explore the web-based AWS console. Let's make a start on our Guestbook deployment.

Deploying the workload

Even though we are our cluster is running on AWS using Amazon EKS, we are going to be using the same set of **YAML Ain't Markup Language** (**YAML**) files we used to launch our workload in GKE. Follow these next steps:

1. As before, our first step is launching the `Redis Leader` deployment and service using the following commands:

    ```
    $ kubectl apply -f https://raw.githubusercontent.com/
    GoogleCloudPlatform/kubernetes-engine-samples/master/
    guestbook/redis-leader-deployment.yaml
    ```

    ```
    $ kubectl apply -f https://raw.githubusercontent.com/
    GoogleCloudPlatform/kubernetes-engine-samples/master/
    guestbook/redis-leader-service.yaml
    ```

2. Once the `Redis Leader` deployment and service have been created, we need to launch the `Redis Follower` deployment, as follows:

```
$ kubectl apply -f https://raw.githubusercontent.com/
GoogleCloudPlatform/kubernetes-engine-samples/master/
guestbook/redis-follower-deployment.yaml
$ kubectl apply -f https://raw.githubusercontent.com/
GoogleCloudPlatform/kubernetes-engine-samples/master/
guestbook/redis-follower-service.yaml
```

3. Once Redis is up and running, it's time to launch the frontend deployment and service using the following commands:

```
$ kubectl apply -f https://raw.githubusercontent.com/
GoogleCloudPlatform/kubernetes-engine-samples/master/
guestbook/frontend-deployment.yaml
$ kubectl apply -f https://raw.githubusercontent.com/
GoogleCloudPlatform/kubernetes-engine-samples/master/
guestbook/frontend-service.yaml
```

4. After a few minutes, we will be able to run the following command to get information on the service we have just launched, which should include details on where to access our workload:

```
$ kubectl get service frontend
```

You will notice that this time, the output is slightly different from the output we got when running the workload on GKE, as we can see in the following screenshot:

```
      kubectl get service frontend
NAME          TYPE            CLUSTER-IP       EXTERNAL-IP
                              PORT(S)          AGE
frontend      LoadBalancer    10.100.119.218   ab5c61c5b581d4cb4970445d5b5b783d-5785634
25.us-east-1.elb.amazonaws.com  80:30555/TCP   65s
```

Figure 15.6 – Getting information on the frontend service

As you can see, rather than an **Internet Protocol (IP)** address, we get a **Uniform Resource Locator (URL)**. Copy that into your browser.

Once you have opened the URL, given that we have used exactly the same work, you won't be surprised to see the Guestbook application, as shown in the following screenshot:

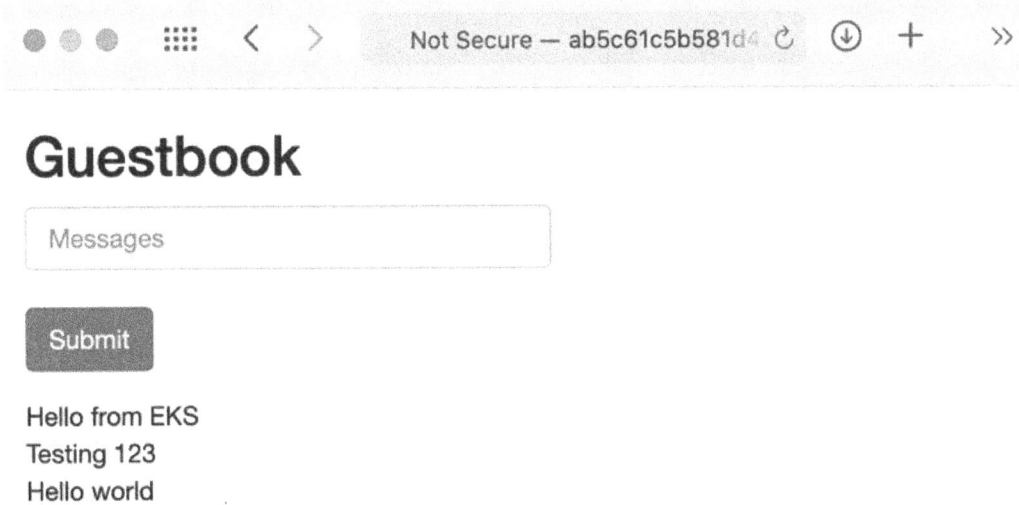

Figure 15.7 – The Guestbook application with a few test messages

Now we have our workload up and running, let's take a look at what we can see through the AWS console.

Exploring the AWS console

In this section, we are going to try to take a look at our newly deployed workload using the AWS console. First, we need to log in, as follows:

1. Open `https://console.aws.amazon.com/`. If you are still logged in from when we generated the IAM credentials, you will need to log out as we will not be using our main AWS user for this part of the chapter.

2. Once logged out, open the CSV file that we downloaded during the *AWS CLI configuration* section of the chapter and then copy and paste the console login link into your browser. Enter the credentials for the `ekscluster` user we created earlier in the *AWS CLI configuration* section of this chapter. If you selected the option to reset the password when the user first logs in, then you will be prompted to do so now.

> **Important note**
>
> You may be thinking to yourself: *Why are we using a different user to log in to the AWS console, as my main user has full access?* There is a good reason for this—when `eksctl` launched our cluster, there was a line that said `adding identity`. This granted permissions to the user that were used to create a cluster to be able to interact with the cluster using AWS services. This means that we will be able to view workloads and the like within the AWS console. If you were to use your regular user, then you would not have permission to see any of the details we are about to look at within the AWS console, due to the cluster not knowing about your main user.

3. Next, make sure that the region we have launched our cluster in is selected. The region selector can be found in the top right next to the **Support** drop-down menu—make sure that **US East (N. Virginia) us-east-1** is selected.

4. Once you have the region selected, enter `EKS` into the **Services** menu and you will be taken to the **Amazon Container Services** page, as illustrated in the following screenshot:

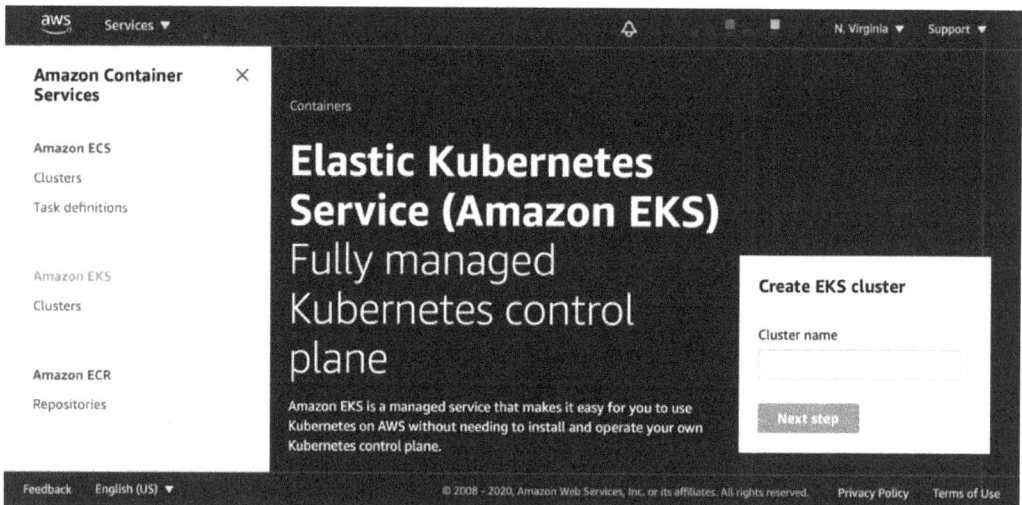

Figure 15.8 – The Amazon Container Services page

5. Next, click on the **Clusters** link underneath the **Amazon EKS** entry in the left-hand side menu, as illustrated in the following screenshot:

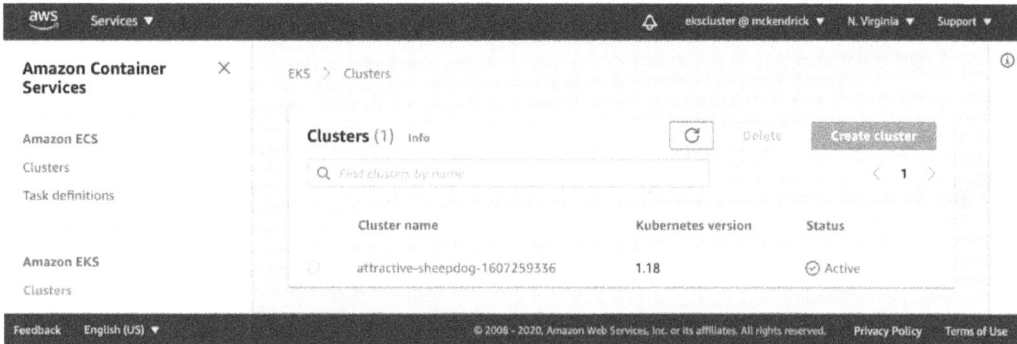

Figure 15.9 – Listing the Amazon EKS clusters

Clicking on the cluster name, which in my case is `attractive-sheepdog-1607259336`, will take you to the cluster view.

6. The initial **Overview** tab will show you the nodes within the cluster. The next option is **Workloads**—clicking this will take you to the workloads view.

7. By default, it will show you all of the Kubernetes namespaces, including the system ones—selecting the `default` namespace in the **All Namespaces** dropdown will display just the workloads we have launched, as illustrated in the following screenshot:

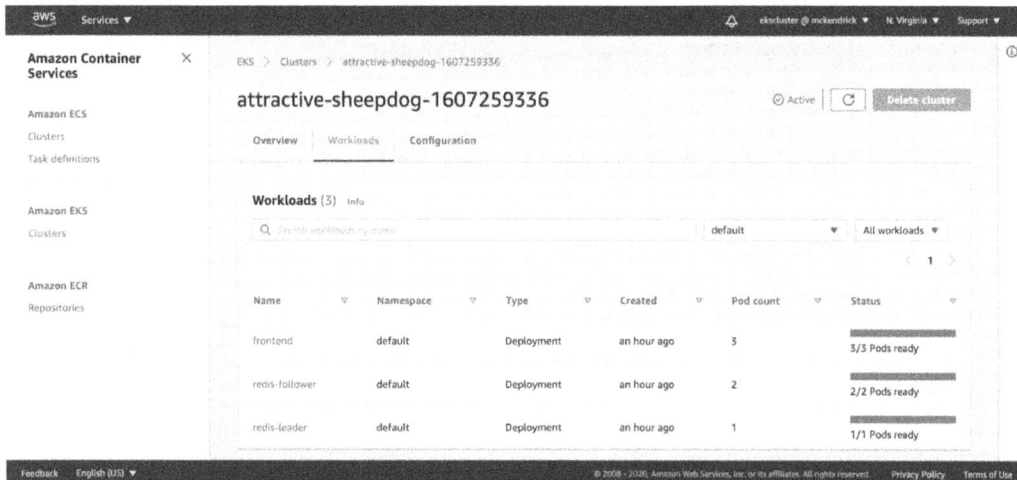

Figure 15.10 – Viewing the workloads in the default namespace

8. Clicking on one of the deployments will give you more information on the deployment—this includes details of the pods, configuration, and so on. However, as you click around, you will find that all you can really do is view information on the services; there are no graphs, logging output, or anything that gives more than a basic overview of our workloads.

9. The final section is **Configuration**; again, this gives basic information on our cluster and not a lot more.

Moving away from the EKS service page and going to the EC2 service section of the AWS console will display the two nodes, as illustrated in the following screenshot:

Figure 15.11 – Viewing the EC2 instances

Here, you will be able to drill down and find out more information on the instance, including its **central processing unit (CPU)**, **random-access memory (RAM)**, and network utilization; however, this is only for the instance itself and not for our Kubernetes workload.

10. Selecting **Load Balancers** from the **Load Balancing** section of the left-hand side menu will show you the elastic load balancer that was launched and configured when we applied the frontend service, as illustrated in the following screenshot:

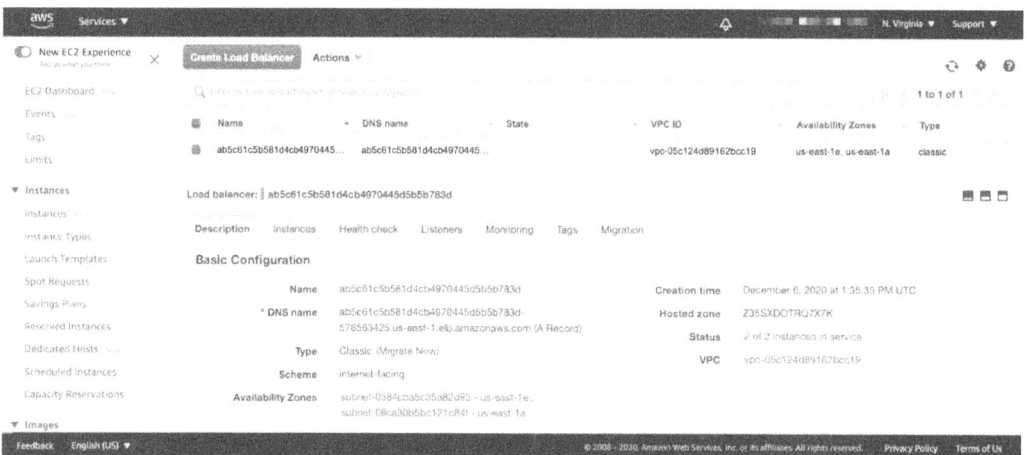

Figure 15.12 – Viewing the elastic load balancer

11. The final AWS service we are using is AWS CloudFormation, so entering CloudFormation in the **Services** menu and clicking on the link will take you to the **CloudFormation** service page.

Here, you will see two stacks, one for the EKS nodes—these are our two EC2 instances—and then one for the EKS cluster, which is our Kubernetes management plane. These stacks are illustrated in the following screenshot:

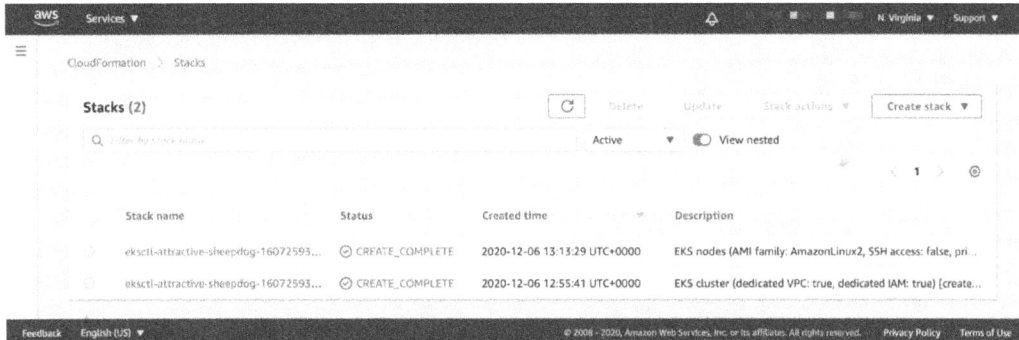

Figure 15.13 – Viewing the AWS CloudFormation stacks

12. Selecting one of the stacks will give you details on what happened when the stack was launched. It will list all of the many resources that were created during the launch of the Amazon EKS cluster using `eksctl`.

13. If you select a template and then view it in the designer, you can even see the template that was generated by `eksctl` with a visual representation. The following screenshot gives a view of the simpler of the two stacks:

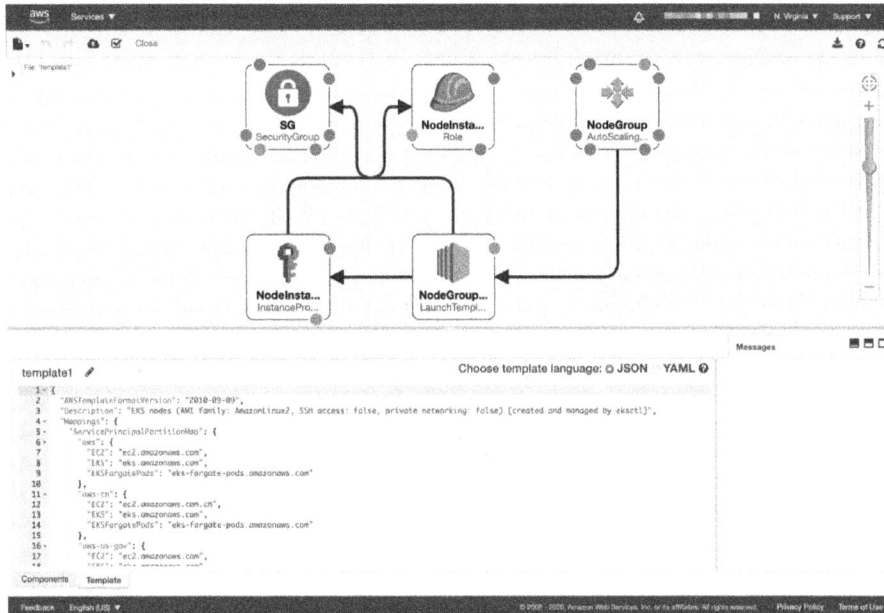

Figure 15.14 – Viewing the AWS CloudFormation template in the designer

That is about all we can see in the AWS console. As we have seen, while Amazon EKS is relatively simple to launch using `eksctl`, its level of integration with the AWS console is somewhat lacking compared to the GKE cluster we launched in the previous chapter.

While we were able to explore and view our workload, we could not interact with too much; also, the feedback on the cluster is tied into the basic monitoring offered by the Amazon EC2 service.

Once you have finished your Amazon EKS cluster, you can delete it.

Deleting your Amazon EKS cluster

You can delete your cluster by running the following command, making sure to replace the cluster name with that of your own:

```
$ eksctl delete cluster --name attractive-sheepdog-1607259336
```

Deleting the cluster does not take as long to run as when we launched it; in fact, it takes around 5 minutes. As before, `eksctl` gives you details on what it is doing as it deletes the resources, starting with the same details on when we launched the cluster, as illustrated in the following code snippet:

```
[i]  eksctl version 0.33.0
[i]  using region us-east-1
```

Then, it provides details on the cluster, as follows:

```
[i]  deleting EKS cluster "attractive-sheepdog-1607259336"
[i]  deleted 0 Fargate profile(s)
```

The first thing that is updated is the local `kubectl` configuration, as we can see here:

```
[✓]  kubeconfig has been updated
```

Then, any resources that have been launched as part of deploying workloads onto our cluster are removed, as follows:

```
[i]  cleaning up AWS load balancers created by Kubernetes
objects of Kind Service or Ingress
```

Then, the two AWS CloudFormation stacks are removed, which in turn removes all of the resources they created and configured, as illustrated in the following code snippet:

```
[i]   2 sequential tasks: { delete nodegroup "ng-6cd00965",
delete cluster control plane "attractive-sheepdog-1607259336"
[async] }
[i]   will delete stack "eksctl-attractive-sheepdog-1607259336-
nodegroup-ng-6cd00965"
[i]   waiting for stack "eksctl-attractive-sheepdog-1607259336-
nodegroup-ng-6cd00965" to get deleted
[i]   will delete stack "eksctl-attractive-sheepdog-1607259336-
cluster"
[√]   all cluster resources were deleted
```

At this point, our cluster has been completely deleted.

> **Important note**
> Please double-check the **EC2**, **EKS**, and **CloudFormation** sections in the AWS console to ensure that all services have been correctly deleted as you will be charged for any orphaned or idle resources that have been left behind. While this is an unlikely scenario, it is best to double-check now rather than receive an unexpected bill at the end of the month.

So, how much would our Amazon EKS cluster have cost us to run for a month?

There are two sets of costs that we need to take into account—the first is for the Amazon EKS cluster itself. It is **US Dollars (USD)** $0.10 per hour for each Amazon EKS cluster you create; however, each Amazon EKS cluster can run multiple node groups, so you shouldn't have to launch more than one per region—so, the cluster itself comes in around $72 per month.

Now, added to the EC2 cluster nodes, which in our case would have cost around $70 each, this means that the total cost to run our cluster for a month would be around $212—and I say *around* because there are then charges for bandwidth and also for the AWS **Elastic Load Balancing (ELB)** service, which will increase the cost of our workload further.

Summary

In this chapter, we discussed the origins of AWS and also Amazon EKS before walking through how to sign up for an account and how to install and configure the two command-line tools required to easily launch an Amazon EKS cluster.

Once our cluster was up and running, we deployed the same workload as when we launched our GKE cluster. We did not have to make any allowances for the workload running on a different cloud provider—it just worked, even deploying a load balancer using the AWS native load balancing service.

We did, however, find that EKS is not as integrated with the AWS console as the Google service we looked at, and also learned that we had to install a second command-line tool to easily launch our cluster due to the complications of trying to do so using the AWS CLI. This would have been around eight steps, and that assumes that the Amazon VPC configuration and IAM roles had been created and deployed.

Personally, this lack of integration and complexity when it comes to launching and maintaining clusters compared to other providers would put me off running my Kubernetes workloads on Amazon EKS.

In the next chapter, we are going to look at launching an **Azure Kubernetes Service** (**AKS**) cluster on Microsoft Azure, which will be the last of the three public providers we will be covering.

Further reading

Here are links to more information on some of the topics and tools we have covered in this chapter:

- AWS: `https://aws.amazon.com/`
- Amazon EKS: `https://aws.amazon.com/eks/`
- The AWS CLI: `https://aws.amazon.com/cli/`
- eksctl: `https://eksctl.io/`
- Weaveworks: `https://www.weave.works/`
- Official documentation: `https://docs.aws.amazon.com/eks/latest/userguide/what-is-eks.html`

16
Kubernetes Clusters on Microsoft Azure with Azure Kubernetes Service

The last of the three public cloud Kubernetes services we are going to look at is **Azure Kubernetes Service** (**AKS**), which is hosted on what most people consider to be one of the "big three" public cloud providers, Microsoft Azure.

By the end of this chapter, you will have configured your local environment with the tools needed to interact with your Microsoft Azure account and launch your AKS cluster. From there we will launch the same workload we launched in the previous two chapters, and then explore the level of integration that your AKS cluster has with the Microsoft Azure portal.

To do this, we will be covering the following topics:

- What are Microsoft Azure and AKS?

- Preparing your local environment

- Launching your AKS cluster

- Deploying a workload and interacting with your cluster

- Deleting your AKS cluster

Technical requirements

To following along with this chapter, you will need a Microsoft Azure account with a valid payment method attached to it.

> **Important note**
>
> Following the instructions in this chapter will incur a cost and it is important that you terminate any resources you launch once you have finished with them to avoid unwanted expenses.

All prices quoted in this chapter are correct at the time of print and we recommend that you review the current costs before you launch any resources.

What are Microsoft Azure and AKS?

Before we start to look at installing the supporting tools, let's quickly discuss the origins of the services we'll be looking at, starting with Microsoft Azure.

Microsoft Azure

In 2008, Microsoft announced a new service called Windows Azure, which it had been working on since 2004. This service was part of a project known internally as Project Red Dog. This project focused on delivering data center services using core Windows components. The five core components that Microsoft announced at their 2008 developer conference were as follows:

- **Microsoft SQL Data Services**: This was a cloud version of the Microsoft SQL Database service running as a **Platform as a Service (PaaS)**, which aimed to remove the complexity of hosting your own SQL services.

- **Microsoft .NET Services**: Another PaaS service that allowed developers to deploy their .NET-based applications into a Microsoft-managed .NET runtime.

- **Microsoft SharePoint**: A **Software as a Service** (**SaaS**) version of the popular intranet product.

- **Microsoft Dynamics**: A **SaaS** version of Microsoft's CRM product.

- **Windows Azure**: An **Infrastructure as a Service** (**IaaS**) offering similar to other cloud providers that allowed users to spin up virtual machines, storage, and the networking services needed to support their compute workloads.

All of the services that Microsoft announced were built on top of the Red Dog operating system, which was a specialized operating system with a cloud layer built in.

In 2014, Windows Azure was renamed Microsoft Azure, which reflected both the name of the underlying operating system powering the cloud services and also the fact that Azure was running a large number of Linux-based workloads – in fact, in 2020, it was revealed that more than 50% of VM cores are running Linux workloads, as well as 60% of the Azure Marketplace images now being Linux-based.

This is largely due to Microsoft embracing both Linux and open source projects such as Kubernetes.

AKS

Microsoft provided its own container-based service called **Azure Container Service** (**ACS**). This allowed you to deploy container workloads backed by three different orchestrators: Docker Swarm, DC/OS, and Kubernetes.

It soon became apparent that Kubernetes was the most popular of the three orchestrators, so ACS was gradually replaced by AKS, which was a CNCF-compliant, purely Kubernetes-based service. The transition took about 2 years, with AKS becoming generally available in 2018 and ACS being retired in early 2020.

The AKS service is closely integrated with Azure Active Directory, Policies, and other key Microsoft Azure services.

Alongside AKS, Microsoft also offer a pure container service called **Azure Container Instances** (**ACI**) as well as the ability to launch container workloads in Azure App Services.

Rather than carrying on discussing the background of the services, I have always found it much easier to roll up your sleeves and get hands-on with a service, so without further delay, let's look at getting the tools installed we will need to launch and manage our AKS cluster.

Preparing your local environment

Before we look at launching our cluster, there are a few tasks we need to complete. First of all, if you don't already have one, you will need to sign up for an Azure account.

If you don't already have an account, then you can head to `https://azure.microsoft.com/free/`. You will be taken to the following page where you can sign up for a free account:

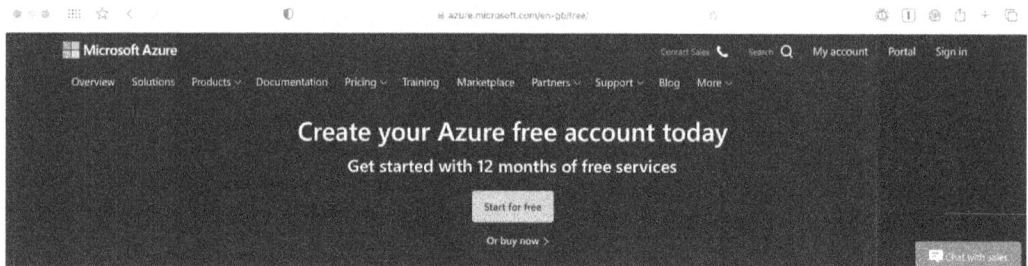

Figure 16.1 – Creating your Azure free account

At the time of writing, your free account includes 12 months of popular services, $200 of credit that can be used to explore and test the different Azure services, and access to over 25 services that will always be free.

Click on the **Start for free** button and follow the onscreen instructions. The sign-up process will take about 15 minutes and you will need to provide valid credit or debit card information to complete the process and gain access to your free account.

Once you have access to your account, the next step is to install the Azure CLI.

The Azure CLI

Microsoft provides a powerful cross-platform command-line tool for managing your Microsoft Azure resources. Installing it on the three main operating systems couldn't be simpler.

Installing on macOS

If you have been following along with the previous two chapters, then you may have already guessed that we will be using Homebrew to install the Azure CLI on macOS. To do this, simply run the following command:

```
$ brew install azure-cli
```

Once installed, run the following command:

```
$ az -version
```

This should return something similar to the following screenshot:

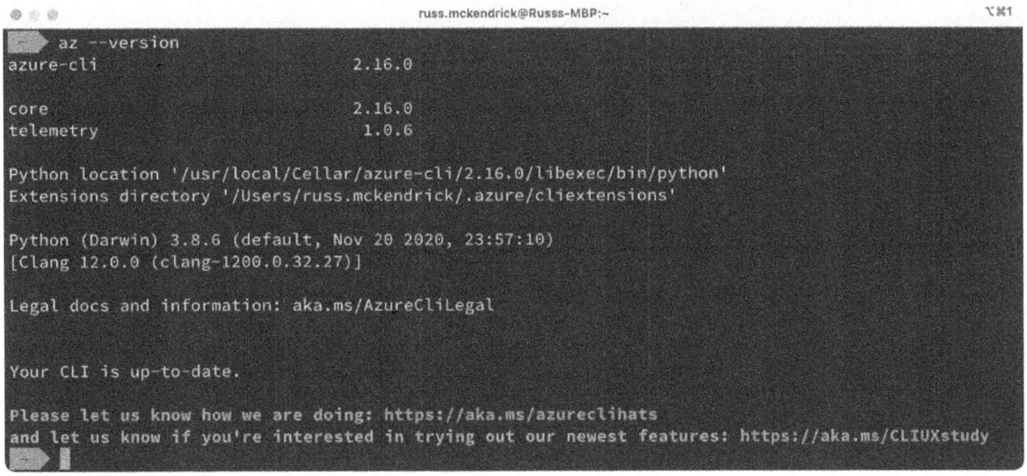

```
       az --version
azure-cli                          2.16.0

core                               2.16.0
telemetry                           1.0.6

Python location '/usr/local/Cellar/azure-cli/2.16.0/libexec/bin/python'
Extensions directory '/Users/russ.mckendrick/.azure/cliextensions'

Python (Darwin) 3.8.6 (default, Nov 20 2020, 23:57:10)
[Clang 12.0.0 (clang-1200.0.32.27)]

Legal docs and information: aka.ms/AzureCliLegal


Your CLI is up-to-date.

Please let us know how we are doing: https://aka.ms/azureclihats
and let us know if you're interested in trying out our newest features: https://aka.ms/CLIUXstudy
```

Figure 16.2 – Checking the Azure CLI version on macOS

Once installed, you can move on to the *Configuring the Azure CLI* section of this chapter.

Installing on Windows

There are few ways you can install the Azure CLI on Windows:

- The first is to download a copy of the installer from `https://aka.ms/installazurecliwindows` and then execute it by double-clicking on it.

- The next option is to use the following PowerShell command, which will download the installer from the preceding URL and install it:

```
$ Invoke-WebRequest -Uri https://aka.ms/
installazurecliwindows -OutFile .\AzureCLI.msi; Start-
Process msiexec.exe -Wait -ArgumentList '/I AzureCLI.msi
/quiet'; rm .\AzureCLI.msi
```

- The third option is to use the Chocolatey package manager and run the following command:

```
$ choco install azure-cli
```

Whichever way you choose to install the package, run the following command once it has been installed to find out the version number:

```
$ az -version
```

You should see something like the following screen:

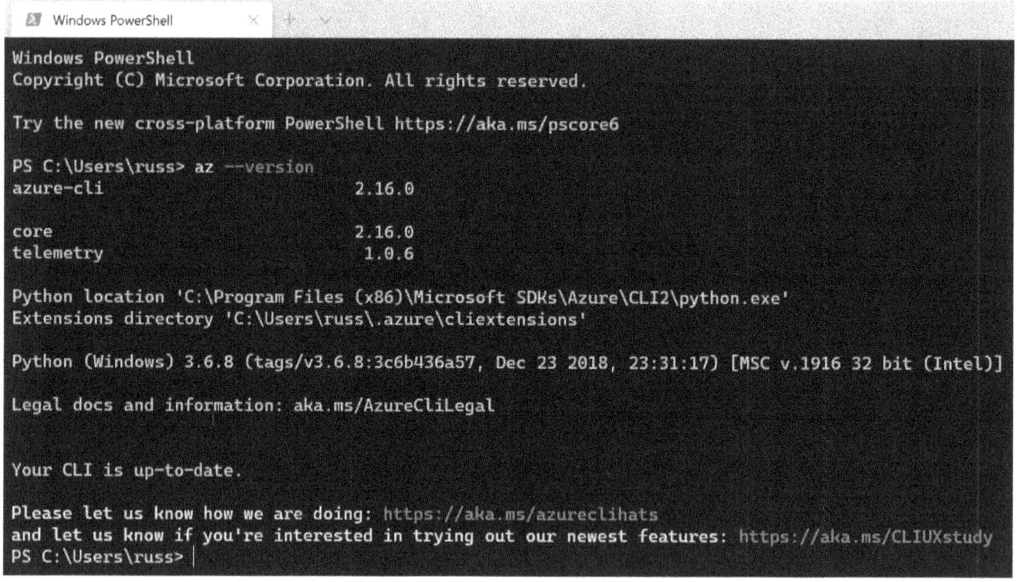

Figure 16.3 – Checking the Azure CLI version on Windows

Once installed, you can move on to the *Configuring the Azure CLI* section.

Installing on Linux

Microsoft provides an installation script that covers the most common Linux distributions. To run the script, use the following command:

```
$ curl -L https://aka.ms/InstallAzureCli | bash
```

Once installed, you will need to restart your session. You can do this by logging out and then back in, or on some distributions by running the following command:

```
$ source ~/.profile
```

Once you have restarted your session, run the following command:

```
$ az -version
```

This will return information on the version of the Azure CLI installed:

```
ubuntu@primary:~$ az --version
azure-cli                  2.16.0

core                       2.16.0
telemetry                  1.0.6

Python location '/home/ubuntu/lib/azure-cli/bin/python'
Extensions directory '/home/ubuntu/.azure/cliextensions'

Python (Linux) 3.8.5 (default, Jul 28 2020, 12:59:40)
[GCC 9.3.0]

Legal docs and information: aka.ms/AzureCliLegal


Your CLI is up-to-date.

Please let us know how we are doing: https://aka.ms/azureclihats
and let us know if you're interested in trying out our newest features: https://aka.ms/CLIUXstudy
ubuntu@primary:~$
```

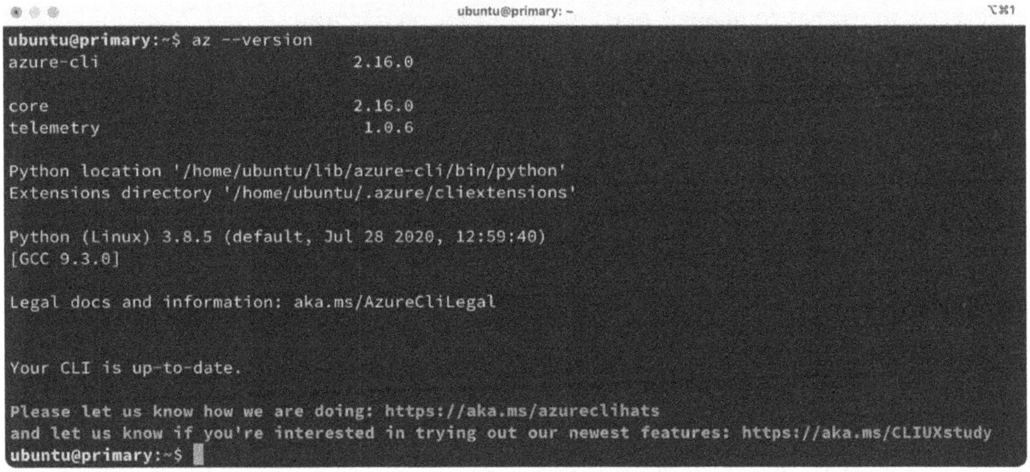

Figure 16.4 – Checking the Azure CLI version on Linux

Now we have the Azure CLI installed, we can configure it.

Configuring the Azure CLI

Configuring the Azure CLI is a really straightforward process; you just need to run the following command:

```
$ az login
```

This will open up your default browser where you will be asked to log in. If you are having problems or running a command-line-only installation of the Azure CLI (on a remote Linux server for example), then running the following command will give you a URL and unique sign-in code to use:

```
$ az login –use-device-code
```

Once logged in, your command-line session should return some information on your Azure account. You can view this again by using the following command:

```
$ az account show
```

If, for any reason, you are not able to install the Azure CLI locally, all is not lost, as there is a web-based terminal with the Azure CLI you can use in the Azure portal. We are going to look at this next.

Accessing Azure Cloud Shell

To get access to Azure Cloud Shell, open `https://portal.azure.com/` and log in with your credentials. Once logged in, you can click on the Cloud Shell icon in the menu bar, which can be found at the top of the page – it is the first icon next to the central search box.

If you have previously opened a Cloud Shell session, then it will launch straight away. If you receive a message saying **You have no storage mounted**, resolve this by making sure that your subscription is selected correctly from the drop-down box:

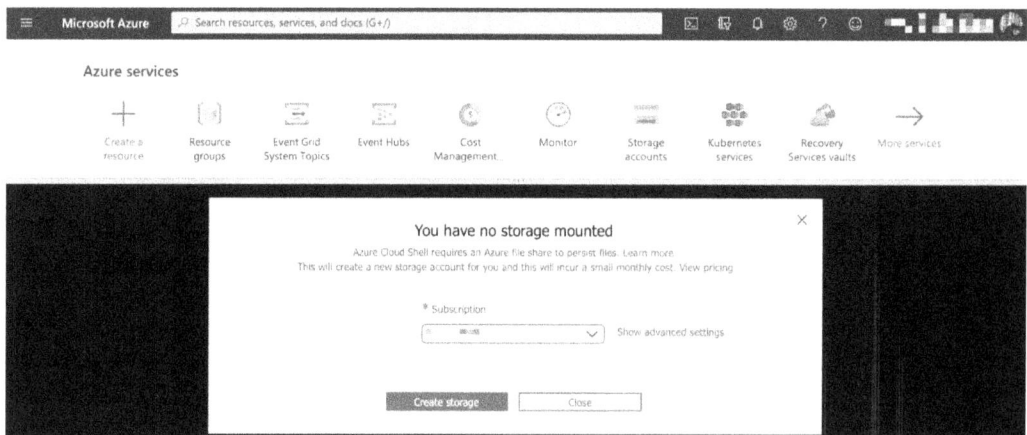

Figure 16.5 – Setting up Cloud Shell

Then, click on the **Create storage** button. After about a minute, your Cloud Shell should open, and you will be presented with a command prompt:

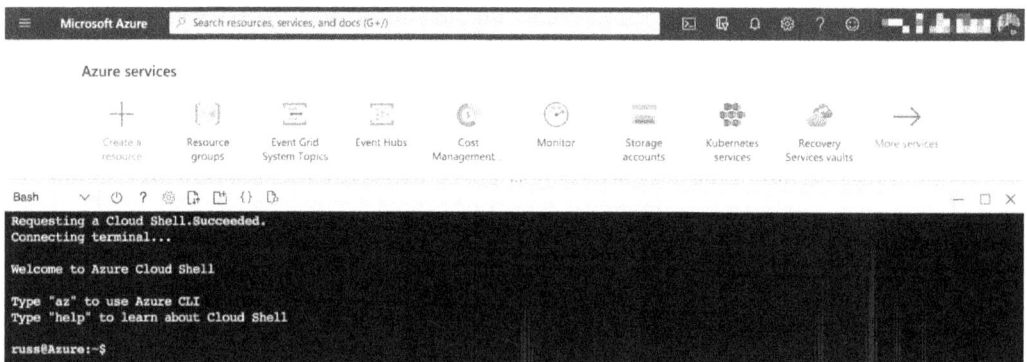

Figure 16.6 – Your Cloud Shell command prompt

Now that you have a command prompt, running the following command, as we did on the local Azure CLI installation, will give you information on the version of the Azure CLI installed:

```
$ az --version
```

You will not need to run the `az login` command as the Azure portal took care of that for you in the background when your Cloud Shell instance launched.

Now that you have access to a configured Azure CLI, in one form or another, we can look at launching our AKS cluster.

Launching your AKS cluster

Launching your AKS cluster requires two commands. The first command creates an Azure resource group:

```
$ az group create --name k8sforbeginners-rg --location eastus
-o table
```

In the preceding command, we are creating a resource group called `k8sforbeginners-rg` in the `eastus` region and setting the output to be formatted as a table rather than JSON, which is the default output type for the Azure CLI.

> **Information**
>
> A resource group is a logical container used to group related Azure resources. Services launched within the resource group can inherit settings such as role-based access controls, locks, and even the location.

Once the resource group has been created, you should see something like the following output:

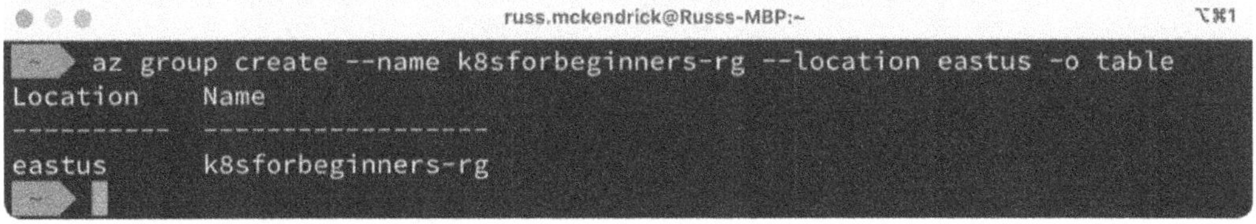

Figure 16.7 – Creating the resource group

Now that we have our resource group, we can launch our AKS cluster by running the following command:

```
$ az aks create --resource-group k8sforbeginners-rg --name
k8sforbeginners-aks --node-count 2 --enable-addons monitoring
--generate-ssh-keys -o yaml
```

Launching the cluster will take about 5 minutes, so while that is running I will work through the options we passed to the preceding `az aks create` command:

- `--resource-group` is the name of the resource group you want to launch your AKS cluster in. The cluster will inherit the resource group's location. In our example, we are using the `k8sforbeginners-rg` resource group we created in the command before last.

- `--name` is the name of the cluster you are launching. We are calling ours `k8sforbeginners-aks`.

- `--node-count` is the number of nodes you want to launch. We are launching 2. At the time of writing, the default instance type for nodes is `Standard_DS2_v2`, meaning that each node will have 2 vCPUs and 7 GB of RAM.

- `--enable-addons` is used to supply a list of add-ons to enable while the cluster is being launched – we are just enabling the `monitoring` add-on.

- `--generate-ssh-keys` will generate SSH public and private key files for the cluster.

- `-o` determines the output. At this time, we are outputting the results returned when we run the command as `yaml` because the output is more readable than both the JSON and table options.

Once your cluster has launched, you should see something like the following output:

```
● ● ●                          russ.mckendrick@Russs-MBP:~                              ⌥⌘1
      az aks create --resource-group k8sforbeginners-rg --name k8sforbeginners-aks --node-count 2 --e
nable-addons monitoring --generate-ssh-keys -o yaml
AAD role propagation done[#########################################]  100.0000%aadProfile: null
addonProfiles:
  KubeDashboard:
    config: null
    enabled: false
    identity: null
  omsagent:
    config:
      logAnalyticsWorkspaceResourceID: /subscriptions/▪▪ ▪▪ ▪ ▪▪ ▪▪▪ ▪ ▪▪▪▪▪ ▪▪ ▪ ▪▪ ▪ ▪ ▪/resourceg
roups/defaultresourcegroup-eus/providers/microsoft.operationalinsights/workspaces/defaultworkspace-c
▪ ▪▪▪ ▪ ▪ ▪ ▪▪▪ ▪▪ ▪▪ ▪▪ ▪ ▪▪▪ ▪ ▪ ▪ ▪-eus
    enabled: true
    identity:
      clientId: 7f1c1d2a-7b42-48dd-ad0e-7079ab0c5640
      objectId: 47722444-0ee8-4257-a7ab-09604305b575
      resourceId: /subscriptions/ ▪▪ ▪ ▪▪▪ ▪ ▪▪ ▪ ▪ ▪ ▪ ▪ ▪ ▪ ▪ ▪ ▪ ▪/resourcegroups/MC_k8sforbeginn
ers-rg_k8sforbeginners-aks_eastus/providers/Microsoft.ManagedIdentity/userAssignedIdentities/omsagen
t-k8sforbeginners-aks
agentPoolProfiles:
- availabilityZones: null
  count: 2
  enableAutoScaling: null
  enableNodePublicIp: false
  maxCount: null
  maxPods: 110
  minCount: null
  mode: System
  name: nodepool1
  nodeImageVersion: AKSUbuntu-1804-2020.12.01
  nodeLabels: {}
  nodeTaints: null
```

Figure 16.8 – Viewing the output of the cluster launch

As you can see, there is a lot of information. We are not going to worry about any of this though, as we will be using the Azure CLI and portal to interact with the cluster.

Now that our cluster has launched, we need to configure our kubectl client so that it can interact with the cluster. To do this, run the following command:

```
$ az aks get-credentials --resource-group k8sforbeginners-rg
--name k8sforbeginners-aks
```

Once run, you should see something like the following:

```
● ● ●                          russ.mckendrick@Russs-MBP:~                              ⌥⌘1
      az aks get-credentials --resource-group k8sforbeginners-rg --name k8sforbeginners-aks
Merged "k8sforbeginners-aks" as current context in /Users/russ.mckendrick/.kube/config
```

Figure 16.9 – Grabbing the credentials and configuring kubectl

This now means that you can start to interact with your cluster, for example, by running the following command:

```
$ kubectl get nodes
```

This will return the nodes within the cluster, as seen in the following screenshot:

```
russ.mckendrick@Russs-MBP:~                                              ⌥⌘1
~    ❯ kubectl get nodes
NAME                                      STATUS    ROLES    AGE     VERSION
aks-nodepool1-10663470-vmss000000         Ready     agent    7m16s   v1.18.10
aks-nodepool1-10663470-vmss000001         Ready     agent    6m45s   v1.18.10
~
```

Figure 16.10 – Viewing the nodes in the cluster

Now that we have our cluster launched and our local `kubectl` client configured with the cluster details and credentials, we can launch our test workload.

Deploying a workload and interacting with your cluster

We are going to be using the same workload we launched in *Chapter 14, Kubernetes Clusters on Google Kubernetes Engine*, and *Chapter 15, Launching a Kubernetes Cluster on Amazon Web Services with Amazon Elastic Kubernetes Service*, so I am not going to go into detail here other than to cover the commands.

Launching the workload

As per the previous chapters, there are three main steps to launching our workload:

1. Create the Redis leader deployment and service:

   ```
   $ kubectl apply -f https://raw.githubusercontent.com/
   GoogleCloudPlatform/kubernetes-engine-samples/master/
   guestbook/redis-leader-deployment.yaml
   $ kubectl apply -f https://raw.githubusercontent.com/
   GoogleCloudPlatform/kubernetes-engine-samples/master/
   guestbook/redis-leader-service.yaml
   ```

2. Create the Redis follower deployment and service:

   ```
   $ kubectl apply -f https://raw.githubusercontent.com/
   GoogleCloudPlatform/kubernetes-engine-samples/master/
   guestbook/redis-follower-deployment.yaml
   $ kubectl apply -f https://raw.githubusercontent.com/
   GoogleCloudPlatform/kubernetes-engine-samples/master/
   guestbook/redis-follower-service.yaml
   ```

3. Create the Guestbook frontend deployment and service:

```
$ kubectl apply -f https://raw.githubusercontent.com/
GoogleCloudPlatform/kubernetes-engine-samples/master/
guestbook/frontend-deployment.yaml
$ kubectl apply -f https://raw.githubusercontent.com/
GoogleCloudPlatform/kubernetes-engine-samples/master/
guestbook/frontend-service.yaml
```

4. The final command to run gets the address of the frontend service:

```
$ kubectl get service frontend
```

This should return something like the following screenshot:

Figure 16.11 – Getting the external IP address of the frontend service

Direct your browser to the IP address and you should be greeted with the Guestbook application. As before, I have added a few test entries:

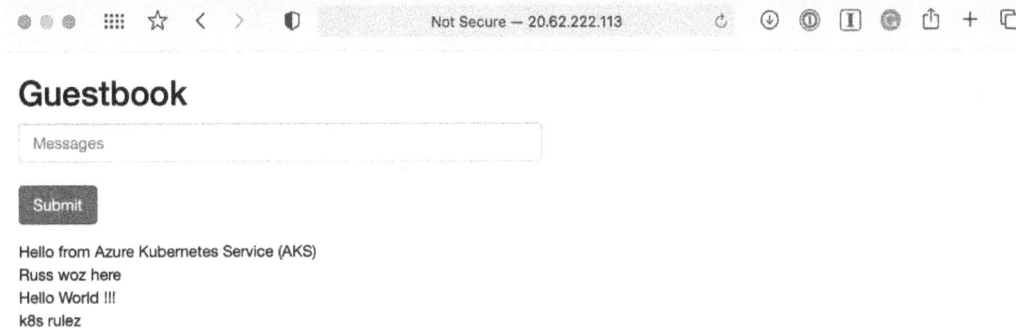

Figure 16.12 – Viewing the Guestbook application

Now that the workload is up and running, we can move to the Azure portal.

Exploring the Azure portal

If you haven't already, log in to the Azure portal found at `https://portal.azure.com/`. Once logged in, enter `kube` into the **Search resources, services, and docs** search box at the very top of the page.

Under the list of services, you will see **Kubernetes services** – click on this service and you will be be presented with a list of Kubernetes services you have launched within your subscription.

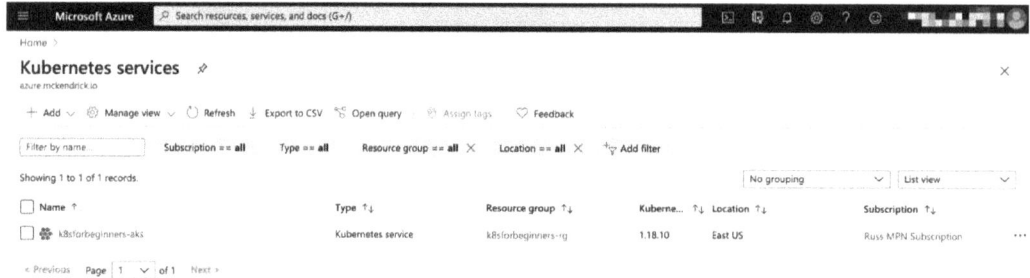

Figure 16.13 – Listing the Kubernetes services

Clicking on **k8sforbeginners-aks** will take you to an overview page. This will be our jumping-off point for viewing information about our workload and cluster.

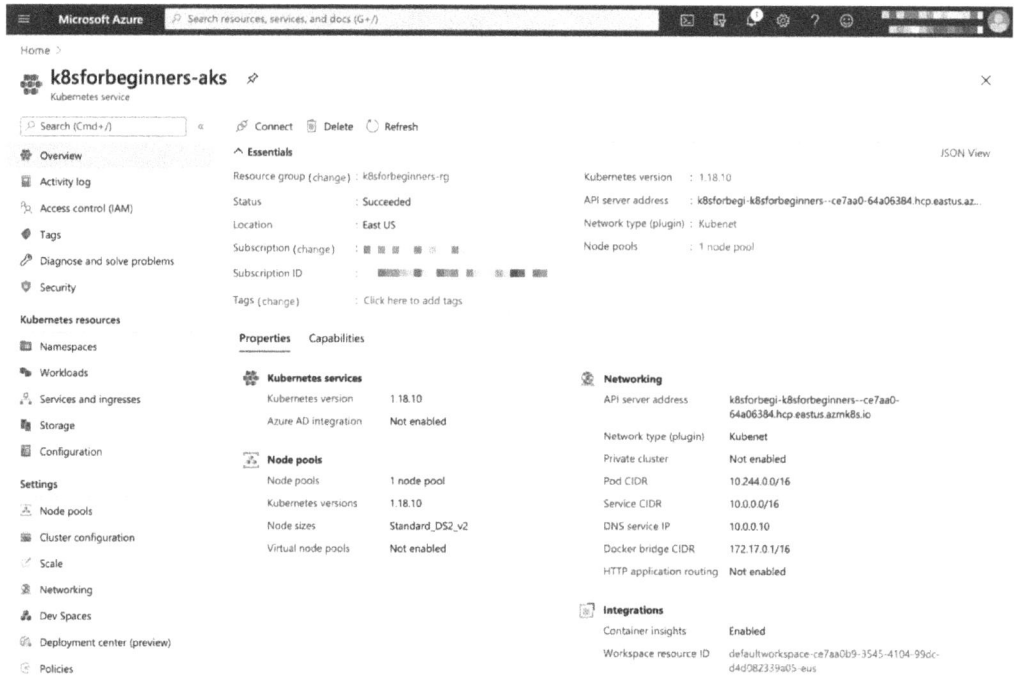

Figure 16.14 – The cluster overview

Under the Kubernetes resources menu on the left, you will see several options. Let's work through them one by one:

- **Namespaces**: Here you will find all the namespaces active within the cluster. As we didn't define a namespace when we launched our workload, our deployments and services will be listed under the `default` namespace.

  As well as the `default` namespace, there are also the ones deployed as part of the cluster, namely, `kube-node-lease`, `kube-public`, and `kube-system`. I would recommend leaving these alone.

  If you were to click on the default namespace, you would be presented with the **Overview** page, where you can edit the **YAML** that defines the namespaces, along with an **Events** log.

- **Workloads | Deployments**: Here, as you may have already guessed, is where you can view information on your workloads. In the following screenshot, I have filtered the list to only show the workloads in the **default** namespace:

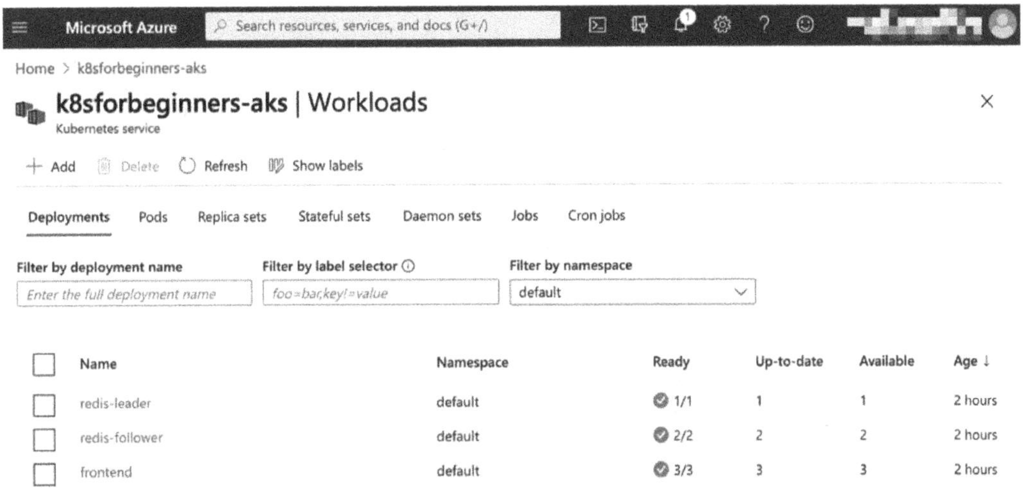

Figure 16.15 – Viewing the deployments

Clicking on one of the deployments will give you a more detailed view of the deployment. For example, selecting the **frontend** deployment shows the following:

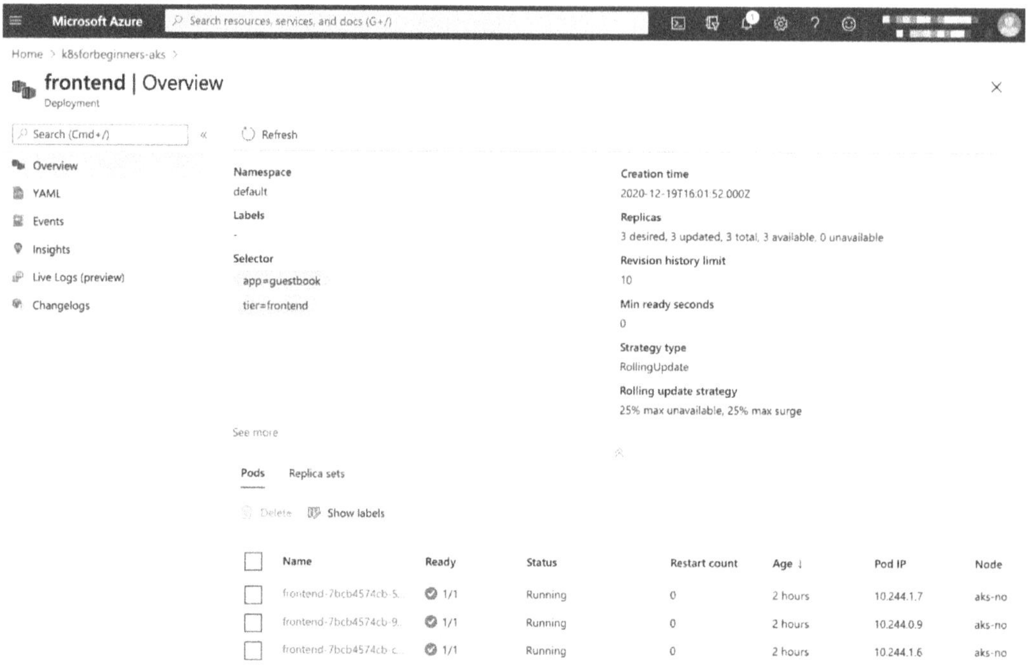

Figure 16.16 – An overview of the frontend deployment

As you can see from the menu on the left, there are a few additions to the options: as well as **YAML** and **Events**, we also have the option to view **Insights**. We will take a closer look at this at the end of this section when we cover Insights in more detail.

The next option is **Live Logs**, which, at the time of writing, is in preview. Here, you can select one of the deployed pods and stream the logs in real time.

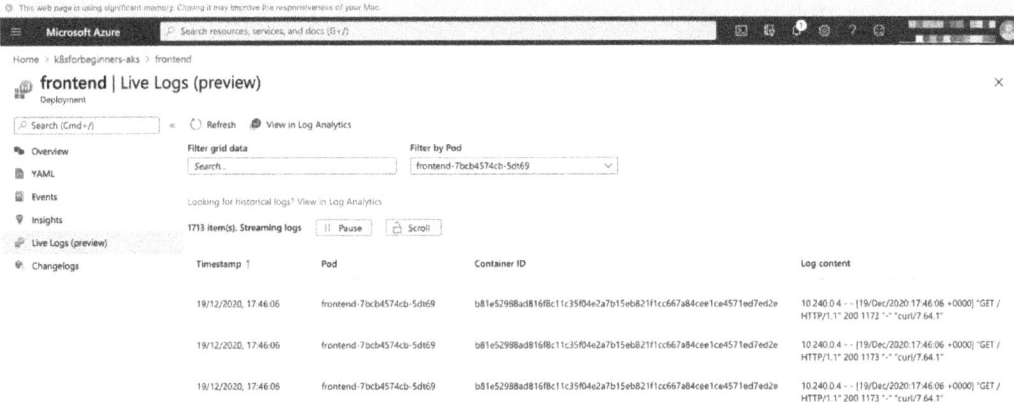

Figure 16.17 – Viewing the pod logs in real time

The **Changelogs** option displays any changelogs available as part of your deployment.

- **Workloads | Pods**: Here, you can view a list of the pods that make up your workload. The IP address along with the node the pod is active on are listed. This is a useful view to get a quick overview of all of your running pods.

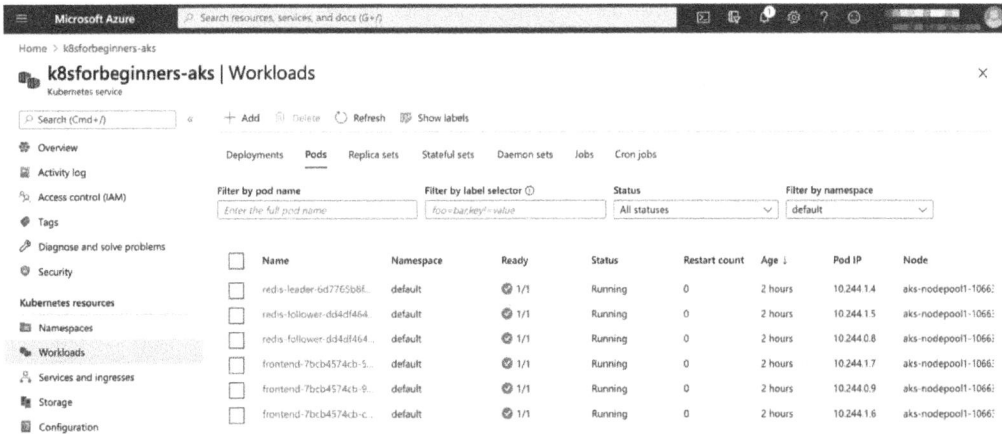

Figure 16.18 – Viewing all of your pods

Clicking on one of the pods will give you an **Overview**, and show you the **YAML** and also any **Events** for the pod.

- **Workloads | Replica sets** is another useful view for getting an at-a-glance view of the replica sets deployed as part of your workload.

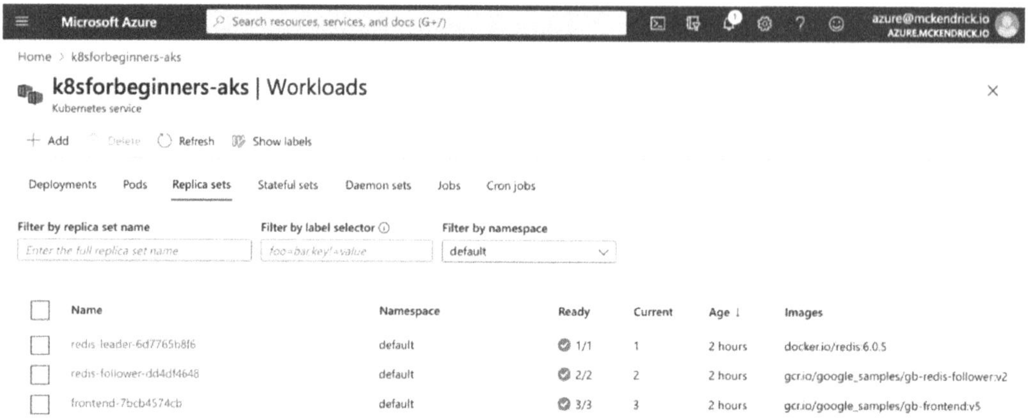

Figure 16.19 – Viewing the replica sets

Clicking on one of the replica sets gives you the now-familiar options of **Overview**, **YAML**, and **Events**.

- **Workloads | Stateful sets**: While we don't have any stateful sets deployed as part of our workload, clicking to view all of the namespaces will show the stateful sets deployed by Microsoft as part of the cluster. These are the Azure OMS Agent (`omsagent`) used for monitoring our cluster – there is one of these deployed per node at all times – and the Proxy service (`kube-proxy`).

- **Workloads | Jobs** and **Cron jobs**: Under these tabs, you will find details of any jobs and cron jobs you have deployed within the cluster.

- **Services and ingresses**: Here, you will be able to find a list of all of the services you have deployed in your cluster. As you can see from the following screenshot, you can get an overview of the **Cluster IP** used for the service along with any **External IP** you have configured:

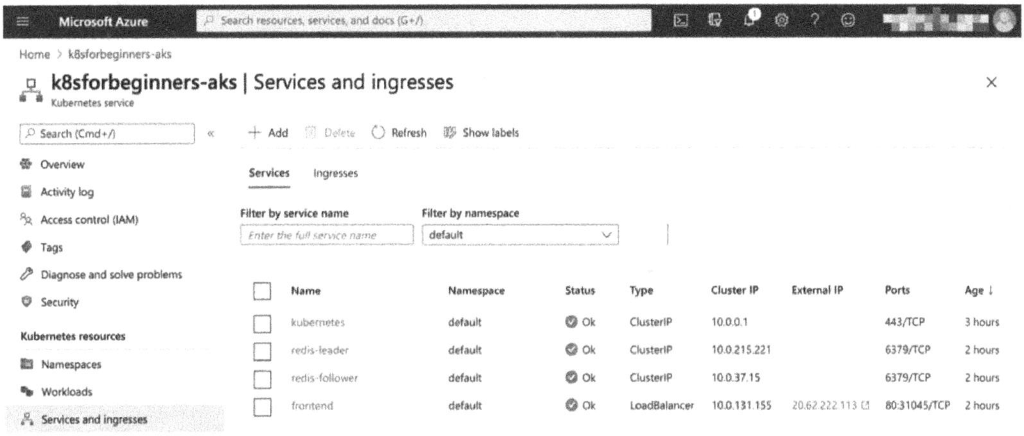

Figure 16.20 – Viewing the services

Clicking on one of the services listed will provide the now-familiar view and allow you to drill deeper into the configuration of the services.

- **Storage**: If we had any persistent storage configured within the cluster, you would be able to view the details here.

- **Configuration**: The final option under the Kubernetes resources menu allows you to view and edit any **Config maps** or **Secrets** you have configured within the cluster. As we don't have any of these configured in our workload, the ones listed are for the cluster itself, so I wouldn't recommend making any changes to the items present.

As you can see, there is a wealth of options available to get information about all of the parts that make up your Kubernetes workload.

The next section in the left-hand menu covers the settings of the cluster itself. Here is a quick overview of what you can find:

- **Node pools**: Here you will find details of the node pool you have, along with the options to **Upgrade** the Kubernetes version running within the pool. This option is only available if you upgrade the version of Kubernetes running on the control plane.

  You can also **Scale** the pool and have the option to **Add a node pool**. In the following screenshot, we can see what the scale option looks like:

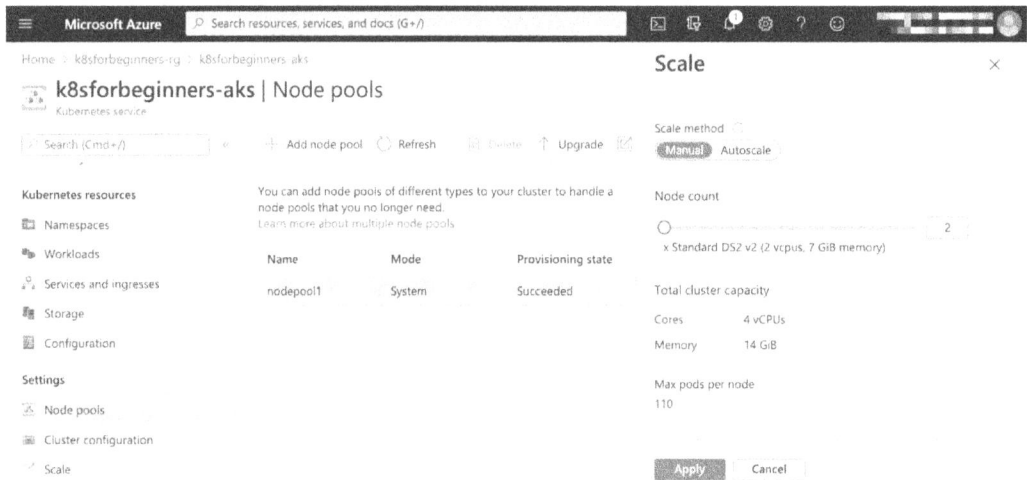

Figure 16.21 – Node pool scale options

- **Cluster configuration**: In the last point, I mentioned that you can only upgrade the Kubernetes version running within your node pools if you upgrade the control plane, and this option is where you do that. The control plane is managed by Microsoft and is separate to your node pools. The Kubernetes control plane offers backward compatibility by up to three releases, so normally you would only be able to upgrade within three releases of the version you are currently running.

- **Scale**: While, at the time of writing, this option is still listed as its own option, the page does mention that the scale functions have been moved to the **Node pool** section and that this page will be removed at some point.

- **Networking**: Here you will find both information and configuration options for the networking of your cluster. This is quite a dense subject and I recommend following the link on the page for more information on the options available when it comes to networking and your AKS cluster.

- **Dev Spaces**: Like the **Scale** option, this has again been deprecated. Microsoft at one point offered an IDE based on Visual Studio Code called Dev Spaces, but this service is due to be retired and Microsoft are now recommending you use Bridge to Kubernetes in Visual Studio and Visual Studio Code.

- **Deployment center**: This service is still in preview, but here you can add and configure an Azure DevOps pipeline to deploy your application straight into your cluster.

- **Policies**: One of the big selling points of Microsoft Azure is the centralized means it offers for managing policies. Enabling this add-on will extend the Azure policy service into your AKS cluster, giving you a way to centrally manage and report on policies.

- **Properties**: This provides a quick overview of your AKS cluster.

- **Locks**: Here, you can add a resource lock that will protect your cluster from accidental deletion or from any configuration changes that may affect the running of your AKS cluster.

The last part of the Azure portal we are going to look at is the **Insights** option found under the **Monitoring** menu in the cluster view. As you may recall, when we deployed our cluster, we enabled the monitoring add-on using the `--enable-addons monitoring` flag.

What this did was enable the OMS Agent stateful set, which we saw when we looked at the **Stateful set** page under the **Workloads** section. The OMS Agent is the tool used by Microsoft to ship data from a resource to the Azure Log Analytics service. Once the data has been shipped to this service, Microsoft then presents this information back to you, most commonly as Insights. Most Azure services have an **Insights** option and the data here can be used by Azure Monitor to create and generate alerts.

Clicking on **Insights** will bring up a page similar to the following:

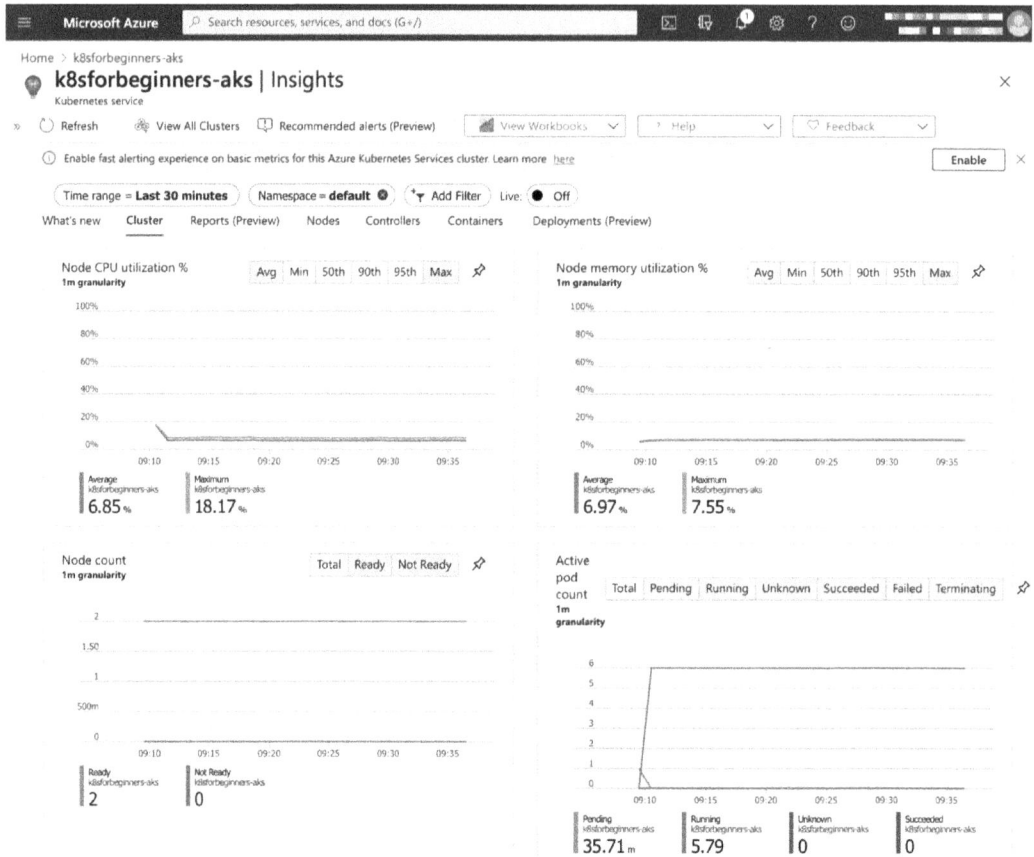

Figure 16.22 – Viewing some of the cluster insights

There are a few options under **Insights** – let's take a look:

- **Cluster**: This is shown in the preceding screenshot and gives you a quick view of the CPU and memory utilization across the whole cluster. It also shows both Node and pod counts.

- **Reports**: At the time of writing, this is in preview. Here you can find pre-written reports around **Node monitoring** (performance), **Resource monitoring** (availability), **Billing**, and also **Networking**. As the service matures, more reports will be added.

- **Nodes**: Here you can get an overview of your Nodes.

- **Controllers**: This is where you will find details on the controllers launched within your cluster – for example, the Replica and Daemon sets.

- **Containers**: Here you can find details of all the containers running on the pods you have deployed.

- **Deployments**: This option shows a list of your deployments.

Now you may think there is a lot of repetition in the preceding sections – and there is a little bit of that; however, if you need to see what is going on within your cluster quickly, you now have a way of getting that information without having to drill through a lot of pages to find it.

I recommend having a look around and clicking on as many options as you can to explore the level of integration of your cluster and the Azure portal. Once you have finished, it is time for us to remove the cluster.

Deleting your AKS cluster

The final thing we are going to look at is how to delete the cluster. Moving back to the Azure CLI, all we need to run to delete the cluster is the following command:

```
$ az aks delete --resource-group k8sforbeginners-rg --name
k8sforbeginners-aks
```

You will be asked if you are sure – answering yes will proceed to delete the cluster.

The process takes about 5 minutes. The preceding command only deletes the cluster itself and not the resource group. To delete the latter, run the following command:

```
$ az group delete --name k8sforbeginners-rg
```

Again, you will be asked if you want to delete the group – just answer yes.

So how much would our cluster cost to run?

Unlike the other two cloud services we looked at in the previous two chapters, there is no charge for cluster management, and all that you need to pay for is the compute resource.

So, in our case, 2 x Standard_DS2_v2 instances in US East would cost around $213 per month. There are other options, such as new generations of instances that could give us a similar-sized cluster for less money. For example, we could launch a different cluster using the following commands:

```
$ az group create --name k8sforbeginners-rg --location eastus
-o table
$ az aks create --resource-group k8sforbeginners-rg --name
k8sforbeginners-aks --node-count 2 --enable-addons monitoring
--generate-ssh-keys --node-vm-size standard_ds3_v2 -o yaml
```

This will give us a two-node cluster with a total of 4 vCPUs and 16 GB of RAM for around $140 per month.

Summary

In this chapter, we looked at how Microsoft Azure came to be, along with a little bit of the history behind some of the container services offered by Microsoft, and how they eventually settled on AKS.

We then signed up for an Azure account and installed and configured the Azure CLI before launching our own AKS cluster. Once launched, we deployed the same workload we deployed to our Google Kubernetes Engine and Amazon Elastic Kubernetes Service clusters.

Once the workload was deployed, we moved onto the Azure portal and looked at the options for gaining insights into our workload and cluster as well as some of the cluster management options.

We then finally deleted the resources we launched, and discussed how much the cluster would cost to run.

Out of the three public cloud services we have looked at over the last three chapters, I personally believe that Microsoft has come up with the most rounded and feature-rich offering.

I would put Google's offering, which we discussed in *Chapter 14, Kubernetes Clusters on Google Kubernetes Engine*, at a close second, as technically it is good, but their pricing makes it difficult to recommend. This leaves Amazon's service, covered in *Chapter 15, Launching a Kubernetes Cluster on Amazon Web Services with Amazon Elastic Kubernetes Service*. AWS comes in as my least recommend service – it simply doesn't feel as polished as the offerings by Microsoft and Google in that it should feel like you are launching a service to complement other services offered by the cloud provider, but instead it really feels like you just happen to be running a Kubernetes cluster in AWS.

Personal opinion aside, the key takeaway from having launched a Kubernetes cluster in three very different public cloud services is that once you have your cluster up and running, and your kubectl client configured to interact with it, the experience is pretty much the same and your workload really doesn't care where it is launched. You also don't have to take into consideration the differences between the three providers – only just a few years ago, this seemed like an unthinkable situation, and truly cloud-agnostic workloads were just a pipe dream.

In the next section of this book, we are going to look at some more advanced Kubernetes usage, starting with Helm charts – all of which can be applied to clusters launched in the public clouds we have discussed in this and the previous two chapters.

Further reading

Here are links to more information on some of the topics and tools covered in this chapter:

- Microsoft Azure: https://azure.microsoft.com/
- AKS: https://azure.microsoft.com/services/kubernetes-service/
- AKS official documentation: https://docs.microsoft.com/en-us/azure/aks/
- Azure DevOps: https://azure.microsoft.com/services/devops/
- Bridge to Kubernetes documentation: https://code.visualstudio.com/docs/containers/bridge-to-kubernetes
- Bridge to Kubernetes (Visual Studio): https://marketplace.visualstudio.com/items?itemName=ms-azuretools.mindaro
- Bridge to Kubernetes (Visual Studio Code): https://marketplace.visualstudio.com/items?itemName=mindaro.mindaro

Section 5: Advanced Kubernetes

There is still a lot to say about Kubernetes. In this section, we are going to discuss some other advanced Kubernetes topics. These topics are not forcibly linked together, but it's nice to discuss all of them. We will discover the Helm package manager, advanced techniques for scheduling, the Ingress resource, and more.

This part of the book comprises the following chapters:

17
Working with Helm Charts

As part of the application development process, you need to think about application *redistribution* and *dependency management*. You may want this as part of your final product offering – the application needs to be easily downloaded and installed by the customers. But you may also want to redistribute the application or a component internally for other teams working on the same product. In the non-container world, you have a variety of **package management systems** (or **package managers**). If you work on **Ubuntu**, you can use the **Advanced Package Tool** (**APT**) to install software. On **Windows**, you can use **Chocolatey** (`https://chocolatey.org/`), and if you are interested in libraries or applications specific to **JavaScript**, you can use **npm**.

Kubernetes is no exception, and it has its own dedicated package managers implemented by the community, the most popular of which is currently **Helm** (`https://helm.sh/`). In general, Helm is currently regarded as the industry standard for internal and external redistribution of your Kubernetes applications, with the largest repository of stable **Helm charts** (currently available as part of **Artifact Hub**: `https://artifacthub.io/`). You can use Helm charts to quickly deploy different components in your Kubernetes cluster. For example, if you need to have a scalable **Elasticsearch** cluster for your logs, no problem – you can deploy a fully functional cluster in a matter of minutes!

By the end of this chapter, you will know the most important principles behind Helm and Helm chart development, and additionally, we will demonstrate how you can install some popular components and solutions using the official stable Helm charts. With this knowledge, you will be able to quickly set up your Kubernetes development environment or even plan for redistribution of your Kubernetes application as a dedicated Helm chart.

In this chapter, we will cover the following topics:

- Understanding Helm
- Releasing software to Kubernetes using Helm
- Helm chart anatomy
- Installing popular solutions using Helm charts

Technical requirements

For this chapter, you will need the following:

- A Kubernetes cluster deployed. We recommend using a *multi-node*, cloud-based Kubernetes cluster. It is possible to run Helm on minikube, but you may encounter limitations as not all functionalities of real clusters are supported.
- The Kubernetes **command-line interface (CLI)** (kubectl) installed on your local machine and configured to manage your Kubernetes cluster.

Basic Kubernetes cluster deployment (local and cloud-based) and kubectl installation have been covered in *Chapter 3, Installing Your First Kubernetes Cluster*.

The following previous chapters can give you an overview of how to deploy a fully functional Kubernetes cluster on different cloud platforms:

- *Chapter 14, Kubernetes Clusters on Google Kubernetes Engine*
- *Chapter 15, Launching a Kubernetes Cluster on Amazon Web Services with the Amazon Elastic Kubernetes Service*
- *Chapter 16, Kubernetes Cluster on Microsoft Azure with Azure Kubernetes Service*

You can download the latest code samples for this chapter from the official GitHub repository: https://github.com/PacktPublishing/The-Kubernetes-Bible/tree/master/Chapter17.

Understanding Helm

The simplest way to distribute your Kubernetes application to others so that they can deploy it on their cluster is by sharing the Kubernetes objects' YAML manifest files, for example using a public source repository such as `https://github.com/`.
This approach is often used as a basic showcase of how you can run a given application as a container on Kubernetes. However, sharing raw YAML manifests has quite a few disadvantages:

- All values in YAML templates are *hardcoded*. This means that if you want to change the number of replicas of a Service object or a value stored in the ConfigMap object, you need to go through the manifest files, find the values you want to configure, and then edit them. Similarly, if you want to deploy the manifests to a different namespace in the cluster than the creators intended, you need to edit all YAML files. On top of that, you do not really know which values in the YAML templates are intended to be configurable by the creator unless they document this.

- The Deployment process may be different for each application. There is no *standardization* in which YAML manifests the creator would provide and which components you need to deploy yourself.

- No *dependency management*. For example, if your application requires a **MySQL** server running as a StatefulSet in the cluster, you either need to deploy it yourself or rely on the creator of the application to provide YAML manifests for the MySQL server.

This is a bit similar to what you see with desktop applications in the Windows ecosystem if you do not use **Windows Store** or a package manager such as Chocolatey. Some applications that you download will come with an installer as a `.exe` file, some as a `.msi`, and others will be just `.zip` files that you need to extract and configure yourself.

In Kubernetes, you can use Helm, which is the most popular package manager for Kubernetes applications and services. If you are familiar with popular package managers such as APT, yum, npm, or Chocolatey, you will find many concepts in Helm similar and easy to understand. The following are the three most important concepts in Helm:

- A **chart** is a Helm *package*. This is what you *install* when you use the Helm **CLI**. A Helm chart contains all Kubernetes YAML manifests required to deploy the application on the cluster. Please note that these YAML manifests may be *parametrized*, so that you can easily inject configuration values provided by the user that installs the chart.

- A **repository** is a storage location for Helm charts that is used for collecting and sharing charts. They can be public or private – there are multiple public repositories that are available, which you can browse on Artifact Hub (`https://artifacthub.io/packages/search?page=1&kind=0`). Up until recently, there was one main, official Helm charts repository (`https://github.com/helm/charts`). This gathered all stable and incubating charts and was maintained by **Google**. This repository is now deprecated and using Artifact Hub is the recommended way to discover charts and repositories. Private repositories are usually used for distributing components running on Kubernetes between teams working on the same product.

- A **release** is an *instance* of a Helm chart that was installed and is running in a Kubernetes cluster. This is what you manage with the Helm CLI, for example by upgrading or uninstalling it. You can install one chart many times on the same cluster and have multiple releases of it that are identified uniquely by release names.

> **Tip**
> In short, Helm charts contain parametrizable YAML manifests that you store in a Helm repository for distribution. When you install a Helm chart, a Helm release is created in your cluster that you can further manage.

Let's quickly summarize the most common use cases for Helm:

- Deploying popular software to your Kubernetes cluster. This makes *development* on Kubernetes much easier – you can deploy third-party components to the cluster in a matter of seconds. The same approach may be used in *production* clusters. You do not need to rely on writing your own YAML manifest for such third-party components.

- Helm charts provide *dependency management* capabilities. If chart A requires chart B to be installed first with specific parameters, Helm supports syntax for this out of the box.

- Sharing your own applications as Helm charts. This can include packaging a product for consumption by the end users or using Helm as an internal package and dependency manager for microservices in your product.

- Ensuring that the applications receive proper upgrades. Helm has its own process for upgrading Helm releases.

- Configuring software deployments for your needs. Helm charts are basically generic YAML templates for Kubernetes object manifests that can be *parametrized*. Helm uses **Go** templates (`https://godoc.org/text/template`) for parametrization. If you are familiar with Go then you will be at home; if not, you will find it pretty similar to other templating systems, such as **Mustache** (`https://mustache.github.io/`).

Currently, Helm is distributed as a binary client (library) that has a CLI similar to `kubectl`. All operations that you perform using Helm do not require any additional components to be installed on the Kubernetes cluster. Please note that this has changed with the release of version `3.0.0` of Helm. Previously, the architecture of Helm was different, and it required a special, dedicated service running on Kubernetes named **Tiller**. This was causing various problems such as security around **role-based access control (RBAC)** and elevated-privilege Pods running inside the cluster. You can read more about the differences between the latest major version of Helm and previous ones in the official FAQ: `https://helm.sh/docs/faq/#changes-since-helm-2`. This is useful to know if you find any online guides that still mention Tiller – they are most likely intended for older versions of Helm.

Now, we are going to install Helm and deploy a simple Helm chart from Artifact Hub to verify that it works correctly on your cluster.

Releasing software to Kubernetes using Helm

In this section, you will learn how to install Helm and how to test the installation by deploying an example Helm chart. Helm is provided as binary releases (`https://github.com/helm/helm/releases`) available for multiple platforms. You can use them or refer to the following guides for installation using a package manager on your desired operating system.

Installing Helm on Ubuntu

To install Helm on Ubuntu, you need to first add the official APT repository using the following commands:

```
$ curl https://baltocdn.com/helm/signing.asc | sudo apt-key add
-
$ sudo apt-get install apt-transport-https –yes
$ echo "deb https://baltocdn.com/helm/stable/debian/ all main"
| sudo tee /etc/apt/sources.list.d/helm-stable-debian.list
$ sudo apt-get update
```

After that, the installation is straightforward using APT – use the following command:

```
$ sudo apt-get install helm
```

Verify that the installation was successful by trying to get the Helm version from the command line:

```
$ helm version
version.BuildInfo{Version:"v3.5.3",
GitCommit:"041ce5a2c17a58be0fcd5f5e16fb3e7e95fea622",
GitTreeState:"dirty", GoVersion:"go1.15.8"}
```

Once installed, you can move on to *Deploying an example chart* in this section.

Installing Helm on Windows

To install Helm on Windows, the easiest way is to use the Chocolatey package manager. If you have not used Chocolatey before, you can find more details and the installation guide in the official documentation: https://chocolatey.org/install.

Execute the following command to install Helm:

```
$ choco install kubernetes-helm
```

Verify that the installation was successful by trying to get the Helm version from the command line:

```
$ helm version
version.BuildInfo{Version:"v3.5.3",
GitCommit:"041ce5a2c17a58be0fcd5f5e16fb3e7e95fea622",
GitTreeState:"dirty", GoVersion:"go1.15.8"}
```

Once installed, you can move on to *Deploying an example chart* in this section.

Installing Helm on macOS

To install Helm on macOS, you can use the standard **Homebrew** package manager. Use the following command to install the Helm formula:

```
$ brew install helm
```

Verify that the installation was successful by trying to get the Helm version from the command line:

```
$ helm version
version.BuildInfo{Version:"v3.5.3",
GitCommit:"041ce5a2c17a58be0fcd5f5e16fb3e7e95fea622",
GitTreeState:"dirty", GoVersion:"go1.15.8"}
```

Once installed, we can deploy an example chart to verify that Helm works properly on your Kubernetes cluster.

Deploying an example chart

By default, Helm comes with no repositories configured. One possibility, which is no longer recommended, is to add the `stable` repository so that you can browse the most popular Helm charts:

```
$ helm repo add stable https://charts.helm.sh/stable
```

Please note that most charts are now in a process of deprecation as they are moved to different Helm repositories where they will be maintained by the original creators. You can see this if you try to search for available Helm charts using the `helm search repo` command:

```
$ $ helm search repo stable
NAME                                        CHART VERSION    APP
VERSION                  DESCRIPTION
stable/acs-engine-autoscaler                2.2.2            2.1.1
DEPRECATED Scales worker nodes within agent pools
stable/aerospike                            0.3.5
v4.5.0.5                 DEPRECATED A Helm chart for Aerospike
in Kubern...
stable/airflow                              7.13.3           1.10.12
DEPRECATED - please use: https://github.com/air...
...
```

Instead, the new recommended way is to use the `helm search hub` command, which allows you to browse the Artifact Hub directly from the CLI:

```
$ helm search hub
URL                                                     CHART
VERSION    APP VERSION    DESCRIPTION
https://artifacthub.io/packages/helm/gabibbo97/...  0.1.0
fedora-32     389 Directory Server
https://artifacthub.io/packages/helm/aad-pod-id...  3.0.3
1.7.4         Deploy components for aad-pod-identity
https://artifacthub.io/packages/helm/arhatdev/a...  0.1.0
latest        Network Manager Living at Edge
...
```

Now, let's try searching for one of the most popular Helm charts that we can use for testing our installation – we would like to deploy **WordPress** on our Kubernetes cluster. First, let's check what the available charts are for WordPress on Artifact Hub:

```
$ helm search hub wordpress
URL                                                       CHART
VERSION     APP VERSION      DESCRIPTION
https://artifacthub.io/packages/helm/groundhog2...        0.3.0
5.7.0-apache    A Helm chart for Wordpress on Kubernetes
https://artifacthub.io/packages/helm/bitnami/wo...        10.7.1
5.7.0           Web publishing platform for building blogs and
...
https://artifacthub.io/packages/helm/seccurecod...        2.5.2
4.0             Insecure & Outdated Wordpress Instance: Never
e...
...
```

Similarly, you can directly use the Artifact Hub web UI (`https://artifacthub.io/packages/search?page=1&ts_query_web=wordpress&kind=0`) and search for WordPress Helm charts:

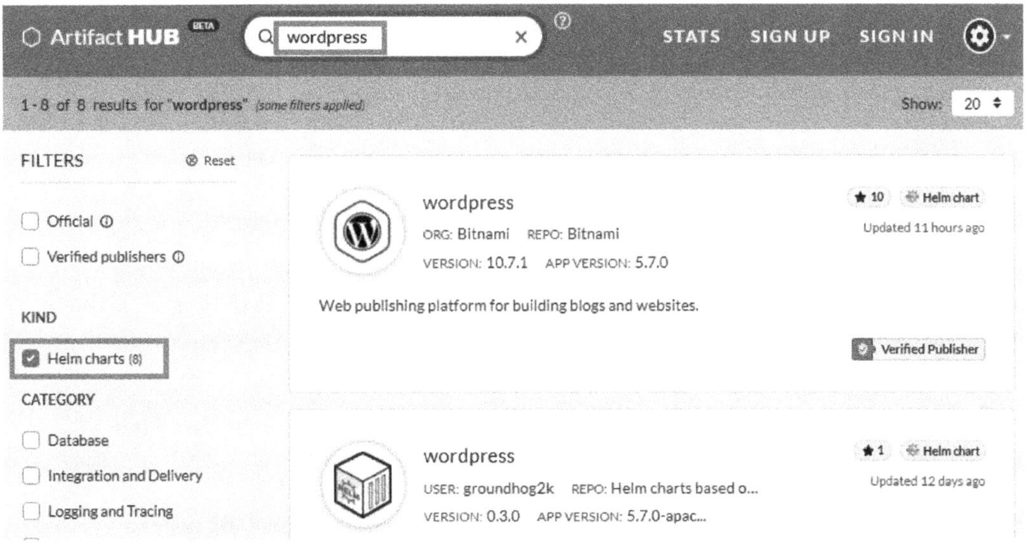

Figure 17.1 – Artifact Hub search results for WordPress Helm charts

We recommend using the Helm chart provided and maintained by **Bitnami** (https://bitnami.com/stacks/helm), a company specializing in distributing open source software on various platforms, such as Kubernetes. They were also maintainers of multiple charts in the deprecated `stable` repository. If you navigate to the search result for WordPress charts by Bitnami (https://artifacthub.io/packages/helm/bitnami/wordpress) you will see the following:

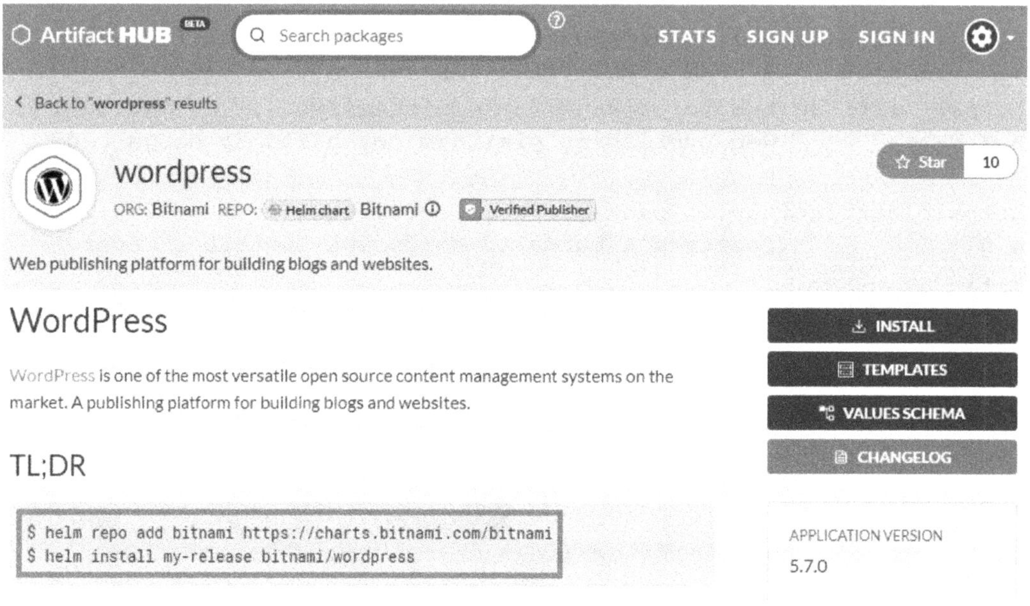

Figure 17.2 – Bitnami WordPress Helm chart on Artifact Hub with install instructions

The page gives you detailed information about how you can add the `bitnami` repository and how to install the Helm chart for WordPress. Additionally, you will find a lot of details about available configuration, known limitations, and troubleshooting. You can also navigate to the home page of each of the charts in order to see the YAML templates that make up the chart (https://github.com/bitnami/charts/tree/master/bitnami/wordpress).

We can now do the installation by following the instructions on the web page. First, add the `bitnami` repository to your Helm installation:

```
$ helm repo add bitnami https://charts.bitnami.com/bitnami
"bitnami" has been added to your repositories
```

With the repository present, we can install the `bitnami/wordpress` Helm chart as a `wordpress-test-release` Helm release with the default configuration:

```
$ helm install wordpress-test-release bitnami/wordpress
...
** Please be patient while the chart is being deployed **
Your WordPress site can be accessed through the following DNS
name from within your cluster:
    wordpress-test-release.default.svc.cluster.local (port 80)
To access your WordPress site from outside the cluster follow
the steps below:
1. Get the WordPress URL by running these commands:
  NOTE: It may take a few minutes for the LoadBalancer IP to be
available.
         Watch the status with: 'kubectl get svc --namespace
default -w wordpress-test-release'
    export SERVICE_IP=$(kubectl get svc --namespace default
wordpress-test-release --template "{{ range (index .status.
loadBalancer.ingress 0) }}{{.}}{{ end }}")
    echo "WordPress URL: http://$SERVICE_IP/"
    echo "WordPress Admin URL: http://$SERVICE_IP/admin"
2. Open a browser and access WordPress using the obtained URL.
3. Login with the following credentials below to see your blog:
  echo Username: user
  echo Password: $(kubectl get secret --namespace default
wordpress-test-release -o jsonpath="{.data.wordpress-password}"
 | base64 --decode)
```

After a while, you will be provided with all the information required to monitor your Deployment of WordPress and how to log in to the WordPress instance. This is the beauty of Helm – you have executed a single `helm install` command and you are presented with a detailed guide of how to use the deployed component on *your* cluster. And meanwhile, the WordPress instance deploys without any intervention from you!

> **Tip**
> It is a good practice to first inspect what will be the Kubernetes objects'
> YAML manifests that were produced by Helm. You can do that by running
> the `helm install` command with additional flags: `helm install`
> `wordpress-test-release bitnami/wordpress --dry-run`
> `--debug`. The output will contain joint output of YAML manifests, and they
> will not be applied to the cluster.

Let's now follow the instructions from the Helm chart installation output:

1. Wait for the `wordpress-test-release` Service object (of the `LoadBalancer`
   type) to acquire external IP:

   ```
   $ kubectl get svc --namespace default -w wordpress-test-
   release
   NAME                      TYPE            CLUSTER-IP
   EXTERNAL-IP     PORT(S)                      AGE
   wordpress-test-release    LoadBalancer    10.0.62.91
   52.226.146.38    80:32049/TCP,443:30351/TCP    9m18s
   ```

   In our case, this will be `52.226.146.3`. You may additionally verify that Pods are
   ready by checking the output of the `kubectl get pods` command.

2. You can also use further commands in the instruction to get the information in an
   automated way:

   ```
   $ export SERVICE_IP=$(kubectl get svc --namespace default
   wordpress-test-release --template "{{ range (index
   .status.loadBalancer.ingress 0) }}{{.}}{{ end }}")
   $ echo "WordPress URL: http://$SERVICE_IP/"
   WordPress URL: http://52.226.146.38/
   $ echo "WordPress Admin URL: http://$SERVICE_IP/admin"
   WordPress Admin URL: http://52.226.146.38/admin
   ```

3. Now open your web browser and navigate to the WordPress admin URL:

Figure 17.3 – WordPress chart deployed on Kubernetes – admin login page

4. Use the following commands to obtain the credentials that are stored in a dedicated `wordpress-test-release` Secret object deployed as part of the chart:

```
$ echo Username: user
Username: user
$ echo Password: $(kubectl get secret --namespace default
wordpress-test-release -o jsonpath="{.data.wordpress-
password}" | base64 --decode)
Password: <hidden>
```

5. Use the credentials to log in as the WordPress admin:

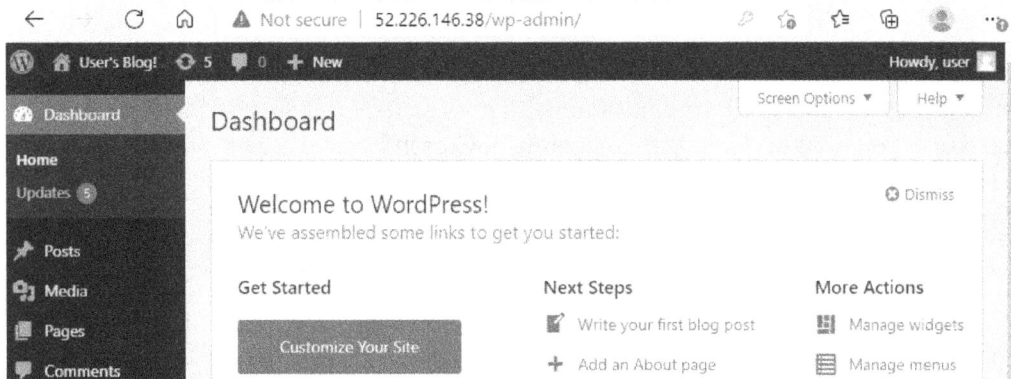

Figure 17.4 – WordPress chart deployed on Kubernetes – admin dashboard

You can enjoy your WordPress now, congratulations! If you are interested, you can inspect the Pods, Services, Deployments, and StatefulSets that were deployed as part of this Helm chart. This will give you a lot of insight into what the components of the chart are and how they interact.

> **Tip**
>
> The Helm CLI offers *autocompletion* for various shells. You can run the `helm completion` command to learn more.

If you want to get information about all Helm releases that are deployed in your Kubernetes cluster, use the following command:

```
$ helm list
NAME                          NAMESPACE        REVISION        UPDA
TED                                            STATUS          CHART
APP VERSION
wordpress-test-release  default
1                      2021-03-23 23:47:18.6285679
+0100 CET    deployed             wordpress-10.7.1
5.7.0
```

When you are ready, you can clean up the cluster by uninstalling the Helm release using the following command:

```
$ helm uninstall wordpress-test-release
release "wordpress-test-release" uninstalled
```

This will delete all Kubernetes objects that the release has created. Please note though that PersistentVolumes and PersistentVolumeClaims created by Helm chart will not be cleaned up – you need to clean them up manually. Currently there is a proposal for Helm v4 to include an option of deleting PersistentVolumeClaims while executing the `helm uninstall` command.

We will now take a closer look at how Helm charts are structured internally.

Helm chart anatomy

As an example, we will take the WordPress Helm chart by Bitnami (`https://github.com/bitnami/charts/tree/master/bitnami/wordpress`) that we have just used to perform a test Deployment in the cluster. Helm charts are simply directories with a specific structure (convention) that can live either in your local filesystem or in a Git repository. The directory name is at the same time the name of the chart, in this case, `wordpress`. The structure of files in the chart directory is as follows:

- `Chart.yaml`: YAML file that contains metadata about the chart such as version, keywords, and references to dependent charts that must be installed.

- `LICENSE`: Optional, plain-text file with license information.

- `README.md`: End user README file that will be visible on Artifact Hub.

- `values.yaml`: The default configuration values for the chart that will be used as YAML template parameters. These values can be overridden by the Helm user, either one by one in the CLI or as a separate YAML file with values.

- `values.schema.json`: Optionally, you can provide a JSON schema that `values.yaml` must follow.

- `charts/`: Optional directory with additional, dependent charts.

- `crds/`: Optional Kubernetes custom resource definitions.

- `templates/`: The most important directory that contains all YAML *templates* for generating Kubernetes YAML manifest files. The YAML templates will be combined with provided *values*. The resulting YAML manifest files will be applied to the cluster.

- `templates/NOTES.txt`: Optional file with short usage notes.

For example, if you inspect `Chart.yaml` in the WordPress Helm chart, you can see that it depends on the **MariaDB** chart by Bitnami, if an appropriate value of `mariadb.enabled` is set to `true` in the provided values:

```
...
appVersion: 5.7.0
dependencies:
  - condition: mariadb.enabled
    name: mariadb
    repository: https://charts.bitnami.com/bitnami
    version: 9.x.x
...
```

Now, if you take a look at the values.yaml file with the default values, which is quite verbose, you can see that by default MariaDB is enabled:

```
...
##
## MariaDB chart configuration
## ref: https://github.com/bitnami/charts/blob/master/bitnami/
mariadb/values.yaml
##
mariadb:
  ## Whether to deploy a mariadb server to satisfy the
applications database requirements. To use an external database
set this to false and configure the externalDatabase parameters
  ##
  enabled: true
...
```

And lastly, let's check what one of the YAML templates looks like – open the deployment.yaml file (https://github.com/bitnami/charts/blob/master/bitnami/wordpress/templates/deployment.yaml), which is a template for the Kubernetes Deployment object for Pods with WordPress containers. For example, you can see how the number of replicas is referenced from the provided values:

```
kind: Deployment
...
spec:
...
  replicas: {{ .Values.replicaCount }}
...
```

This will be replaced by the replicaCount value (for which the default value of 1 you can find in the values.yaml file). The details about how to use Go templates can be found at https://pkg.go.dev/text/template. You can also learn by example by analyzing existing Helm charts – most of them are using similar patterns for processing provided values.

> **Tip**
> The detailed documentation on Helm chart structure can be found at
> https://helm.sh/docs/topics/charts/.

In most cases, you will need to override some of the default values from the
`values.yaml` file during the installation of a chart. Let's say we want to set a
`wordpressBlogName` value to `Kubernetes for Beginners Blog`. To do that,
you have two options:

- Use Helm CLI parameters, where the syntax is as follows:

```
$ helm install wordpress-test-release --set
wordpressBlogName="Kubernetes for Beginners Blog"
bitnami/wordpress
```

  You can use the `--set` parameter multiple times to set other values.

- Create your own `values.yaml` file that has exactly the same structure as the
  original one. You need to specify only the values that you override, which in this
  case would be as follows:

```
wordpressBlogName: Kubernetes for Beginners Blog
```

  Then you can use it as a parameter to the `helm install` command:

```
$ helm install wordpress-test-release -f values.yaml
bitnami/wordpress
```

  This approach is much more scalable if you have multiple values to override.
  Additionally, you can do versioning of these files in your own repository.

Now, when you know the most important details about the Helm chart structure, we can
deploy some selected, popular solutions on Kubernetes using Helm charts.

Installing popular solutions using Helm charts

In this section, we will demonstrate how you can quickly install the following software for
your Kubernetes cluster. They can be useful in your development scenarios or as building
blocks of your cloud-native applications:

- **Kubernetes Dashboard**: Provides an overview of the cluster with a nice web UI.

- **Elasticsearch with Kibana**: Popular full-text search engine, commonly used in log
  analytics. Kibana is used as a visualizations UI.

- **Prometheus with Grafana**: Popular monitoring system with a time series database.
  Grafana is used as a visualizations UI.

Let's begin by installing Kubernetes Dashboard.

Kubernetes Dashboard

Kubernetes Dashboard is the official web UI for providing an overview of your cluster. The Helm chart for this component is officially maintained by the Kubernetes community (`https://artifacthub.io/packages/helm/k8s-dashboard/kubernetes-dashboard`). We are going to install it with the default parameters, as there is no need for any customizations at this point. First, add the `kubernetes-dashboard` repository to Helm:

```
$ helm repo add kubernetes-dashboard https://kubernetes.github.io/dashboard/
"kubernetes-dashboard" has been added to your repositories
```

Now, we can install the Helm chart as a `kubernetes-dashboard-test` release in the cluster:

```
$ helm install kubernetes-dashboard-test kubernetes-dashboard/kubernetes-dashboard
...
Get the Kubernetes Dashboard URL by running:
  export POD_NAME=$(kubectl get pods -n default -l "app.kubernetes.io/name=kubernetes-dashboard,app.kubernetes.io/instance=kubernetes-dashboard-test" -o jsonpath="{.items[0].metadata.name}")
  echo https://127.0.0.1:8443/
  kubectl -n default port-forward $POD_NAME 8443:8443
```

Use the following command to verify that the Pod with Kubernetes Dashboard is running and ready:

```
$ kubectl get pods -n default -l "app.kubernetes.io/name=kubernetes-dashboard,app.kubernetes.io/instance=kubernetes-dashboard-test"
NAME                                          READY   STATUS
RESTARTS    AGE
kubernetes-dashboard-test-84fb9b495f-rfpm8    1/1     Running
0           64s
```

Now, you can use the commands for accessing the dashboard. The `kubectl port-forward` command will make it possible to connect to a Pod running inside your cluster from your local machine in a safe way, without exposing the dashboard as a `LoadBalancer` Service. You can find more details in the official documentation: `https://kubernetes.io/docs/tasks/access-application-cluster/port-forward-access-application-cluster/`. Execute the following commands:

```
$ export POD_NAME=$(kubectl get pods -n default -l "app.
kubernetes.io/name=kubernetes-dashboard,app.kubernetes.io/
instance=kubernetes-dashboard-test" -o jsonpath="{.items[0].
metadata.name}")
$ echo https://127.0.0.1:8443/
https://127.0.0.1:8443/
$ kubectl -n default port-forward $POD_NAME 8443:8443
Forwarding from 127.0.0.1:8443 -> 8443
Forwarding from [::1]:8443 -> 8443
```

The `kubectl port-forward` command will do the forwarding as long as it is running. If you terminate the process with *Ctrl + C*, forwarding will be stopped.

Before we can access the dashboard, we need to create a ServiceAccount and ClusterRoleBinding, which we can use for the dashboard. The exact steps are documented in `https://github.com/kubernetes/dashboard/blob/master/docs/user/access-control/creating-sample-user.md`. Follow the given steps:

1. Run the following command to create an `admin-user` service account in the cluster:

```
$ cat <<EOF | kubectl apply -f -
apiVersion: v1
kind: ServiceAccount
metadata:
  name: admin-user
  namespace: default
EOF
```

2. Create `ClusterRoleBinding` between the service account and built into the `cluster-admin` cluster role:

```
$ cat <<EOF | kubectl apply -f -
apiVersion: rbac.authorization.k8s.io/v1
```

```
kind: ClusterRoleBinding
metadata:
  name: admin-user
roleRef:
  apiGroup: rbac.authorization.k8s.io
  kind: ClusterRole
  name: cluster-admin
subjects:
- kind: ServiceAccount
  name: admin-user
  namespace: default
EOF
```

3. Get the bearer token that we can use for logging in to the dashboard:

```
$ kubectl -n default get secret $(kubectl -n default
get sa/admin-user -o jsonpath="{.secrets[0].name}") -o
go-template="{{.data.token | base64decode}}"
```

Now, you can use the pretty long token to log in to the dashboard in the browser.
Navigate to `https://127.0.0.1:8443/`, select **Token**, and copy and paste the
value of your bearer token:

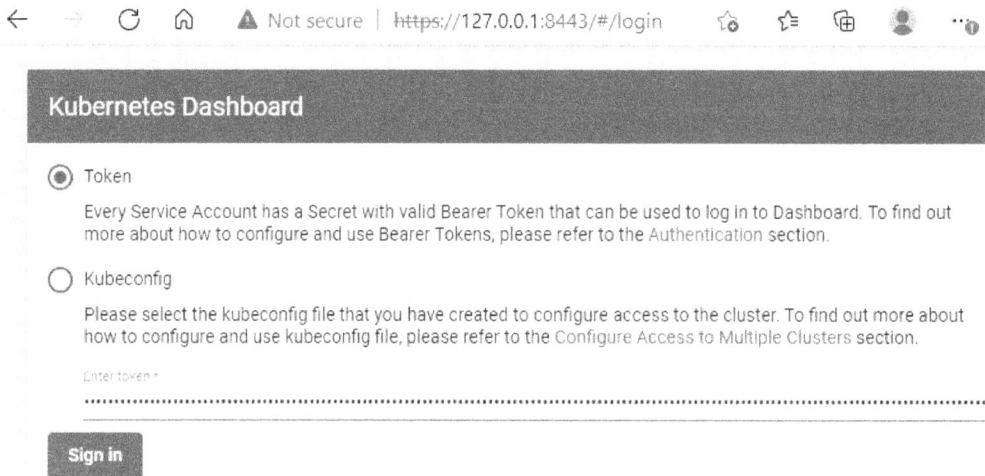

Figure 17.5 – Kubernetes Dashboard chart – login page

At this point, you have access to the dashboard and you can browse its functionalities:

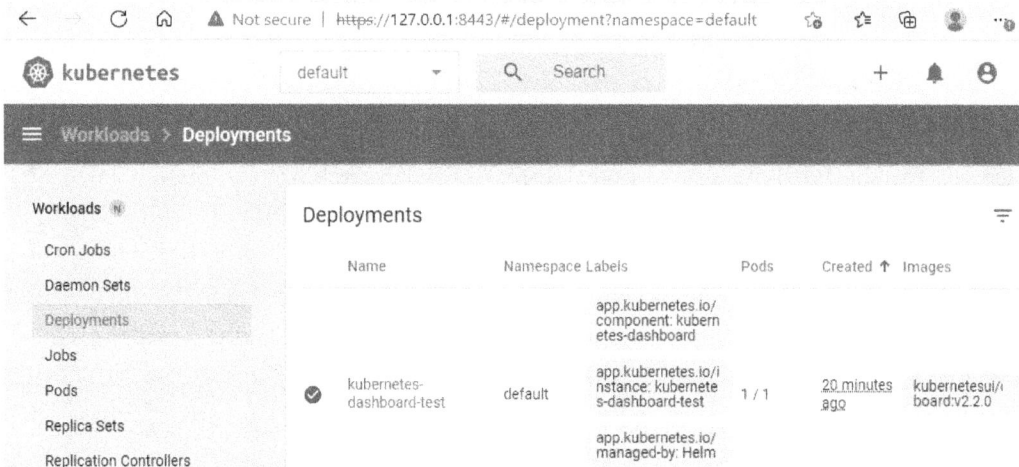

Figure 17.6 – Kubernetes Dashboard chart – Deployments page

The bearer token is for a user with the `cluster-admin` role, so be careful, as you can perform any operations, including deleting resources.

Elasticsearch with Kibana

Elasticsearch is a popular full-text search engine that is commonly used for log indexing and log analytics. **Kibana**, which is part of the Elasticsearch ecosystem, is a visualization UI for the Elasticsearch database. To install this stack, we will need to use two charts, both of which are maintained by Elasticsearch creators:

- Elasticsearch chart (`https://artifacthub.io/packages/helm/elastic/elasticsearch`)

- Kibana chart (`https://artifacthub.io/packages/helm/elastic/kibana`)

First, we need to add the `elastic` repository to Helm:

```
$ helm repo add elastic https://helm.elastic.co
"elastic" has been added to your repositories
```

After that, we can install Elasticsearch in our cluster:

```
$ helm install elasticsearch-test --set replicas=1 elastic/
elasticsearch
```

By default, the chart deploys a three-node cluster of Elasticsearch. Depending on your Deployment, it may not be possible to have three Nodes of Elasticsearch, as each of them needs to run on a separate Kubernetes Node. We adjusted the `replicas` value for the template to 1.

With Elasticsearch being deployed, we can already continue with installing Kibana using the following command:

```
$ helm install kibana-test --set service.type=LoadBalancer
elastic/kibana
```

The chart default values are configured in such a way that by default Kibana will connect to `http://elasticsearch-master:9200`, which is also the instance that we have just deployed using the Helm chart. Additionally, we need to override the `service. type` value to `LoadBalancer` to be able to access Kibana externally.

Wait for the Kibana Service object to get the external IP address:

```
$ kubectl get svc
NAME                          TYPE            CLUSTER-IP
EXTERNAL-IP      PORT(S)         AGE
kibana-test-kibana            LoadBalancer    10.0.28.100
20.62.189.45     5601:32212/TCP   10m
```

In our case, it is `20.62.189.45`.

Now, open the web browser and navigate to `http://20.62.189.45:5601/`. When your Elasticsearch instance and Kibana are ready, you will see the following:

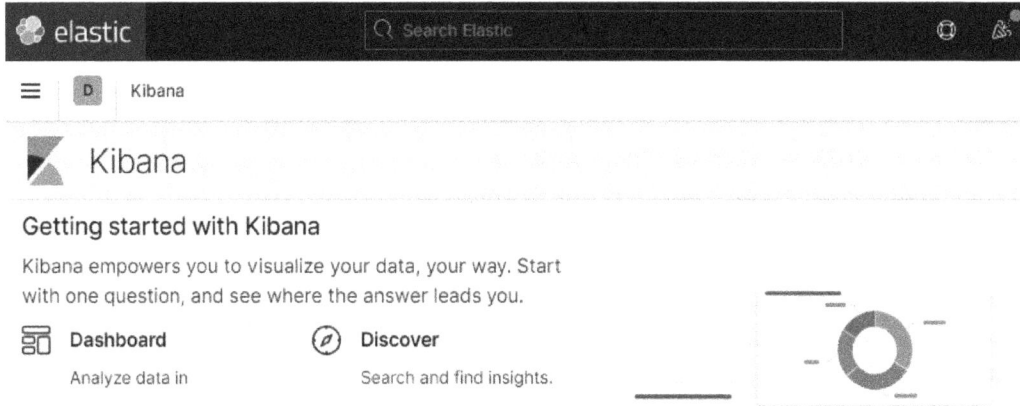

Figure 17.7 – Elasticsearch and Kibana charts – Kibana UI

You can now explore the functionalities of Elasticsearch and Kibana. If you are interested in working on sample data, you can follow the official guide: `https://www.elastic.co/guide/en/kibana/current/get-started.html`.

The next chart that we will take a look at is the Prometheus stack chart, which installs Prometheus together with Grafana.

Prometheus with Grafana

Prometheus is a popular monitoring system with a time series database that is widely used in the Kubernetes community. Grafana is commonly used as a visualizations UI for Prometheus. This combination is so common that the Prometheus community has prepared an official Helm chart (`https://artifacthub.io/packages/helm/prometheus-community/kube-prometheus-stack`) that deploys both Prometheus and Grafana as a single chart, together with a set of useful Prometheus *rules* for basic monitoring.

Let's begin by adding the `prometheus-community` repository to Helm:

```
$ helm repo add prometheus-community https://prometheus-
community.github.io/helm-charts
"prometheus-community" has been added to your repositories
```

Next, the installation is as simple as running this single command:

```
$ helm install prometheus-stack-test prometheus-community/kube-
prometheus-stack
```

You can monitor whether Pods are running using the `kubectl get pods` command:

```
$ kubectl get pods
NAME
READY    STATUS    RESTARTS    AGE
prometheus-prometheus-stack-test-kube-prometheus-0          2/2
Running    1           11m
prometheus-stack-test-grafana-97b7cd8c4-fx561              2/2
Running    0           11m
prometheus-stack-test-kube-operator-5f46876fcc-5dnkp       1/1
Running    0           11m
prometheus-stack-test-kube-state-metrics-b845577bb-tp99s   1/1
Running    0           11m
prometheus-stack-test-prometheus-node-exporter-jg85x       1/1
Running    0           11m
prometheus-stack-test-prometheus-node-exporter-ks5bn       1/1
Running    0           11m
```

To access Prometheus and Grafana, we will use the `kubectl port-forward` command again. In this case, we have not exposed the Pods using the `LoadBalancer` service. Feel free to do so after inspecting the default `values.yaml` file. Execute the following two commands in *separate* shells – the `kubectl` process must be running at the same time as you want to forward the ports:

```
$ kubectl port-forward prometheus-prometheus-stack-test-kube-
prometheus-0 9090
$ kubectl port-forward prometheus-stack-test-grafana-97b7cd8c4-
fx561 3000
```

Of course, you need to use a proper Pod name for Grafana, and in your case it may be different.

Next, let's navigate to the Prometheus UI in a web browser. Open the `http://127.0.0.1:9090/` address:

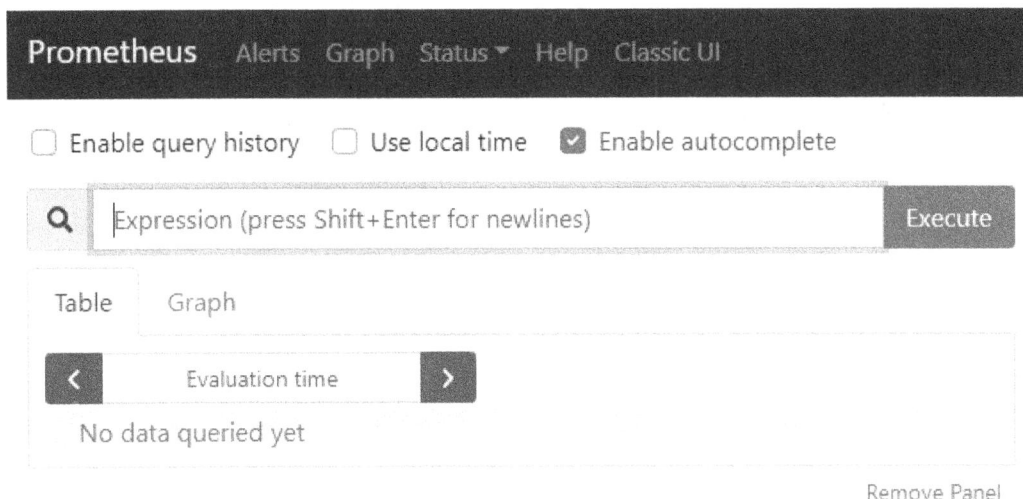

Figure 17.8 – Prometheus stack chart – Prometheus web UI

Similarly, you can navigate to `http://127.0.0.1:3000/` in order to access the Grafana UI:

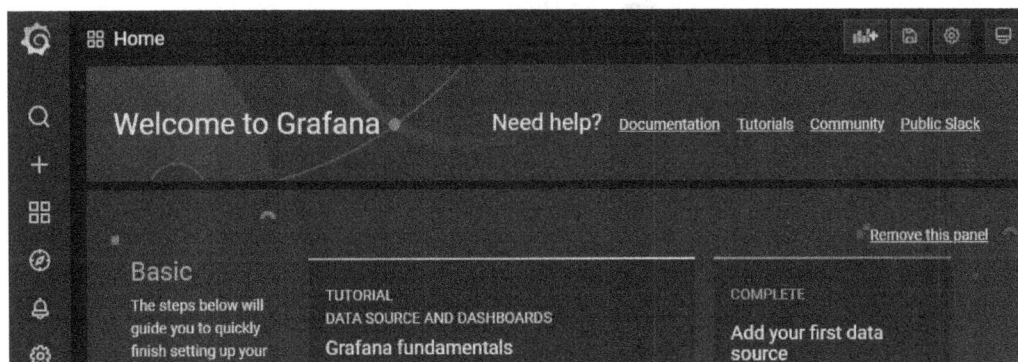

Figure 17.9 – Prometheus stack chart – Grafana web UI

Congratulations, you have successfully deployed the Prometheus stack on your Kubernetes cluster! As you can see, working with Helm charts, even for complex, multi-component solutions, is easy and can provide a lot of benefits for your development and production environments.

Summary

This chapter has covered the details of working with Helm and Helm charts. First, you have learned what the purpose of package management is and how Helm works as a package manager for Kubernetes. We have demonstrated how you can install Helm on your local machine, and how you can deploy the WordPress chart to test the installation. Then, we went through the structure of Helm charts, and we have shown how the YAML templates in charts can be configured using user-provided values. Lastly, we have shown the installation of a few popular solutions on a Kubernetes cluster using Helm. We have installed Kubernetes Dashboard, Elasticsearch together with Kibana, and the Prometheus Stack including Grafana.

In the next chapter, we are going to explore authentication and authorization on Kubernetes. We will dive deeper into RBAC available in Kubernetes – which you got a sneak-peek of in this chapter while creating ServiceAccount for accessing Kubernetes Dashboard!

Further reading

For more information regarding Helm and Helm charts, please refer to the following *Packt Publishing* book:

- *Learn Helm*, by *Andrew Block, Austin Dewey* (`https://www.packtpub.com/product/learn-helm/9781839214295`)

You can learn more about Elasticsearch and Prometheus in the following *Packt Publishing* books:

- *Learning Elasticsearch*, by *Abhishek Andhavarapu* (`https://www.packtpub.com/product/learning-elasticsearch/9781787128453`)

- *Hands-On Infrastructure Monitoring with Prometheus*, by *Joel Bastos, Pedro Araújo* (`https://www.packtpub.com/product/hands-on-infrastructure-monitoring-with-prometheus/9781789612349`)

You can also refer to the official Helm documentation: `https://helm.sh/docs/`.

18
Authentication and Authorization on Kubernetes

In software systems, **authentication** and **authorization** play a significant role in providing security. These terms may seem similar but they are very *different* security processes that work together to provide **identity and access management**. In short, authentication determines whether a given user of a system is indeed who they claim to be – the easiest way to visualize this is using a username and password to complete an authentication process. Authorization, on the other hand, determines what the user can and cannot do or access in the system. This also means that in secure systems, authentication is the *first* step and authorization must always follow authentication. One of the approaches of modeling authorization in systems is using **role-based access control (RBAC)**, where you organize access control and management with roles and privileges. Roles can be assigned to users in the system, which gives them certain privileges and access. In this way, you can achieve very fine-grained access management that can be used to enforce the **principle of least privilege**.

Kubernetes, as a mature and widely adopted container orchestration system, is no exception – it has been designed with security as a first-class citizen. Authentication and authorization in Kubernetes are extensible and can fulfill the needs of enterprise systems.

For authentication, you can use one of the built-in strategies, such as X509 client certificates or OpenID Connect tokens, which are considered industry standards. If there is a need for less common authentication providers such as LDAP, you can use an **authenticating proxy** or **authentication webhook**. We will discuss this in the first section of the chapter and extend this knowledge by demonstrating how you can integrate **Azure Kubernetes Service** (**AKS**) with **Azure Active Directory** (**AAD**) for authentication, and even use Azure RBAC for providing authorization in AKS.

For authorization, Kubernetes comes with a built-in RBAC mechanism that allows you to configure fine-grained sets of permissions and assign them to users, groups, and ServiceAccounts (subjects). In this way, as a cluster administrator, you can control how cluster users (internal and external) interact with the Kubernetes API server, which API resources they can access, and which actions (verbs) they can perform. We will discuss the details in the second section of this chapter.

In this chapter, we will cover the following topics:

- Authentication and user management
- Authentication – AKS and AAD integration
- Authorization – introduction to RBAC

Technical requirements

For this chapter, you will need the following:

- A Kubernetes cluster to be deployed. We recommend using a *multi-node*, cloud-based Kubernetes cluster. To be able to follow the section regarding AKS and, AAD you need to have a managed AKS cluster deployed and the Azure CLI installed.

- The Kubernetes CLI (`kubectl`) installed on your local machine and configured to manage your Kubernetes cluster.

Basic Kubernetes cluster deployment (local and cloud-based) and `kubectl` installation were covered in *Chapter 3*, *Installing Your First Kubernetes Cluster*.

The following chapters can give you an overview of how to deploy a fully functional Kubernetes cluster on different cloud platforms:

- *Chapter 14, Kubernetes Clusters on Google Kubernetes Engine*

- *Chapter 15, Launching a Kubernetes Cluster on Amazon Web Services with the Amazon Elastic Kubernetes Service*

- *Chapter 16, Kubernetes Clusters on Microsoft Azure with the Azure Kubernetes Service.*

You can download the latest code samples for this chapter from the official GitHub repository: `https://github.com/PacktPublishing/The-Kubernetes-Bible/tree/master/Chapter18`.

Authentication and user management

The Kubernetes API server provides RESTful endpoints for managing Kubernetes cluster and acts as the frontend to the shared state of the cluster. All users and all internal components interact with the cluster via the Kubernetes API server. Requests to the API are always one of the following:

- Associated with an external, normal user or a **ServiceAccount** defined in the Kubernetes cluster

- Treated as *anonymous* requests if the cluster has been configured to allow anonymous requests

This is determined in the *authentication* process – the entire HTTP request is used as input to the process, but usually only request hearers or the client certificate is analyzed. Authentication is carried out by authentication modules that depend on the cluster configuration. Your cluster may have multiple authentication modules enabled and then each of them is executed in sequence until one succeeds. If the request fails to authenticate, the API server will either respond with an HTTP status code of `401` (`unauthorized`) or, if anonymous requests are enabled, treat it as anonymous.

> **Tip**
> Anonymous requests are essentially mapped to a special username called `system:anonymous` and a group called `system:unauthenticated`. This means that you can organize your authorization to resources for such requests, just as you can for other users or ServiceAccounts.

Since all operations inside and outside the cluster must go through the Kubernetes API server, this means that all of them must go through the authentication process. This includes the operations of internal cluster components and Pods, which query the API server. For you, as an external user of the cluster, any requests that you make using `kubectl` commands or directly to the Kubernetes API server will also go through the authentication process:

- **Normal users**: Such users are managed *externally*, independent from the Kubernetes cluster. Currently, Kubernetes does not provide any objects to represent such users. The external management of users may be as simple (but *not* recommended) as static user-password files passed to the API server using the `token-auth-file` argument during startup. In real production scenarios, you should use cloud services such as AAD or AWS Identity and Access Management to manage the users and integrate with your Kubernetes cluster using **OpenID Connect** (`https://openid.net/connect/`) tokens to seamlessly authenticate. Note that normal user accounts are global and do not have cluster namespaces.

- **Service accounts**: These are managed by Kubernetes cluster and modeled as ServiceAccount objects. You can create and manage service accounts just like any other resource in Kubernetes; for example, using `kubectl` and YAML manifest files. This type of account is intended for processes in cluster components or running in Pods. The credentials for ServiceAccounts are stored as **Secrets** in the cluster that are mounted into Pods so that the container process can use them to talk to the Kubernetes API server. When a process authenticates using a `ServiceAccount` token, it is seen as a user called `system:serviceaccount:<namespace>:<serviceAccountName>`. Note that ServiceAccounts are namespaced.

As you can see, user management in Kubernetes is a mixture of different approaches that should fit all the needs of different organizations. The key takeaway here is that after the authentication process, the request will be either rejected (optionally treated as anonymous) or will be treated as coming from a particular user. The `username` attribute may be provided by the external user management system, as in the case of normal users, or it will be `system:serviceaccount:<namespace>:<serviceAccountName>` for ServiceAccounts. Additionally, the request will have more attributes associated with it, such as **User ID** (`UID`), **groups**, and **extra fields**. This information is used for authorization processes based on RBAC, which we will explain in the next section.

Now, let's look at the authentication methods that you can use with Kubernetes.

Static token files

This method is the most basic one that Kubernetes offers for managing normal users. The approach somewhat resembles the /etc/shadow and /etc/passwd files, which were used in the early days of Unix systems. Note, however, that it is *not* recommended and is considered *unsecure* for production clusters.

In this method, you define a .csv file where each line has the following format:

```
token,user,uid,"group1,group2,group3"
```

Then, you pass the file when starting the Kubernetes API server process using the token-auth-file parameter. To authenticate against the API server, you need to use a standard HTTP **bearer authentication scheme** for your requests. This means that your requests will need to use an additional header that's in the following form:

```
Authorization: Bearer <token>
```

Based on this request information, the Kubernetes API server will match the token against the static token file and assign user attributes based on the matched record.

When using kubectl, you must modify your kubeconfig. You can do this using the kubectl command:

```
$ kubectl config set-credentials <contextUser> --token=<token>
```

After that, you need to create and use context with this user for your requests using the kubectl config use-context command.

> **Important note**
>
> In Kubernetes versions prior to 1.19, there was a similar authentication method that allowed us to use an HTTP **basic authentication scheme** and a file passed by the basic-auth-file parameter to the API server. This method is no longer supported due to security reasons.

The following diagram visualizes the principles behind this method of authentication:

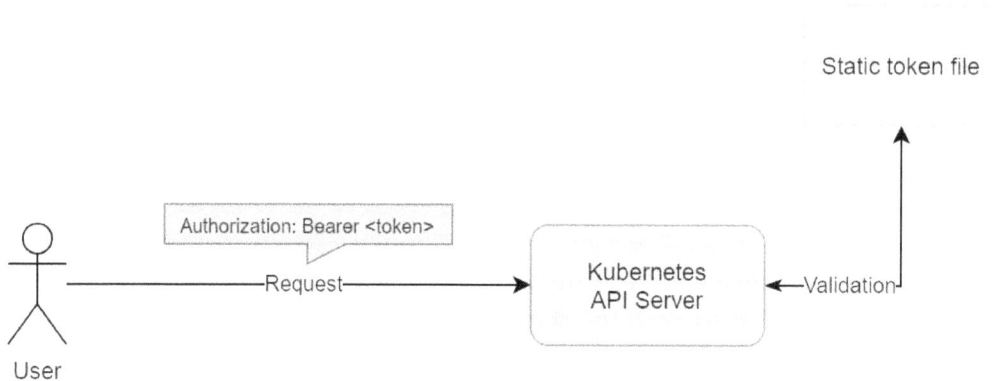

Figure 18.1 – Static token file authentication in Kubernetes

We can now summarize the advantages and disadvantages of using static token file method for authentication.

The advantages are as follows:

- Easy to configure
- Easy to understand

The disadvantages are as follows:

- Unsecure; exposing a token file compromises all cluster users.
- Requires that we manually manage users.
- Adding new users or removing existing ones requires that we restart the Kubernetes API server.
- Rotating any tokens requires that we restart the Kubernetes API server.

In short, this method is good for development environments and learning the principles behind authentication in Kubernetes, but it is not recommended for production use cases. Next, we will take a look at authenticating users using ServiceAccount tokens.

ServiceAccount tokens

As we mentioned in the introduction to this section, ServiceAccounts are meant for in-cluster identities for processes running in Pod containers or for cluster components. However, they can be used for authenticating external requests as well.

ServiceAccounts are Kubernetes objects and can be managed like any other resource in the cluster; that is, by using `kubectl` or raw HTTP requests to the API server. The tokens for ServiceAccounts are **JSON Web Tokens (JWTs)** and are stored as Kubernetes Secret objects. Secrets were covered in *Chapter 6, Configuring Your Pods Using ConfigMaps and Secrets*. Usually, when defining a Pod, you will specify what ServiceAccount should be used for processes running in the containers. You can do this using `.spec.serviceAccountName` in the Pod specification. The JWT token will be injected into the container; then, the process inside can use it in the HTTP bearer authentication scheme to authenticate to the Kubernetes API server. This is only necessary if it interacts with the API server in any way, for example, if it needs to discover other Pods in the cluster. We have summarized this authentication method in the following diagram:

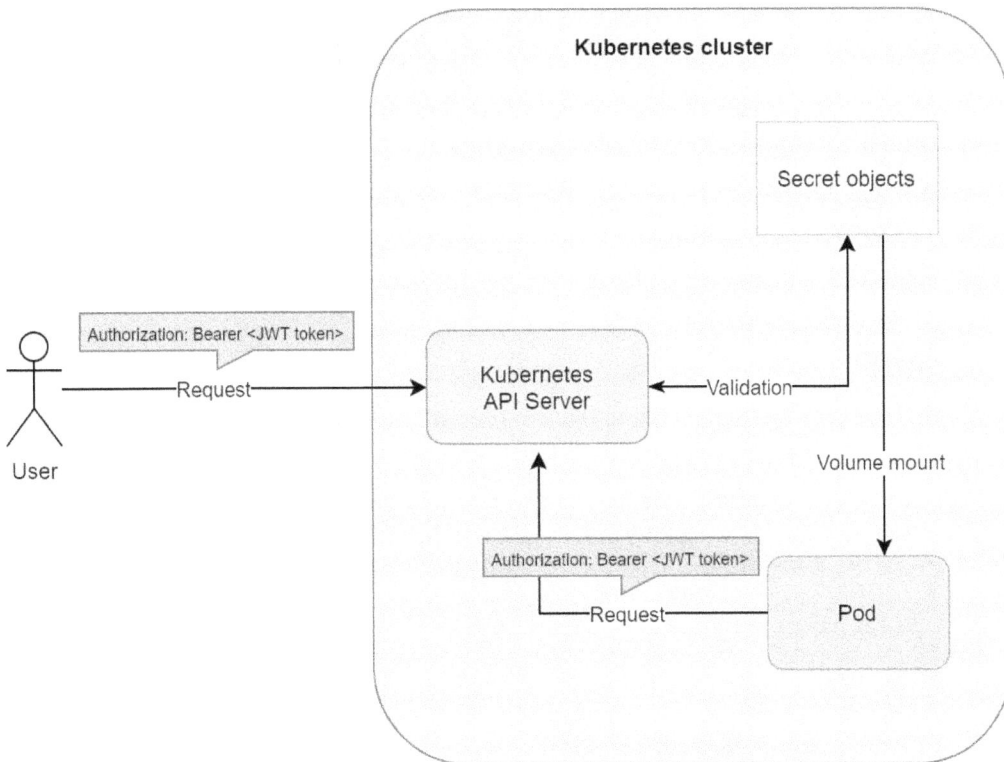

Figure 18.2 – ServiceAccount authentication in Kubernetes

This also shows why ServiceAccount tokens can be used for external requests – the API server does not care about the origin of the request; all it is interested in is the bearer token that comes with the request header. Again, you can use this token in `kubectl` or in raw HTTP requests to the API server. Please note that this is generally not a recommended way to use ServiceAccounts, but it can be used in some scenarios, especially when you are unable to use an external authentication provider for normal users.

We will now demonstrate how you can create and manage ServiceAccounts and how you can use JWT tokens to authenticate when using `kubectl`. This will also give a sneak peek into RBAC, which we are going to look at in more detail in the next section. Please follow the following steps:

1. Create a YAML manifest for a new ServiceAccount named `example-account`. We will configure RBAC for this account so that it can only read Pods in the `default` namespace. The `example-account-serviceaccount.yaml` YAML manifest file has the following contents:

    ```
    apiVersion: v1
    kind: ServiceAccount
    metadata:
      name: example-account
      namespace: default
    ```

    Note that applying the preceding manifest has the same effect as the *imperative* `kubectl create serviceaccount example-account` command.

2. Create a YAML manifest for a `Role` object named `pod-reader` in the `default` namespace. This role will allow you to get, watch, and list Pods in this namespace. The `pod-reader-role.yaml` YAML manifest file has the following contents:

    ```
    apiVersion: rbac.authorization.k8s.io/v1
    kind: Role
    metadata:
      namespace: default
      name: pod-reader
    rules:
    - apiGroups: [""]
      resources: ["pods"]
      verbs: ["get", "watch", "list"]
    ```

3. Create a YAML manifest for `RoleBinding` named `reads-pods`. This is what *associates* the role that we created with our `example-account` ServiceAccount – the account will now have the privilege of read-only access to Pods, and nothing more. The `read-pods-rolebinding.yaml` YAML manifest file has the following contents:

```
apiVersion: rbac.authorization.k8s.io/v1
kind: RoleBinding
metadata:
  name: read-pods
  namespace: default
subjects:
- kind: ServiceAccount
  name: example-account
  namespace: default
roleRef:
  kind: Role
  name: pod-reader
  apiGroup: rbac.authorization.k8s.io
```

4. Now, we can apply all the manifest files to the cluster at once using the `kubectl apply` command:

```
$ kubectl apply -f ./
serviceaccount/example-account created
role.rbac.authorization.k8s.io/pod-reader created
rolebinding.rbac.authorization.k8s.io/read-pods created
```

5. We need to retrieve the JWT for the service account. To do that, use the following command, which retrieves it from the associated Secret object and decodes it from Base64 encoding:

```
$ kubectl -n default get secret $(kubectl -n default get
sa/example-account -o jsonpath="{.secrets[0].name}") -o
go-template="{{.data.token | base64decode}}"
```

6. Copy the token; we will need it for further operations. If you are interested, you can inspect the contents of the JWT using `https://jwt.io/`:

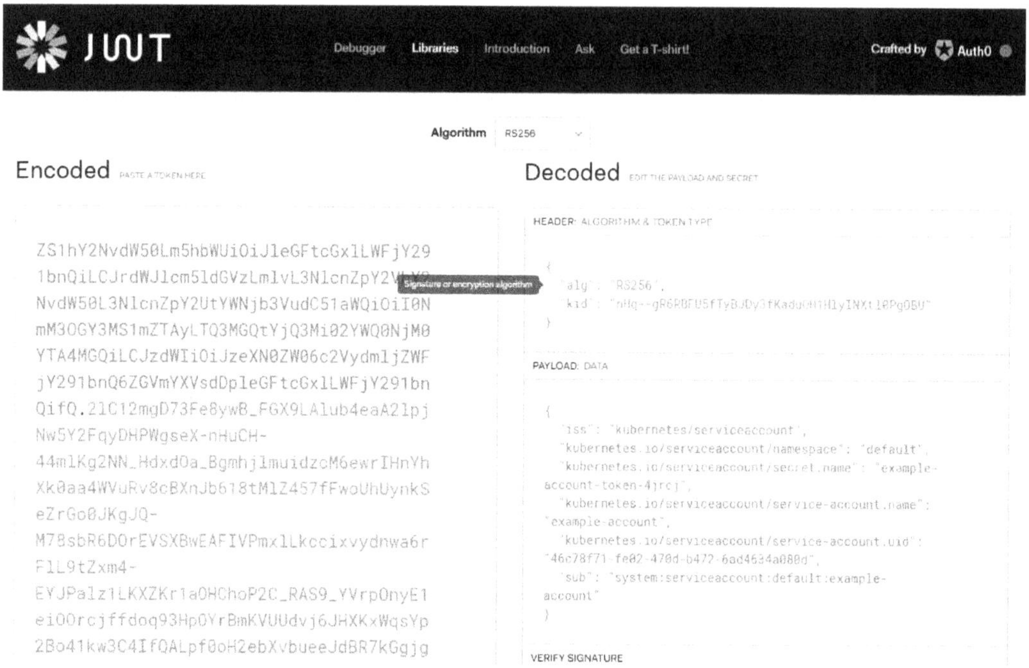

Figure 18.3 – Inspecting a JWT for ServiceAccount

As you can see, the JWT maps to the `example-account` ServiceAccount. Additionally, you can identify that the actual username (marked as `subject` in the payload) it will be mapped to in Kubernetes is `system:serviceaccount:defa ult:example-account`, as we explained previously.

7. With this JWT, we can set up `kubectl` to test it. First, you need to create a user in your `kubeconfig` using the following command:

```
$ kubectl config set-credentials example-account–
token=<jwtToken>
```

8. Create a new context that uses this user. You also need to know the cluster name that you are connecting to right now – you can check it using the `kubectl config view` command. Use the following command to create a new context named `example-account-context`:

```
$ kubectl config set-context example-account-context–
user=example-account–cluster=<clusterName>
```

9. You may want to check the name of the context that you are currently using by using the `kubectl config current-context` command. This will make it easier to go back to your old cluster admin context. Switch to the new context using the following command:

```
$ kubectl config use-context example-account-context
Switched to context "example-account-context".
```

10. We are now ready to verify that our authentication works and that the RBAC roles allow read-only access to Pods in the `default` namespace. First, try getting Pods:

```
$ kubectl get pods
NAME
READY    STATUS    RESTARTS    AGE
alertmanager-prometheus-stack-test-kube-alertmanager-0
2/2      Running   0           47h
elasticsearch-master-0
1/1      Running   0           47h
...
```

11. This has worked as expected! Now, try getting Pods from the `kube-system` namespace:

```
$ kubectl get pods -n kube-system
Error from server (Forbidden): pods is forbidden: User
"system:serviceaccount:default:example-account" cannot
list resource "pods" in API group "" in the namespace
"kube-system"
```

12. We have authenticated correctly, but the action was forbidden by RBAC authorization, which is what we expected. Lastly, let's try getting Service objects:

```
$ kubectl get svc
Error from server (Forbidden): services is forbidden:
User "system:serviceaccount:default:example-account"
cannot list resource "services" in API group "" in the
namespace "default"
```

As you can see, we have successfully used our ServiceAccount JWT as a bearer token for authentication and we have verified that our privileges work correctly. You can now switch back to your old `kubectl` context using the `kubectl config use-context` command.

> **Tip**
>
> The preceding procedure of configuring the `kubectl` context with a bearer token can be used for the static token file authentication method as well.

Let's summarize what are the advantages and disadvantages of using ServiceAccount tokens for authentication are.

The advantages are as follows:

- Easy to configure and use, similar to static token files.

- Entirely managed by the Kubernetes cluster, so there's no need for external authentication providers.

- ServiceAccounts are namespaced.

The disadvantages are as follows:

- ServiceAccounts are intended for processes running in Pod containers to give them identity and let them use Kubernetes RBAC.

- Any Pod that has access to reading Secrets can discover all tokens for ServiceAccounts! This is an important security implication, especially if you are thinking about having high-privilege service accounts in your cluster. This would also violate the **principle of least privilege**.

- **Rotation** of ServiceAccount tokens is cumbersome and there is no automated way to do this out of the box. This makes any mitigations to security incidents much harder.

- `ServiceAccount` tokens do not expire, which is another security concern. There is a design proposal to make this possible, though; you can read more here: `https://github.com/kubernetes/community/blob/master/contributors/design-proposals/auth/bound-service-account-tokens.md`.

In general, using ServiceAccount tokens for external authentication is good for development and test scenarios when you cannot integrate with external authentication providers. But for production clusters, it is not the best option, mainly due to security concerns. Now, let's take a look at using X.509 client certificates.

X.509 client certificates

Using X.509 client certificates is one of the industry standards for authentication processes. There is one important catch, however – you need to have good means of managing certificate signing, revoking, and rotation – otherwise, you may hit very similar security issues as with using ServiceAccount tokens. You can learn more about X.509 certificates and the processes around them here: `https://www.ssl.com/faqs/what-is-an-x-509-certificate/`.

This method works in Kubernetes as follows:

1. The Kubernetes API server is started with the `client-ca-file` argument. This provides **certificate authority (CA)** information to be used to validate client certificates presented to the API server.

2. Users that want to authenticate against the API server need to request an X.509 client certificate from the CA. This should be a secure and audited process. The **subject common name** (the CN attribute in the subject) of the certificate is used as the username attribute when authentication is successful. Note that as of Kubernetes 1.19, you can use the Certificates API to manage signing requests. More information is available in the official documentation: `https://kubernetes.io/docs/reference/access-authn-authz/certificate-signing-requests/`.

3. The user must present the client certificate during authentication to the API server, which validates the certificate against the CA. Based on that, the request goes through the authentication process successfully or is rejected.

While using the `kubectl` commands, you can configure this method of authentication for your user using the `kubectl config set-credentials` command. We have summarized this process in the following diagram:

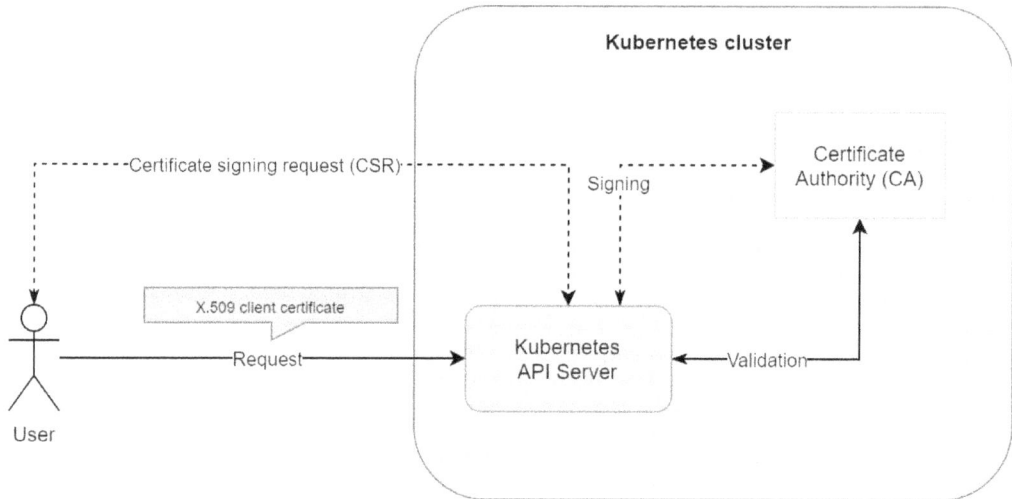

Figure 18.4 – X.509 client certificate authentication in Kubernetes

Please note that this visualizes the case when initial CSR by the user is handled by the Certificate API in a Kubernetes cluster. This does not need to be the case as CA may be external to the cluster, and the Kubernetes API server can rely on a copy of the CA `.pem` file.

We can summarize the advantages of this method as follows:

- It's a much more secure process than using ServiceAccount tokens or static token files.

- Being unable to store certificates in the cluster means that it is not possible to compromise all certificates, as was the case of using ServiceAccount tokens. X.509 client certificates can be used for high-privileged user accounts.

- X.509 client certificates can be revoked on demand. This is very important in case of security incidents.

The disadvantages of X.509 client certificate authentication are as follows:

- Certificates have an expiry date, which means they cannot be valid indefinitely. For simple use cases in development, this is a disadvantage. From a security perspective, in production clusters, this is a huge *advantage*.

- Monitoring certificate expiration, revocation, and rotation must be handled. This should be an automated process so that we can quickly react in the case of security incidents.

- The built-in Certificate API has limited functionality.

- Using client certificates in the browser for authentication is troublesome, for example, when you would like to authenticate to Kubernetes Dashboard.

The key takeaway is that using X.509 client certificates is secure but requires sophisticated certificate management so that we have all the benefits. Now, we will take a look at OpenID Connect tokens, which is the recommended method for cloud environments.

OpenID Connect tokens

Using **OpenID Connect** (**OIDC**), you can achieve a **single sign-on** (**SSO**) experience for your Kubernetes cluster (and possibly other resources in your organization). OIDC is an authentication layer that's created on top of OAuth 2.0, which allows third-party applications to verify the identity of the end user and obtain basic user profile information. OIDC uses JWTs, which you can obtain using flows that conform to the OAuth 2.0 specifications. The most significant issue with using OIDC for authenticating in Kubernetes is the limited availability of OpenID **providers**. But if you are deploying in a cloud environment, all tier 1 cloud service providers such as Microsoft Azure, Amazon Web Services, and Google Cloud Platform have their versions of OpenID providers. The beauty of **managed** Kubernetes cluster deployments in the cloud, such as, AKS Amazon EKS, and Google Kubernetes Engine, is that they provide *integration* with their native OpenID provider out of the box or by a simple flip of a configuration switch. In other words, you do not need to worry about reconfiguring the Kubernetes API server and making it work with your chosen OpenID provider – you get it alongside the managed solution. In the last section of this chapter, we will demonstrate how you can do that for AKS throughout.

If you are interested in learning more about the OpenID Connect protocol, you can refer to the official web page: `https://openid.net/connect/`. For more details and more specific flows, such as in the context of AAD please take a look here: `https://docs.microsoft.com/en-us/azure/active-directory/develop/v2-protocols-oidc`. In the following diagram, you can see the basics of the OIDC authentication flow on Kubernetes:

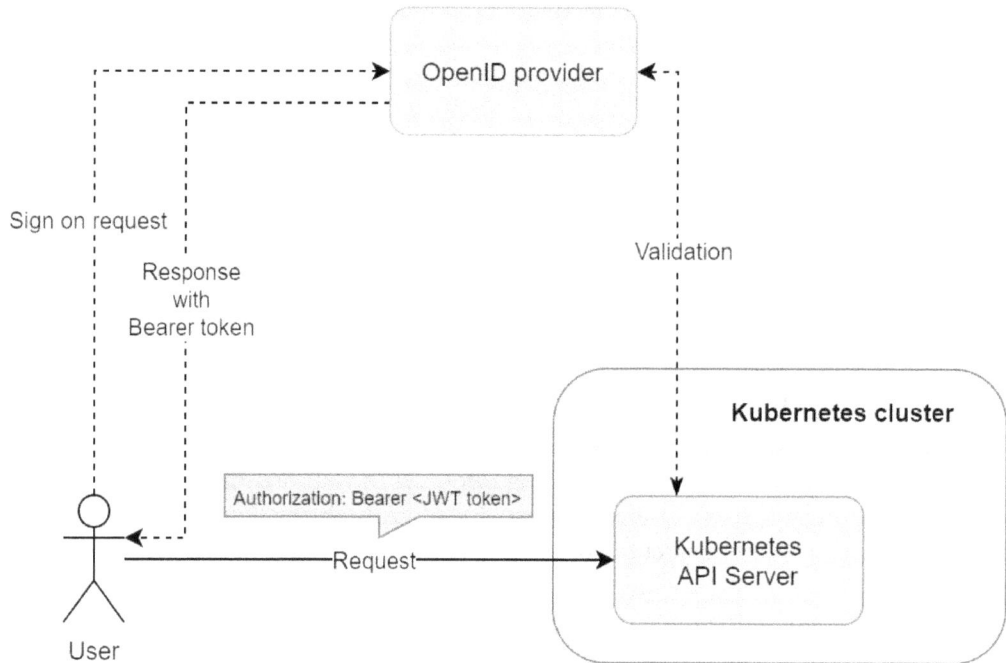

Figure 18.5 – OpenID Connect authentication in Kubernetes

The most important thing is that the OpenID provider is responsible for the SSO experience and managing the bearer tokens. Additionally, the Kubernetes API server must validate the bearer token that's received against the OpenID provider.

Using OIDC has the following advantages:

- You get SSO experience, which you can use with other services in your organization.

- Tier 1 cloud service providers have their own OpenID providers that easily integrate with their managed Kubernetes offerings.

- It can be also used with other OpenID providers and non-cloud deployments – this requires a bit more configuration though.

- It's a secure and scalable solution.

The disadvantages of this approach can be summarized as follows:

- Kubernetes has no web interface where you can trigger the authentication process. This means that you need to get the credentials by manually requesting them from the identity provider. In managed cloud Kubernetes offerings, this is often solved by additional simple tooling to generate `kubeconfig` with credentials.

- Tokens cannot be revoked, so they are set to expire in a short time. This requires the tokens to be frequently renewed.

The key takeaway about OIDC is that this is your best bet when configuring authentication for Kubernetes, especially if you are deploying production clusters in the cloud. Lastly, let's take a quick look at the other available authentication methods.

Other methods

Kubernetes offers a few other authentication methods that you can use. They are mainly intended for advanced use cases, such as integrating with LDAP or Kerberos. The first one is **authenticating proxy**.

When you use authenticating proxy in front of the Kubernetes API server, you can configure the API server to use certain HTTP headers to extract authentication user information from them. In other words, your authenticating proxy is doing the job of authenticating the user and passing down this information alongside the request in the form of additional headers.

You can find more information in the official documentation: `https://kubernetes.io/docs/reference/access-authn-authz/authentication/#authenticating-proxy`.

Another approach is known as **webhook token authentication**, where the Kubernetes API server uses an external service to verify the bearer tokens. The external service receives the information in the form of a TokenReview object from the API server via an HTTP POST request, performs verification, and sends back a TokenReview object with additional information about the result.

You can find more information in the official documentation: `https://kubernetes.io/docs/reference/access-authn-authz/authentication/#webhook-token-authentication`.

In general, you only need these two methods in special cases where you want to integrate with existing identity providers in your organization that are not supported by Kubernetes out of the box.

In the next section, we will look at RBAC.

Authorization and introduction to RBAC

While authentication is about determining whether a given user of a system is indeed who they claim to be, **authorization** determines what the user can and cannot do or access. As such, authorization usually complements authentication – these two processes are used together to provide security for the system. Authentication is the first step in determining the identity of the user, whereas authorization is the next step when verifying if the user can perform the action they want to.

In the Kubernetes API server, authenticating a request results in a set of additional request attributes such as user, group, API request verb, or HTTP request verb. These are then passed further to authorization modules that, based on these attributes, answer whether the user is allowed to do the action or not. If the request is denied by any of the modules, the user will be presented with an HTTP status code of 403 (Forbidden).

> **Tip**
>
> This is an important difference between HTTP status codes. If you receive 401 (Unauthorized), this means that you have been not recognized by the system; for example, you have provided incorrect credentials or the user does not exist. If you receive 403 (Forbidden), this means that authentication has been successful, you have been recognized, but you are not *allowed* to do the action you requested. This is useful when debugging issues regarding access to a Kubernetes cluster.

Kubernetes has a few authorization modes available that can be enabled by using the authorization-mode argument when starting the Kubernetes API server:

- **RBAC**: Allows you to organize access control and management with roles and privileges. RBAC is one of the industry standards for access management, also outside of Kubernetes. Roles can be assigned to users in the system, which gives them certain privileges and access. In this way, you can achieve very fine-grained access management that can be used to enforce the **principle of least privilege**. For example, you can define a role in the system that allows you to access certain files on a network share. Then, you can assign such roles to individual users on groups in the system to allow them to access these files. This can be done by associating the user with a role – in Kubernetes, you model this using the RoleBinding and ClusterRoleBinding objects. In this way, multiple users can be assigned a role and a single user can have multiple roles assigned. Please note that in Kubernetes, RBAC is *permissive*, which means that there are no *deny* rules. Everything is denied by default, and you have to define *allow* rules instead.

- **Attribute-Based Access Control** (**ABAC**): This is part of the access control paradigm, used not only in Kubernetes, which uses policies based on the attributes of the user, resource, and environment. This is a very fine-grained access control approach – you can, for example, define that the user can access a given file, but only if the user has clearance to access confidential data (user attribute), the owner of the file is Mike (resource attribute), and the user tries to access the file from an internal network (environment attribute). So, policies are sets of attributes that must be present together for the action to be performed. In Kubernetes, this is modeled using Policy objects. For example, you can define that the authenticated user, `mike`, can read any Pods in the `default` Namespace. If you want to give the same access to user `bob`, then you need to create a new Policy for user `bob`.

- **Node**: A special-purpose authorization mode used for authorizing API requests made by `kubelet` in the cluster.

- **Webhook**: This mode is similar to webhooks for authentication. You can define an external service that needs to handle HTTP POST requests with an SubjectAccessReview object that's sent by the Kubernetes API server. This service must process the request and determine if the request should be allowed or denied. The response from the service should contain `SubjectAccessReview`, along with information, whether the subject is allowed the access. Based on that, the Kubernetes API server will either proceed with the request or reject it with an HTTP status code of `403`. This approach is useful when you are integrating with existing access control solutions in the organization.

Currently, RBAC is considered an industry standard in Kubernetes due to its flexibility and ease of management. For this reason, RBAC is the only authentication mode we are going to describe in more detail.

RBAC mode in Kubernetes

Using RBAC in Kubernetes involves two types of API resources that belong to the `rbac.authorization.k8s.io` API group:

- Role and ClusterRole: They define a set of permissions. Each `rule` in Role says which verb(s) are allowed for which API resource(s). The only difference between Role and ClusterRole is that Role is namespace-scoped, whereas ClusterRole is global.

- RoleBinding and ClusterRoleBinding: They associate users or a set of users (alternatively, groups or ServiceAccounts) with a given Role. Similarly, RoleBinding is namespace-scoped, while ClusterRoleBinding is cluster-wide. Please note that ClusterRoleBinding works with ClusterRole, but RoleBinding works with both ClusterRole and Role.

All these Kubernetes objects can be managed using `kubectl` and YAML manifests, just as you do with Pods, Services, and so on.

We will now demonstrate this in practice. In the previous section, we showed a basic RBAC configuration for a service account that was being used for authentication using `kubectl`. The example that we are going to use here will be a bit different and will involve creating a Pod that runs under a *dedicated* service account and periodically queries the Kubernetes API server for a list of Pods. In general, having dedicated service accounts for running your Pods is a good practice and makes it possible to ensure the principle of least privilege. For example, if your Pod needs to get the list of Pods in the cluster but does not need to create a new Pod, the ServiceAccount for this Pod should have a role assigned that allows you to list read-only Pods, nothing more. Please follow these steps to configure this example:

1. Begin by creating a dedicated ServiceAccount named `pod-logger`. Create a YAML manifest named `pod-logger-serviceaccount.yaml`:

```
apiVersion: v1
kind: ServiceAccount
metadata:
  name: pod-logger
  namespace: default
```

2. Apply the manifest to the cluster using the following command:

```
$ kubectl apply -f ./pod-logger-serviceaccount.yaml
```

3. Create a role named `pod-reader`. This role will only allow the `get`, `watch`, and `list` verbs on `pods` resources in the Kubernetes RESTful API. In other words, this translates into an `/api/v1/namespaces/default/pods` endpoint in the API. Note that `apiGroups` specified as `""` mean the `core` API group. The structure of the `pod-reader-role.yaml` YAML manifest file is as follows:

```
apiVersion: rbac.authorization.k8s.io/v1
kind: Role
metadata:
  namespace: default
  name: pod-reader
rules:
- apiGroups: [""]
  resources: ["pods"]
  verbs: ["get", "watch", "list"]
```

4. Apply the manifest to the cluster using the following command:

```
$ kubectl apply -f ./pod-reader-role.yaml
```

5. Now, we would normally create a RoleBinding object to associate the service account with the role. But to make this demonstration more interesting, we willcreate a Pod that's running under the `pod-logger` service account. This will essentially make the Pod unable to query the API for Pods because it will be *unauthorized* (remember that everything is denied by default in RBAC). Create a YAML manifest named `pod-logger-static-pod.yaml` for a static Pod called `pod-logger-static`, running without any additional controllers:

```
apiVersion: v1
kind: Pod
metadata:
  name: pod-logger-static
spec:
  serviceAccountName: pod-logger
  containers:
  - name: logger
    image: radial/busyboxplus:curl
    command:
    - /bin/sh
    - -c
```

```
      - |
        SERVICEACCOUNT=/var/run/secrets/kubernetes.io/
    serviceaccount
        TOKEN=$(cat ${SERVICEACCOUNT}/token)
        while true
        do
            echo "Querying Kubernetes API Server for Pods in
    default namespace..."
            curl-cacert $SERVICEACCOUNT/ca.crt-header
    "Authorization: Bearer $TOKEN" -X GET https://
    kubernetes/api/v1/namespaces/default/pods
            sleep 10
        done
```

Here, the most important fields are .spec.serviceAccountName, which specifies the service account that the Pod should run under, and command in the container definition, which we have overridden to periodically query the Kubernetes API. Assigning the pod-logger service account, as explained in the previous section, will result in a Secret with a bearer JWT for this account to be mounted in the container filesystem under /var/run/secrets/kubernetes.io/serviceaccount/token. The overridden commands run an infinite loop in a Bourne shell in 10-second intervals. In each iteration, we query the Kubernetes API endpoint (https://kubernetes/api/v1/namespaces/default/pods) for Pods in the default namespace with the HTTP GET method using the curl command. To properly authenticate, we need to pass the contents of /var/run/secrets/kubernetes.io/serviceaccount/token as a **bearer** token in the Authorization header for the request. Additionally, we need to pass a CA certificate path to verify the remote server using the cacert argument. The certificate is injected into /var/run/secrets/kubernetes.io/serviceaccount/ca.crt by the Kubernetes runtime.

6. When you create this Pod and inspect its logs, you should expect to see just a bunch of messages with an HTTP status code of 403 (Forbidden). This is because the ServiceAccount does not have a RoleBinding type that associates it with the pod-reader Role yet. First, apply the manifest to the cluster:

```
$ kubectl apply -f ./pod-logger-static-pod.yaml
```

7. Start following the logs of the pod-logger-static Pod using the following command:

```
$ kubectl logs pod-logger-static -f
Querying Kubernetes API Server for Pods in default
namespace...
...
{
  "kind": "Status",
  "apiVersion": "v1",
  "metadata": {

  },
  "status": "Failure",
  "message": "pods is forbidden: User \"system:serviceac
count:default:pod-logger\" cannot list resource \"pods\"
in API group \"\" in the namespace \"default\"",
  "reason": "Forbidden",
  "details": {
  "kind": "pods"
  },
  "code": 403
}
```

8. In a new console window, we will create and apply a RoleBinding that *associates* the ServiceAccount with the pod-reader Role. Create a YAML manifest named read-pods-rolebinding.yaml that contains the following contents:

```
apiVersion: rbac.authorization.k8s.io/v1
kind: RoleBinding
metadata:
  name: read-pods
  namespace: default
subjects:
- kind: ServiceAccount
  name: pod-logger
  namespace: default
roleRef:
  kind: Role
```

```
name: pod-reader
apiGroup: rbac.authorization.k8s.io
```

There are three key components in the RoleBinding manifest: `name`, which is used to identify the user, `subjects`, which reference the users, groups, or service accounts, and `roleRef`, which references the role.

9. Apply the manifest file using the following command:

```
$ kubectl apply -f ./read-pods-rolebinding.yaml
```

10. In the previous console window, which still follows the logs of the Pod, you will see that the Pod was able to successfully retrieve the list of Pods in the cluster. In other words, the request was successfully authorized:

```
$ kubectl logs pod-logger-static -f
...
Querying Kubernetes API Server for Pods in default
namespace...
...
{
"kind": "PodList",
"apiVersion": "v1",
"metadata": {
"selfLink": "/api/v1/namespaces/default/pods",
"resourceVersion": "4052324"
},
"items": [
    {
"metadata": {
"name": "alertmanager-prometheus-stack-test-kube-
alertmanager-0",
"generateName": "alertmanager-prometheus-stack-test-kube-
alertmanager-",
"namespace": "default",
...
```

11. Lastly, you can delete the RoleBinding type using the following command:

```
$ kubectl delete rolebinding read-pods
```

12. Now, if you inspect the logs of the Pod again, you will see that the requests are denied with an HTTP status code of 403 again.

Congratulations! You have successfully used RBAC in Kubernetes to be able to read the Pods in the cluster for a Pod running under ServiceAccount. Next, we will take a look at how to practically integrate AKS with AAD for authentication and authorization.

Azure Kubernetes Service and Azure Active Directory integration

Tier 1 cloud service providers such as Microsoft Azure, Google Cloud Platform, and Amazon Web Services have their own **managed** Kubernetes cluster offerings. We covered the Kubernetes deployments for these three cloud platforms in the previous chapters. What is important here is that managed Kubernetes clusters come with a lot of additional integrations with other cloud services. In this section, we will show you how to use **AAD** integrations for **AKS** to provide authentication using OpenID Connect and authorization using Azure RBAC. This approach unifies user management and access control across Azure resources, AKS, and Kubernetes resources.

> **Important note**
>
> At the time of writing, integration with AAD for authentication in AKS is in *general availability* and may be enabled on demand. Azure RBAC for Kubernetes authorization is currently in *preview* and can be enabled only when creating a new cluster. It will be possible to enable it on demand when the feature reaches *general availability*. For this reason, we will demonstrate these two features by deploying a new cluster from scratch.

Let's begin this demonstration by taking care of the prerequisites.

Prerequisites

First, we need to ensure that the prerequisites have been installed and enabled (these steps must be fulfilled when the feature is still in preview; otherwise, a standard Azure CLI installation should be sufficient):

1. Using the Azure CLI, register the `EnableAzureRBACPreview` feature flag using the `az feature register` command:

    ```
    $ az feature register—namespace "Microsoft.
    ContainerService"—name "EnableAzureRBACPreview"
    ```

2. Wait for the flag to be registered; this can take a few minutes. You can query the status using the following command:

    ```
    $ az feature list -o table | grep EnableAzureRBACPreview
    Microsoft.ContainerService/EnableAzureRBACPreview
    Registered
    ```

3. When the status turns into `Registered`, perform a registration refresh of the `Microsoft.ContainerService` resource provider:

    ```
    $ az provider register—namespace Microsoft.
    ContainerService
    ```

4. Install the `aks-preview` CLI extension and update it to the latest version (`0.4.55` or higher, if required):

    ```
    $ az extension add—name aks-preview
    $ az extension update—name aks-preview
    ```

With all the prerequisites ready, we can deploy the managed AKS cluster with AAD integration and Azure RBAC integration.

Deploying a managed AKS cluster with AAD and Azure RBAC integration

To deploy the cluster, follow these steps:

1. If you haven't created a resource group named `k8sforbeginners-rg` yet, you need to create it using the following command:

    ```
    $ az group create—name k8sforbeginners-rg—location eastus
    ```

2. Start provisioning a cluster named `k8sforbeginners-aks-aad` with the AAD and Azure RBAC integration features:

```
$ az aks create—resource-group k8sforbeginners-rg—name
k8sforbeginners-aks-aad—node-count 2 --enable-aad—enable-
azure-rbac
```

3. This will take a few minutes. In the end, you should see that the following section is present in the response body:

```
"aadProfile": {
    "adminGroupObjectIds": null,
    "clientAppId": null,
    "enableAzureRbac": true,
    "managed": true,
    "serverAppId": null,
    "serverAppSecret": null,
    "tenantId": ...
},
```

We now have a managed AKS cluster with AAD and Azure RBAC integration ready. Next, we are going to access the cluster using `kubectl` to verify the AAD integration.

Accessing the AKS cluster with AAD integration enabled

In the previous sections, we explained what the available authentication modes in Kubernetes are. One of them is **OpenID Connect** integration, which is an identity layer built on top of the OAuth 2.0 protocol to provide **single sign-on** (**SSO**) capabilities. AKS cluster with AAD integration internally relies on this authentication mode. The most important benefit of such an integration is that you can manage users and groups in AAD, just like with any Azure service. Your AKS cluster will seamlessly use AAD to authenticate normal users! This means you can build RBAC policies on top of that.

To access the newly deployed AKS cluster with `kubectl`, please follow these steps:

1. The AAD user that you use for the Azure CLI needs to have the `Azure Kubernetes Service Cluster User` role (`https://docs.microsoft.com/en-us/azure/role-based-access-control/built-in-roles#azure-kubernetes-service-cluster-user-role`). Of course, if you are the owner of the subscription, this is enough.

2. Execute the following Azure CLI command, which will generate `kubeconfig` for accessing the cluster. If you are presented with any SSO instructions, please follow them:

```
$ az aks get-credentials—resource-group
k8sforbeginners-rg—name k8sforbeginners-aks-aad
```

3. Now, attempt to get Pods in the cluster:

```
$ kubectl get pods
To sign in, use a web browser to open the page https://
microsoft.com/devicelogin and enter the code ... to
authenticate.
```

4. You need to complete the SSO process using the provided code. Eventually, you will see the following in your browser (provided that you have the correct AAD role assigned):

■■ Microsoft

Azure Kubernetes Service AAD Client

You have signed in to the Azure Kubernetes Service AAD Client application on your device. You may now close this window.

Figure 18.6 – Signing into AKS integrated with AAD

5. However, you will be presented with, maybe surprisingly, a message stating that listing Pods is forbidden:

```
$ kubectl get pods
...
Error from server (Forbidden): pods is forbidden:
User "9b3fde3b-4059-40fa-9e93-4147cc93164d" cannot
```

```
list resource "pods" in API group "" in the namespace
"default": User does not have access to the resource in
Azure. Update role assignment to allow access.
```

This is *expected*. The message is coming from the Kubernetes API server authorization module, which means that authentication using AAD SSO was successful! The reason that we have been forbidden access is that we are also using Azure RBAC integration and by default, our user does not have any Kubernetes Roles assigned. Take note of the AAD user principal ID (in this example, `9b3fde3b-4059-40fa-9e93-4147cc93164d`) as it will be needed in the next steps. Alternatively, you can use the `az ad signed-in-user show` command and check `objectId`.

> Tip
> AAD offers various solutions for managing access to resources. You can read the following documentation if you are interested in providing conditional access and **just in time** (JIT) access to an AKS cluster: `https://docs.microsoft.com/en-us/azure/aks/managed-aad#use-conditional-access-with-azure-ad-and-aks`, `https://docs.microsoft.com/en-us/azure/aks/managed-aad#configure-just-in-time-cluster-access-with-azure-ad-and-aks`. JIT access is the most secure way to allow elevated access to the cluster for a *limited time* and with *full auditing* capabilities. This is regarded as an industry standard for securely managing production clusters.

Now, let's learn how to work with Azure RBAC integration for AKS.

Using Azure RBAC for an AKS cluster

To make it possible for our AAD user to list and manage the Pods in the cluster, we will do two things. First, we are going to use the built-in `Azure Kubernetes Service RBAC Admin` role in Azure RBAC, which is essentially a **superuser** administration role that allows you to perform any action on any resource. This role should rarely be used as it violates the principle of least privilege. If you are going to use such a highly privileged role for production systems, you need to consider using JIT cluster access. The alternative way of going about this would be to create AAD custom roles, where you can create your own role that allows you to manage Pods in the `default` namespace. Such a role is defined in Azure RBAC, but because we have Azure RBAC integration turned on for an AKS cluster, this will be effective for Kubernetes resources.

To use the built-in `Azure Kubernetes Service RBAC Admin` role, follow these steps:

1. Get the resource ID of your AKS cluster using the following command and store it in the `AKS_ID` environment variable:

```
$ AKS_ID=$(az aks show-resource-group k8sforbeginners-rg-
name k8sforbeginners-aks-aad-query id -o tsv)
$ echo $AKS_ID
/subscriptions/.../resourcegroups/k8sforbeginners-rg/
providers/Microsoft.ContainerService/managedClusters/
k8sforbeginners-aks-aad
```

2. Using your user principal ID from the previous steps (in our case, `9b3fde3b-4059-40fa-9e93-4147cc93164d`), create an Azure RBAC role assignment for your user and the `Azure Kubernetes Service RBAC Admin` role:

```
$ az role assignment create-role "Azure Kubernetes
Service RBAC Admin"-assignee 9b3fde3b-4059-40fa-9e93-
4147cc93164d-scope $AKS_ID
```

3. After a while, attempt to get the Pods from all the namespaces using the `kubectl` command:

```
$ kubectl get pods -A
NAMESPACE      NAME                                   READY
STATUS    RESTARTS    AGE
kube-system    coredns-748cdb7bf4-kjbxb               1/1
Running    0            58m
kube-system    coredns-748cdb7bf4-ww4gg               1/1
Running    0            60m
...
```

4. Since we want to demonstrate using an Azure RBAC custom role, delete the role assignment. First, you need to get the ID of the assignment and then pass it to the second command, like so:

```
$ az role assignment list-scope $AKS_ID-query [].id -o
tsv
/subscriptions/.../resourcegroups/k8sforbeginners-rg/
providers/Microsoft.ContainerService/managedClusters/
k8sforbeginners-aks-aad/providers/Microsoft.
Authorization/roleAssignments/9d67b507-7f87-44dc-a3f1-
7fd053b308f6
```

```
$ az role assignment delete-ids /subscriptions/
cc9a8166-829e-401e-a004-76d1e3733b8e/resourcegroups/
k8sforbeginners-rg/providers/Microsoft.ContainerService/
managedClusters/k8sforbeginners-aks-aad/providers/
Microsoft.Authorization/roleAssignments/9d67b507-7f87-
44dc-a3f1-7fd053b308f6
```

5. Lastly, you can verify that `kubectl get pods` returns `forbidden` again.

The alternative solution would be to create an Azure RBAC custom role that allows you to manage Pods. You can check the full list of available actions for roles in the official documentation: `https://docs.microsoft.com/en-us/azure/role-based-access-control/resource-provider-operations#microsoftcontainer service`. We are interested in the following actions:

- `Microsoft.ContainerService/managedClusters/pods/read`

- `Microsoft.ContainerService/managedClusters/pods/write`

- `Microsoft.ContainerService/managedClusters/pods/delete`

To create the custom role and assign it to your AAD user, follow these steps:

1. Determine your subscription ID using the `az account show` command. The subscription ID will be present under the `id` property in the output.

2. Create an `aks-pod-writer.json` file that contains a role definition that allows you to read, write, and delete Pods in the AKS cluster. Replace `<subscriptionId>` with your subscription ID:

```
{
     "Name": "AKS Pods Writer",
     "Description": "Allows read-write management of Pods
in cluster/namespace.",
     "Actions": [],
     "NotActions": [],
     "DataActions": [
          "Microsoft.ContainerService/managedClusters/pods/
read",
          "Microsoft.ContainerService/managedClusters/pods/
write",
          "Microsoft.ContainerService/managedClusters/pods/
delete"
     ],
```

```
    "NotDataActions": [],
    "assignableScopes": [
        "/subscriptions/<subscriptionId>"
    ]
}
```

3. Use the following command to create a custom role definition:

```
$ az role definition create—role-definition @aks-pod-
writer.json
```

4. Now, you can assign the new role to your user (in our example, 9b3fde3b-4059-
40fa-9e93-4147cc93164d):

```
$ az role assignment create—role "AKS Pods Writer"—
assignee 9b3fde3b-4059-40fa-9e93-4147cc93164d—scope $AKS_
ID
```

5. After a while, you can check the Pods in the cluster using the kubectl get pods
-A command. Note that the new role assignments can take up to 5 minutes to
propagate:

```
$ kubectl get pods -A
NAMESPACE       NAME                                   READY
STATUS      RESTARTS     AGE
kube-system     coredns-748cdb7bf4-kjbxb               1/1
Running     0            89m
kube-system     coredns-748cdb7bf4-ww4gg               1/1
Running     0            90m
```

6. Now, try to list all the deployments in the cluster – you will get a Forbidden
result, as expected:

```
$ kubectl get deploy -A
Error from server (Forbidden): deployments.apps is
forbidden: User "9b3fde3b-4059-40fa-9e93-4147cc93164d"
cannot list resource "deployments" in API group "apps"
at the cluster scope: User does not have access to the
resource in Azure. Update role assignment to allow
access.
```

7. Lastly, we need to verify whether we can create Pods. Run a simple static Pod with the `busybox` container image:

```
$ kubectl run -i—tty busybox—image=busybox:1.28 --rm—
restart=Never—sh
If you don't see a command prompt, try pressing enter.
/ #
```

As you can see, we have successfully used Azure RBAC roles to manage authorization in an AKS cluster. What's more, we haven't manually managed the Role and RoleBinding objects in the Kubernetes cluster itself. Please note that most of the actions that we have executed in the Azure CLI can be also done in the Azure portal.

> **Important note**
>
> If you do not need the AKS resources, remember to clean them up to avoid any unnecessary costs.

Congratulations! You have successfully deployed a managed AKS cluster with AAD and Azure RBAC integration. Now, let's summarize what you have learned in this chapter.

Summary

This chapter covered *authentication* and *authorization* in Kubernetes. First, we provided an overview of the available authentication methods in Kubernetes and explained how you can use ServiceAccount tokens for external user authentication. Next, we focused on RBAC in Kubernetes. You learned how to use Roles, ClusterRoles, RoleBindings, and ClusterRoleBindings to manage authorization in your cluster. We demonstrated a practical use case of RBAC for ServiceAccounts by creating a Pod that can list Pods in the cluster using the Kubernetes API (respecting the principle of least privilege). Finally, we provided an overview of how easily you can integrate your AKS with AAD for single sign-on authentication and Azure RBAC for authorization.

In the next chapter, we are going to dive deep into advanced techniques for scheduling Pods.

Further reading

For more information regarding authorization and authentication in Kubernetes, please refer to the following PacktPub books:

- *The Complete Kubernetes Guide*, by *Jonathan Baier, Gigi Sayfan, Jesse White* (`https://www.packtpub.com/virtualization-and-cloud/complete-kubernetes-guide`)

- *Getting Started with Kubernetes – Third Edition*, by *Jonathan Baier, Jesse White* (`https://www.packtpub.com/virtualization-and-cloud/getting-started-kubernetes-third-edition`)

- *Kubernetes for Developers*, by *Joseph Heck* (`https://www.packtpub.com/virtualization-and-cloud/kubernetes-developers`)

You can also refer to the official documentation:

- The Kubernetes documentation (`https://kubernetes.io/docs/home/`), which is always the most up-to-date source of knowledge about Kubernetes in general.

- AKS authentication and authorization best practices are available in the official Microsoft documentation: `https://docs.microsoft.com/en-us/azure/aks/operator-best-practices-identity`.

- Details about Azure RBAC for Kubernetes are documented at `https://docs.microsoft.com/en-us/azure/aks/manage-azure-rbac`.

- More advanced use cases for Azure RBAC for Kubernetes are covered in this guide: `https://docs.microsoft.com/en-us/azure/aks/azure-ad-rbac`.

19
Advanced Techniques for Scheduling Pods

At the beginning of the book, in *Chapter 2*, *Kubernetes Architecture – From Docker Images to Running Pods*, we explained the principles behind the **Kubernetes scheduler (kube-scheduler)** control plane component and its crucial role in the cluster. In short, its responsibility is to schedule container workloads (Kubernetes Pods) and assign them to healthy worker **Nodes** that fulfill the *criteria* required for running a particular workload.

This chapter will cover how you can control the criteria for scheduling Pods in the cluster. We will especially dive deeper into Node **affinity**, **taints**, and **tolerations** for Pods. We will also take a closer look at **scheduling policies**, which give kube-scheduler flexibility in how it prioritizes Pod workloads. You will find all of these concepts important in running production clusters at cloud scale.

In this chapter, we will cover the following topics:

- Refresher – What is kube-scheduler?
- Managing Node affinity
- Using Node taints and tolerations
- Scheduling policies

Technical requirements

For this chapter, you will need the following:

- Kubernetes cluster deployed. We recommend using a *multi-node*, cloud-based Kubernetes cluster. Having a multi-node cluster will make understanding Node affinity, taints, and tolerations much easier.
- Kubernetes CLI (`kubectl`) installed on your local machine and configured to manage your Kubernetes cluster.

Basic Kubernetes cluster deployment (local and cloud-based) and `kubectl` installation have been covered in *Chapter 3, Installing Your First Kubernetes Cluster*.

The following previous chapters can give you an overview of how to deploy a fully functional Kubernetes cluster on different cloud platforms:

- *Chapter 14, Kubernetes Clusters on Google Kubernetes Engine*
- *Chapter 15, Launching a Kubernetes Cluster on Amazon Web Services with the Amazon Elastic Kubernetes Service*
- *Chapter 16, Kubernetes Clusters on Microsoft Azure with the Azure Kubernetes Service*

You can download the latest code samples for this chapter from the official GitHub repository: `https://github.com/PacktPublishing/The-Kubernetes-Bible/tree/master/Chapter19`.

Refresher – What is kube-scheduler?

In Kubernetes clusters, kube-scheduler is a component of the control plane that runs on **Master Nodes**. The main responsibility of this component is scheduling container workloads (Pods) and **assigning** them to healthy worker Nodes that fulfill the criteria required for running a particular workload. To recap, a Pod is a group of one or more containers with a shared network and storage and is the smallest **deployment unit** in the Kubernetes system. You usually use different Kubernetes controllers, such as Deployment objects and StatefulSet objects, to manage your Pods, but it is kube-scheduler that eventually assigns the created Pods to particular Nodes in the cluster.

> **Important note**
>
> For **managed** Kubernetes clusters in the cloud, such as the managed **Azure Kubernetes Service** or the **Amazon Elastic Kubernetes Service**, you normally do not have access to the Master Nodes, as they are managed by the cloud service provider for you. This means you will not have access to kube-scheduler itself, and usually, you cannot control its configuration, such as scheduling policies. But you can control all parameters for Pods that influence their scheduling.

Kube-scheduler queries the **Kubernetes API Server (kube-apiserver)** at a regular interval in order to list the Pods that have not been *scheduled*. At creation, Pods are marked as *not* scheduled – this means no worker Node was elected to run them. A Pod that is not scheduled will be registered in the `etcd` cluster state but without any worker Node assigned to it, and thus, no running kubelet will be aware of this Pod. Ultimately, no container described in the Pod specification will run at this point.

Internally, the Pod object, as it is stored in `etcd`, has a property called `nodeName`. As the name suggests, this property should contain the name of the worker Node that will host the Pod. When this property is set, we say the Pod is in a `scheduled` state, otherwise, the Pod is in a `pending` state.

We need to find a way to fill this value, and this is the role of the kube-scheduler. For this, the kube-scheduler poll continues the kube-apiserver at a regular interval. It looks for Pod resources with an empty `nodeName` property. Once it finds such Pods, it will execute an algorithm to elect a worker Node and will update the `nodeName` property in the Pod object, by issuing an HTTP request to the kube-apiserver. When electing a worker Node, the kube-scheduler will take into account its internal scheduling policies and criteria that you defined for the Pods. Finally, the kubelet which is responsible for running Pods on the selected worker Node will notice that there is a new Pod in the `scheduled` state for the Node and will attempt starting the Pod. These principles have been visualized in the following diagram:

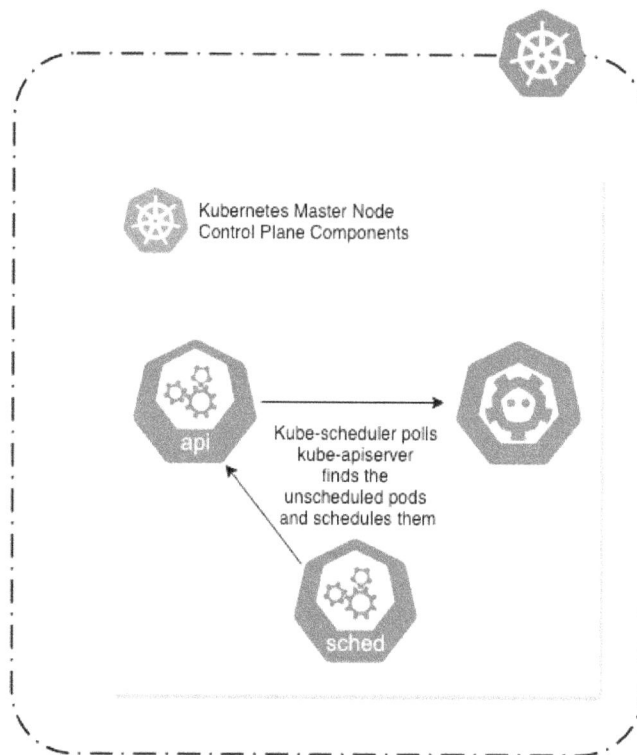

Figure 19.1 – Interactions of kube-scheduler and Kubernetes API server

The scheduling process for a Pod is performed in two phases:

- **Filtering**: Kube-scheduler determines the set of Nodes that are capable of running a given Pod. This includes checking the actual state of the Nodes and verifying any resource requirements and criteria specified by the Pod definition. At this point, if there are no Nodes that can run a given Pod, the Pod cannot be scheduled and remains pending.

- **Scoring**: Kube-scheduler assigns scores for each Node based on a set of **scheduling policies**. Then, the Pod is assigned by the scheduler to the Node with the highest score. We will cover scheduling policies in the last section of this chapter.

The kube-scheduler will consider criteria and configuration values you can optionally pass in the Pod specification. By using these configurations, you can control precisely how the kube-scheduler will elect a worker Node.

> **Important note**
>
> The decisions of kube-scheduler are valid *precisely* at the point in time of scheduling the Pod. When the Pod gets scheduled and is running, kube-scheduler will *not* do any rescheduling operations while it is running (which can be days or even months). So even if this Pod no longer matches the Node according to your rules, it will remain running. Rescheduling will only happen if the Pod is terminated and a new Pod needs to be scheduled.

In the next sections, we will discuss the following configurations to control the scheduling of Pods:

- **Node name** and **Node selector**, which are the simplest forms of static scheduling.
- **Node affinity** and **inter-Pod affinity/anti-affinity**.
- **Taints** and **tolerations**.

Let's first take a look at Node affinity, together with Node name and Node selector.

Managing Node affinity

To better understand how Node **affinity** works in Kubernetes, we need first to take a look at the most basic scheduling options, which are using Node name and Node selector for Pods.

Pod Node name

As we mentioned before, each Pod object has a nodeName field which is usually controlled by the kube-scheduler. Nevertheless, it is possible to set this property directly in the YAML manifest when you create a Pod or create a controller that uses a Pod template. This is the simplest form of statically scheduling Pods on a given Node and is generally *not recommended* – it is not flexible and does not scale at all. The names of Nodes can change over time and you risk running out of resources on the Node.

> **Tip**
>
> You may find setting `nodeName` explicitly useful in debugging scenarios when you want to run a Pod on a specific Node.

We are going to demonstrate all scheduling principles on an example Deployment object that we introduced in *Chapter 11, Deployment – Deploying Stateless Applications*. This is a simple Deployment that manages *five* Pod replicas of an `nginx` webserver running in a container. Create the following YAML manifest named `nginx-deployment.yaml`:

```yaml
apiVersion: apps/v1
kind: Deployment
metadata:
  name: nginx-deployment-example
spec:
  replicas: 5
  selector:
    matchLabels:
      app: nginx
      environment: test
  template:
    metadata:
      labels:
        app: nginx
        environment: test
    spec:
      containers:
      - name: nginx
        image: nginx:1.17
        ports:
        - containerPort: 80
```

At this point, the Pod template in `.spec.template.spec` does not contain any configuration that affects the scheduling of the Pod replicas. Before we apply the manifest to the cluster, we need to know what Nodes we have in the cluster so that we can understand how they are scheduled and how we can influence the scheduling of Pods. You can get the list of Nodes using the `kubectl get nodes` command:

```
$ kubectl get nodes
NAME                              STATUS   ROLES   AGE
VERSION
aks-nodepool1-77120516-vmss000000 Ready    agent   1d
v1.18.14
aks-nodepool1-77120516-vmss000001 Ready    agent   1d
v1.18.14
aks-nodepool1-77120516-vmss000002 Ready    agent   1d
v1.18.14
```

In our example, we are running a three-Node cluster. For simplicity, we will refer to `aks-nodepool1-77120516-vmss000000` as Node 0, `aks-nodepool1-77120516-vmss000001` as Node 1, and `aks-nodepool1-77120516-vmss000002` as Node 2.

Now, let's apply the `nginx-deployment.yaml` YAML manifest to the cluster:

```
$ kubectl apply -f ./nginx-deployment.yaml
deployment.apps/nginx-deployment-example created
```

The Deployment object will create five Pod replicas. You can get their statuses, together with the Node names that they were scheduled for, using the following command:

```
$ kubectl get pods --namespace default --output=custom-
columns="NAME:.metadata.name,STATUS:.status.phase,NODE:.spec.
nodeName"
NAME                                                  STATUS     NODE
nginx-deployment-example-5549875c78-nndb4    Running
aks-nodepool1-77120516-vmss000001
nginx-deployment-example-5549875c78-ps7pd    Running
aks-nodepool1-77120516-vmss000000
nginx-deployment-example-5549875c78-s824f    Running
aks-nodepool1-77120516-vmss000002
nginx-deployment-example-5549875c78-xfbkj    Running
aks-nodepool1-77120516-vmss000002
nginx-deployment-example-5549875c78-zg2w7    Running
aks-nodepool1-77120516-vmss000000
```

As you can see, by default the Pods have been distributed uniformly – Node 0 has received two Pods, Node 1 one Pod, and Node 2 two Pods. This is a result of the default scheduling policies enabled in the kube-scheduler for filtering and scoring.

> Tip
> If you are running a **non-managed** Kubernetes cluster, you can inspect the logs for the kube-scheduler Pod using the `kubectl logs` command, or even directly at the master Nodes in `/var/log/kube-scheduler.log`. This may also require increased verbosity of logs for the kube-scheduler process. You can read more at `https://kubernetes.io/docs/reference/command-line-tools-reference/kube-scheduler/`.

We will now **forcefully** assign all Pods in the Deployment to Node 0 in the cluster using the `nodeName` field in the Pod template. Change the `nginx-deployment.yaml` YAML manifest so that it has this property set with the correct Node name for *your* cluster:

```
apiVersion: apps/v1
kind: Deployment
metadata:
  name: nginx-deployment-example
```

```
spec:
...
  template:
...
    spec:
      nodeName: aks-nodepool1-77120516-vmss000000
...
```

Apply the manifest to the cluster using the `kubectl apply -f ./nginx-deployment.yaml` command and inspect the Pod status and Node assignment again:

```
$ kubectl get pods --namespace default --output=custom-
columns="NAME:.metadata.name,STATUS:.status.phase,NODE:.spec.
nodeName"
```

NAME	STATUS	NODE
nginx-deployment-example-6977595df5-95sfh aks-nodepool1-77120516-vmss000000	Running	
nginx-deployment-example-6977595df5-cxgqb aks-nodepool1-77120516-vmss000000	Running	
nginx-deployment-example-6977595df5-h5wwk aks-nodepool1-77120516-vmss000000	Running	
nginx-deployment-example-6977595df5-pww9g aks-nodepool1-77120516-vmss000000	Running	
nginx-deployment-example-6977595df5-q5xxs aks-nodepool1-77120516-vmss000000	Running	

As expected, *all five* Pods are now running on Node 0. These are all new Pods – when you change the Pod template in the Deployment specification, it causes internally a rollout using a new ReplicaSet object, while the old ReplicaSet object is scaled down, as explained in *Chapter 11, Deployment – Deploying Stateless Applications*.

> **Important note**
>
> In this way, we have actually **bypassed** kube-scheduler. If you inspect events for one of the Pods using the `kubectl describe pod` command, you will see that it lacks any events with `Scheduled` as a reason.

Next, we are going to take a look at another basic method of scheduling Pods, which is the Node selector.

Pod Node selector

Pod specification has a special field, `.spec.nodeSelector`, that gives you the ability to schedule your Pod only on Nodes that have certain label values. This concept is similar to **label selectors** that you know from Deployments or StatefulSets, but the difference is that it allows only simple *equality-based* comparisons for labels. You cannot do advanced *set-based* logic.

A very common use case for scheduling Pods using `nodeSelector` is managing Pods in hybrid Windows/Linux clusters. Every Kubernetes Node comes by default with a set of labels, which include the following:

- `kubernetes.io/arch`: Describes the Node's processor architecture, for example, `amd64` or `arm`. This is also defined as `beta.kubernetes.io/arch`.

- `kubernetes.io/os`: Has a value of `linux` or `windows`. This is also defined as `beta.kubernetes.io/os`.

If you inspect the labels for one of the Nodes, you will see that there are plenty of them – in our case some of them are specific to **Azure Kubernetes Service** (AKS) clusters only:

```
$ kubectl describe node aks-nodepool1-77120516-vmss000000
...
Labels:              agentpool=nodepool1
                     beta.kubernetes.io/arch=amd64
                     beta.kubernetes.io/instance-type=Standard_
DS2_v2
                     beta.kubernetes.io/os=linux
                     failure-domain.beta.kubernetes.io/
region=eastus
                     failure-domain.beta.kubernetes.io/zone=0
                     kubernetes.azure.com/cluster=MC_
k8sforbeginners-rg_k8sforbeginners-aks_eastus
                     kubernetes.azure.com/mode=system
                     kubernetes.azure.com/node-image-
version=AKSUbuntu-1804gen2-2021.02.17
                     kubernetes.azure.com/role=agent
                     kubernetes.io/arch=amd64
                     kubernetes.io/hostname=aks-nodepool1-
77120516-vmss000000
                     kubernetes.io/os=linux
                     kubernetes.io/role=agent
```

```
                        node-role.kubernetes.io/agent=
                        node.kubernetes.io/instance-type=Standard_
DS2_v2
                        storageprofile=managed
                        storagetier=Premium_LRS
                        topology.kubernetes.io/region=eastus
                        topology.kubernetes.io/zone=0
...
```

Of course, you can define your *own* labels for the Nodes and use them to control scheduling. Please note that in general you should use semantic labeling for your resources in Kubernetes, rather than give them special labels just for the purpose of scheduling. Let's demonstrate how to do that by following these steps:

1. Use the kubectl label nodes command to add a node-type label with a superfast value to Node 1 and Node 2 in the cluster:

```
$ kubectl label nodes aks-nodepool1-77120516-vmss000001
node-type=superfast
node/aks-nodepool1-77120516-vmss000001 labeled
$ kubectl label nodes aks-nodepool1-77120516-vmss000002
node-type=superfast
node/aks-nodepool1-77120516-vmss000002 labeled
```

2. Edit the ./nginx-deployment.yaml deployment manifest so that it has nodeSelector in the Pod template set to node-type: superfast (nodeName that we used previously should be removed):

```
apiVersion: apps/v1
kind: Deployment
metadata:
  name: nginx-deployment-example
spec:
...
  template:
...
    spec:
      nodeSelector:
        node-type: superfast
...
```

3. Apply the manifest to the cluster using the `kubectl apply -f ./nginx-deployment.yaml` command and inspect the Pod status and Node assignment again. You may need to wait a while for the deployment rollout to finish:

```
$ kubectl get pods --namespace default --output=custom-
columns="NAME:.metadata.name,STATUS:.status.phase,NODE:.
spec.nodeName"
NAME                                                STATUS
NODE
nginx-deployment-example-8485bc9569-2pm5h           Running
aks-nodepool1-77120516-vmss000001
nginx-deployment-example-8485bc9569-79gn9           Running
aks-nodepool1-77120516-vmss000002
nginx-deployment-example-8485bc9569-df6x8           Running
aks-nodepool1-77120516-vmss000001
nginx-deployment-example-8485bc9569-fd4gv           Running
aks-nodepool1-77120516-vmss000002
nginx-deployment-example-8485bc9569-tlxgl           Running
aks-nodepool1-77120516-vmss000002
```

As you can see, Node 1 has been assigned with two Pods and Node 2 with three Pods. The Pods have been distributed among Nodes that have the `node-type=superfast` label.

4. In contrast, if you change the `./nginx-deployment.yaml` manifest so that it has `nodeSelector` in the Pod template set to `node-type: slow`, which no Node in the cluster has assigned, we will see that Pods could not be scheduled and the deployment will be stuck. Edit the manifest:

```
apiVersion: apps/v1
kind: Deployment
metadata:
  name: nginx-deployment-example
spec:
...
  template:
...
    spec:
      nodeSelector:
        node-type: slow
...
```

5. Apply the manifest to the cluster using the `kubectl apply -f ./nginx-deployment.yaml` command and inspect the Pod status and Node assignment again:

```
$ kubectl get pods --namespace default --output=custom-
columns="NAME:.metadata.name,STATUS:.status.phase,NODE:.
spec.nodeName"
```

NAME NODE	STATUS
nginx-deployment-example-54dbf4699f-jdx42 <none>	Pending
nginx-deployment-example-54dbf4699f-sk2jd <none>	Pending
nginx-deployment-example-54dbf4699f-xjdp2 <none>	Pending
nginx-deployment-example-8485bc9569-2pm5h aks-nodepool1-77120516-vmss000001	Running
nginx-deployment-example-8485bc9569-df6x8 aks-nodepool1-77120516-vmss000001	Running
nginx-deployment-example-8485bc9569-fd4gv aks-nodepool1-77120516-vmss000002	Running
nginx-deployment-example-8485bc9569-tlxgl aks-nodepool1-77120516-vmss000002	Running

The reason why three new Pods are pending and four old Pods are still running is the default configuration of rolling updates in the Deployment object. By default, maxSurge is set to 25% of Pod replicas (absolute number is *rounded up*), so in our case, it is two Pods allowed to be created above the desired number of five Pods. In total, we now have seven Pods. At the same time, maxUnavailable is also 25% of Pod replicas (but absolute number is *rounded down*), so in our case, one Pod out of five can be not available. In other words, four Pods must be Running. And because the new Pending Pods cannot get a Node in the process of scheduling, the Deployment is stuck waiting and not progressing. Normally, in this case you need to either perform a rollback to the previous version for the Deployment or change nodeSelector to one that matches existing Nodes properly. Of course, there is also an alternative of adding a new Node with matching labels or adding missing labels to the existing ones, without performing a rollback.

We will now continue the topic of scheduling Pods with the first of more advanced techniques: Node affinity.

Node affinity configuration for Pods

The concepts of Node affinity expand the `nodeSelector` approach and provide a richer language for defining which Nodes are preferred or avoided for your Pod. In everyday life, the word affinity describes *a natural liking for and understanding of someone or something*, and this best describes the purpose of Node affinity for Pods. That is, you can control which Nodes your Pod will be *attracted* to or *repelled* by.

With Node affinity, represented in `.spec.affinity.nodeAffinity` for the Pod, you get the following enhancements over simple `nodeSelector`:

- You get a richer language for expressing the rules for matching Pods to Nodes. For example, you can use `In`, `NotIn`, `Exists`, `DoesNotExist`, `Gt`, and `Lt` operators for labels.

- Similarly to `nodeAffinity`, it is possible to do scheduling using **inter-Pod** affinity (`podAffinity`) and additionally **anti-affinity** (`podAntiAffinity`). Anti-affinity has an opposite effect to affinity – you can define rules that repel the Pods. In this way, you can make your Pods be attracted to Nodes that *already run* certain Pods. This is especially useful if you want to collocate Pods to decrease latency.

- It is possible to define **soft** affinity and anti-affinity rules that represent a *preference* instead of a **hard** rule. In other words, if the scheduler can still schedule the Pod, even if it cannot match the soft rule. Soft rules are represented by the `preferredDuringSchedulingIgnoredDuringExecution` field in specification, whereas hard rules are represented by the `requiredDuringSchedulingIgnoredDuringExecution` field.

- Soft rules can be **weighted**.

> **Tip**
> Even though there is no Node anti-affinity field provided by a separate field in spec, as in the case of inter-Pod anti-affinity you can still achieve similar results by using the `NotIn` and `DoesNotExist` operators. In this way, you can make Pods be repelled from Nodes with specific labels, also in a soft way.

The use cases and scenarios for defining the Node affinity and inter-Pod affinity/anti-affinity rules are *unlimited*. It is possible to express all kinds of requirements in this way, provided that you have enough labeling on the Nodes. For example, you can model requirements like scheduling the Pod only on a Windows Node with an Intel CPU and premium storage in the West Europe region but currently not running Pods for MySQL, or try not to schedule the Pod in availability Zone 1, but if it is not possible, then availability Zone 1 is still OK.

To demonstrate Node affinity, we will try to model the following requirements for our Deployment: "*Try* to schedule the Pod only on Nodes with a `node-type` label with a `fast` or `superfast` value, but if it this not possible, use any Node but *strictly* not with a `node-type` label with an `extremelyslow` value." For this, we need to use:

- **Soft Node affinity** rule of type `preferredDuringSchedulingIgnoredDuringExecution` to match `fast` and `superfast` Nodes.

- **Hard Node affinity** rule of type `requiredDuringSchedulingIgnoredDuringExecution` to repel the Pod strictly from Nodes with `node-type` as `extremelyslow`. We need to use the `NotIn` operator to get the anti-affinity effect.

In our cluster we are going to first have the following labeling for Nodes:

- Node 0: `slow`

- Node 1: `fast`

- Node 2: `superfast`

As you can see, according to our requirements the Deployment Pods should be scheduled on Node 1 and Node 2, unless there is something preventing them from being allocated there, like a lack of CPU or memory resources. In that case, Node 0 would also be allowed as we use the soft affinity rule.

Next, we will relabel the Nodes in the following way:

- Node 0: `slow`

- Node 1: `extremelyslow`

- Node 2: `extremelyslow`

Subsequently, we will need to redeploy our Deployment (for example, scale it down to zero and up to the original replica count, or use the `kubectl rollout restart` command) to **reschedule** the Pods again. After that, looking at our requirements, kube-scheduler should assign all Pods to Node 0 (because it is still allowed by the soft rule) but avoid *at all costs* Node 1 and Node 2. If by any chance Node 0 has no resources to run the Pod, then the Pods would be stuck in the `Pending` state.

> **Tip**
>
> To solve the issue of rescheduling already running Pods (in other words, to make kube-scheduler consider them again), there is an incubating Kubernetes project named **Descheduler**. You can find out more here: `https://github.com/kubernetes-sigs/descheduler`.

To do the demonstration, please follow these steps:

1. Use the `kubectl label nodes` command to add a `node-type` label with a `slow` value for Node 0, a `fast` value for Node 1, and a `superfast` value for Node 2:

   ```
   $ kubectl label nodes --overwrite aks-nodepool1-77120516-
   vmss000000 node-type=slow
   node/aks-nodepool1-77120516-vmss000000 labeled
   $ kubectl label nodes --overwrite aks-nodepool1-77120516-
   vmss000001 node-type=fast
   node/aks-nodepool1-77120516-vmss000001 labeled
   $ kubectl label nodes --overwrite aks-nodepool1-77120516-
   vmss000002 node-type=superfast
   node/aks-nodepool1-77120516-vmss000002 not labeled  #
   Note that this label was already present with this value
   ```

2. Edit the `./nginx-deployment.yaml` Deployment manifest (the full file is available in the official GitHub repository for the book: `https://github.com/PacktPublishing/Kubernetes-for-Beginners/blob/master/Chapter19/03_affinity/nginx-deployment.yaml`), and remove `nodeSelector`. Instead, define the soft Node affinity rule as follows:

   ```
   apiVersion: apps/v1
   kind: Deployment
   metadata:
     name: nginx-deployment-example
   spec:
   ...
     template:
   ...
       spec:
         affinity:
           nodeAffinity:
   ```

```
        requiredDuringSchedulingIgnoredDuringExecution:
          nodeSelectorTerms:
          - matchExpressions:
            - key: node-type
              operator: NotIn
              values:
              - extremelyslow
        preferredDuringSchedulingIgnoredDuringExecution:
        - weight: 1
          preference:
            matchExpressions:
            - key: node-type
              operator: In
              values:
              - fast
              - superfast
...
```

As you can see, we have used `nodeAffinity`
(not `podAffinity` or `podAntiAffinity`) with
`preferredDuringSchedulingIgnoredDuringExecution` set so that it
has only one soft rule: `node-type` *should* have a `fast` value or a `superfast`
value. This means that if there are no resources on such Nodes, they can still be
scheduled on other Nodes. Additionally, we specify one hard anti-affinity rule
in `requiredDuringSchedulingIgnoredDuringExecution`, which
says that `node-type` *must not* be `extremelyslow`. You can find the full
specification of Pod's `.spec.affinity` in the official documentation: `https://`
`kubernetes.io/docs/reference/generated/kubernetes-api/`
`v1.19/#affinity-v1-core`.

3. Apply the manifest to the cluster using the `kubectl apply -f ./nginx-`
    `deployment.yaml` command and inspect the Pod status and Node assignment
    again. You may need to wait a while for the Deployment rollout to finish:

```
$ kubectl get pods --namespace default --output=custom-
columns="NAME:.metadata.name,STATUS:.status.phase,NODE:.
spec.nodeName"
NAME                                                STATUS
NODE
```

```
nginx-deployment-example-7ff6c65bd4-8z7z5     Running
aks-nodepool1-77120516-vmss000002

nginx-deployment-example-7ff6c65bd4-ps9md     Running
aks-nodepool1-77120516-vmss000002

nginx-deployment-example-7ff6c65bd4-pszkq     Running
aks-nodepool1-77120516-vmss000001

nginx-deployment-example-7ff6c65bd4-qpv5d     Running
aks-nodepool1-77120516-vmss000001

nginx-deployment-example-7ff6c65bd4-vh6dx     Running
aks-nodepool1-77120516-vmss000002
```

Our Node affinity rules were defined to prefer Nodes that have `node-type` set to either `fast` or `superfast`, and indeed the Pods were scheduled for Node 1 and Node 2 only.

Now we will do an experiment to demonstrate how the soft part of Node affinity together with the hard part of Node anti-affinity work. We will relabel the Nodes as described in the introduction, redeploy the Deployment, and observe what happens. Please follow these steps:

1. Use the `kubectl label nodes` command to add a `node-type` label with a `slow` value for Node 0, an `extremelyslow` value for Node 1, and an `extremelyslow` value for Node 2:

    ```
    $ kubectl label nodes --overwrite aks-nodepool1-77120516-
    vmss000000 node-type=slow
    node/aks-nodepool1-77120516-vmss000000 not labeled
    $ kubectl label nodes --overwrite aks-nodepool1-77120516-
    vmss000001 node-type=extremelyslow
    node/aks-nodepool1-77120516-vmss000001 labeled
    $ kubectl label nodes --overwrite aks-nodepool1-77120516-
    vmss000002 node-type=extremelyslow
    node/aks-nodepool1-77120516-vmss000002 labeled
    ```

2. At this point, if you were to check Pods assignments using `kubectl get pods`, there would be no difference. This is because, as we explained before, a Pod's assignment to Nodes is valid only at the time of scheduling, and after that, it is not changed unless they are restarted. To force the restart of Pods, we could scale the Deployment down to zero replicas and then back to five. But there is an easier way, which is to use an imperative `kubectl rollout restart` command. This approach has the benefit of not making the Deployment unavailable, and it performs a rolling restart of Pods without a decrease in the number of available Pods. Execute the following command:

```
$ kubectl rollout restart deploy nginx-deployment-example
deployment.apps/nginx-deployment-example restarted
```

3. Inspect the Pod status and Node assignment again. You may need to wait a while for the Deployment rollout to finish:

```
$ kubectl get pods --namespace default --output=custom-
columns="NAME:.metadata.name,STATUS:.status.phase,NODE:.
spec.nodeName"
```

NAME NODE	STATUS
nginx-deployment-example-6c4fdd447d-4mjfm aks-nodepool1-77120516-vmss000000	Running
nginx-deployment-example-6c4fdd447d-qgqmc aks-nodepool1-77120516-vmss000000	Running
nginx-deployment-example-6c4fdd447d-qhrtf aks-nodepool1-77120516-vmss000000	Running
nginx-deployment-example-6c4fdd447d-tnvpm aks-nodepool1-77120516-vmss000000	Running
nginx-deployment-example-6c4fdd447d-ttfnk aks-nodepool1-77120516-vmss000000	Running

The output shows that, as expected, all Pods have been scheduled to Node 0, which is labeled with `node-type=slow`. We allow such Nodes if there is nothing better, and in this case Node 1 and Node 2 have label `node-type=extremelyslow`, which is prohibited by the hard Node anti-affinity rule.

> **Tip**
>
> To achieve even higher granularity and control of Pod scheduling, you can use **Pod topology spread constraints**. More details are available in the official documentation: `https://kubernetes.io/docs/concepts/workloads/pods/pod-topology-spread-constraints/`.

Congratulations, you have successfully configured Node affinity for our Deployment Pods! We will now explore another way of scheduling Pods – taints and tolerations.

Using Node taints and tolerations

Using the Node and inter-Pod affinity mechanism for scheduling Pods is very powerful, but sometimes you need a simpler way of specifying which Nodes should *repel* Pods. Kubernetes offers a slightly older and simpler feature for this purpose – **taints** and **tolerations**. You apply a taint to a given Node (which describes some kind of limitation) and the Pod must have a specific toleration defined to be schedulable on the tainted Node. If the Pod has a toleration, it does not mean that the taint is *required* on the Node. The real-life definition of *taint* is "a trace of a bad or undesirable substance or quality," and this reflects the idea pretty well – all Pods will *avoid* a Node if there is a taint set for them, but we can instruct Pods to *tolerate* a specific taint.

> **Tip**
>
> If you look closely at how taints and tolerations are described, you can see that you can achieve similar results with Node labels and Node hard and soft affinity rules with the `NotIn` operator. There is one catch – you can define taints with a `NoExecute` effect which will result in the termination of the Pod if it cannot tolerate it. You cannot get similar results with affinity rules unless you restart the Pod manually.

Taints for Nodes have the following structure: `<key>=<value>:<effect>`. The **key** and **value** pair *identifies* the taint and can be used for more granular tolerations, for example tolerating all taints with a given key and any value. This is similar to labels, but please bear in mind that taints are separate properties, and defining a taint does not affect Node labels. In our example demonstration, we will use our own taint with a `machine-check-exception` key and a `memory` value. This is, of course, a theoretical example where we want to indicate that there is a hardware issue with memory on the host, but you could also have a taint with the same key and instead a `cpu` or `disk` value. In general, your taints should *semantically* label the type of issue that the Node is experiencing. There is nothing preventing you from using any keys and values for creating taints, but if they make semantic sense, it is much easier to manage them and define tolerations.

The taint can have different **effects**:

- `NoSchedule` – kube-scheduler *will not schedule* Pods to this Node. Similar behavior can be achieved using a hard Node affinity rule.

- `PreferNoSchedule` – kube-scheduler *will try to not schedule* Pods to this Node. Similar behavior can be achieved using a soft Node affinity rule.

- `NoExecute` – kube-scheduler *will not schedule* Pods to this Node and *evict* (terminate and reschedule) running Pods from this Node. You cannot achieve similar behavior using Node affinity rules. Note that when you define a toleration for a Pod for this type of taint, it is possible to control how long the Pod will tolerate the taint before it gets evicted, using `tolerationSeconds`.

Kubernetes manages quite a few `NoExecute` taints automatically by monitoring the Node hosts. The following taints are built-in and managed by **NodeController** or the kubelet:

- `node.kubernetes.io/not-ready`: Added when NodeCondition `Ready` has a `false` status.

- `node.kubernetes.io/unreachable`: Added when NodeCondition `Ready` has an `Unknown` status. This happens when NodeController cannot reach the Node.

- `node.kubernetes.io/out-of-disk`: Node has no disk available.

- `node.kubernetes.io/memory-pressure`: Node is experiencing memory pressure.

- `node.kubernetes.io/disk-pressure`: Node is experiencing disk pressure.

- `node.kubernetes.io/network-unavailable`: Network is currently down on the Node.

- `node.kubernetes.io/unschedulable`: Node is currently in an `unschedulable` state.

- `node.cloudprovider.kubernetes.io/uninitialized`: Intended for Nodes that are prepared by an external cloud provider. When the Node gets initialized by cloud-controller-manager, this taint is removed.

To add a taint on a Node, you use the `kubectl taint node` command in the following way:

```
$ kubectl taint node <nodeName> <key>=<value>:<effect>
```

So, for example, if we want to use key `machine-check-exception` and a `memory` value with a `NoExecute` effect for Node 1, we will use the following command:

```
$ kubectl taint node aks-nodepool1-77120516-vmss000001 machine-check-exception=memory:NoExecute
```

To remove the same taint, you need to use the following command (bear in mind the - character at the end of the taint definition):

```
$ kubectl taint node aks-nodepool1-77120516-vmss000001 machine-check-exception=memory:NoExecute-
```

You can also remove all taints with a specified key:

```
kubectl taint node aks-nodepool1-77120516-vmss000001 machine-check-exception:NoExecute-
```

To counteract the effect of the taint on a Node for specific Pods, you can define tolerations in their specification. In other words, you can use tolerations to ignore taints and still schedule the Pods to such Nodes. If a Node has multiple taints applied, the Pod must tolerate all of its taints. Tolerations are defined under `.spec.tolerations` in the Pod specification and have the following structure:

```
tolerations:
- key: <key>
  operator: <operatorType>
  value: <value>
  effect: <effect>
```

The operator can be either `Equal` or `Exists`. `Equal` means that both `key` and `value` of taint must match exactly, whereas `Exists` means that just `key` must match and value is not considered. In our example, if we want to ignore the taint, the toleration will need to look like this:

```
tolerations:
- key: machine-check-exception
  operator: Equal
  value: memory
  effect: NoExecute
```

You can define multiple tolerations for a Pod.

In the case of `NoExecute` tolerations, it is possible to define an additional field called `tolerationSeconds`, which specifies how long the Pod will tolerate the taint until it gets evicted. So, this is a way of having partial toleration of taint with a timeout. Please note that if you use `NoExecute` taints, you usually also need to add a `NoSchedule` taint. In this way, you can prevent any **eviction loops** happening when the Pod has a `NoExecute` toleration with `tolerationSeconds` set. This is because the taint has no effect for a specified number of seconds, which also includes *not* preventing the Pod from being scheduled for the tainted Node.

> **Important Note**
>
> When Pods are created in the cluster, Kubernetes automatically adds two `Exists` tolerations for `node.kubernetes.io/not-ready` and `node.kubernetes.io/unreachable` with `tolerationSeconds` set to `300`.

We will now put this knowledge into practice with a few demonstrations. Please follow the next steps:

1. If you have the `nginx-deployment-example` Deployment with Node affinity defined still running from the previous section, it will currently have all Pods running on Node 0. The Node affinity rules are constructed in such a way that the Pods cannot be scheduled on Node 1 and Node 2. Let's see what happens if you taint Node 0 with `machine-check-exception=memory:NoExecute`:

```
$ kubectl taint node aks-nodepool1-77120516-vmss000000
machine-check-exception=memory:NoExecute
node/aks-nodepool1-77120516-vmss000000 tainted
```

2. Check the Pod status and Node assignment:

```
$ kubectl get pods --namespace default --output=custom-
columns="NAME:.metadata.name,STATUS:.status.phase,NODE:.
spec.nodeName"
NAME                                                STATUS
NODE
nginx-deployment-example-6c4fdd447d-c42z2           Pending
<none>
nginx-deployment-example-6c4fdd447d-dstbl           Pending
<none>
nginx-deployment-example-6c4fdd447d-ktfzh           Pending
<none>
```

```
nginx-deployment-example-6c4fdd447d-ptcwc    Pending
<none>
nginx-deployment-example-6c4fdd447d-wdmb9    Pending
<none>
```

All Deployment Pods are now in the Pending state because kube-scheduler is unable to find a Node that can run them.

3. Edit the ./nginx-deployment.yaml Deployment manifest and remove affinity. Instead, define taint toleration for machine-check-exception=memory:NoExecute with a timeout of 60 seconds:

```yaml
apiVersion: apps/v1
kind: Deployment
metadata:
  name: nginx-deployment-example
spec:
...
  template:
...
    spec:
      tolerations:
      - key: machine-check-exception
        operator: Equal
        value: memory
        effect: NoExecute
        tolerationSeconds: 60
...
```

When this manifest is applied to the cluster, the old Node affinity rules which prevented scheduling to Node 1 and Node 2 will be gone. The Pods will be able to schedule on Node 1 and Node 2, but Node 0 has taint machine-check-exception=memory:NoExecute. So, the Pods should *not* be scheduled to Node 0, as NoExecute implies NoSchedule, *right*? Let's check that.

4. Apply the manifest to the cluster using the kubectl apply -f ./nginx-deployment.yaml command and inspect the Pod status and Node assignment again. You may need to wait a while for the Deployment rollout to finish:

```
$ kubectl get pods -o wide
NAME                                   ... AGE   IP
```

```
NODE
nginx-deployment-example-6b774d7f6c-95ttq ...    14s
10.244.1.230    aks-nodepool1-77120516-vmss000000

nginx-deployment-example-6b774d7f6c-hthwj ...    16m
10.244.0.110    aks-nodepool1-77120516-vmss000001

nginx-deployment-example-6b774d7f6c-lskr7 ...    14s
10.244.1.231    aks-nodepool1-77120516-vmss000000

nginx-deployment-example-6b774d7f6c-q94kw ...    16m
10.244.2.19     aks-nodepool1-77120516-vmss000002

nginx-deployment-example-6b774d7f6c-wszfn ...    16m
10.244.0.109    aks-nodepool1-77120516-vmss000001
```

This result may be a bit surprising. As you can see, we got two Pods scheduled on Node 1 and one Pod on Node 2, but at the same time Node 0 has received two Pods, and they are in eviction loop every 60 seconds! The explanation for this is that `tolerationSeconds` for the `NoExecute` taint implies that the whole taint is ignored for 60 seconds. So kube-scheduler can schedule the Pod on Node 0, even though it will get evicted later.

5. Let's fix this behavior by applying a recommendation to use a `NoSchedule` taint whenever you use a `NoExecute` taint. In this way, the evicted Pods will have no chance to be scheduled on the tainted Node again, unless of course they start tolerating this type of taint too. Execute the following command to taint Node 0:

```
$ kubectl taint node aks-nodepool1-77120516-vmss000000
machine-check-exception=memory:NoSchedule
node/aks-nodepool1-77120516-vmss000000 tainted
```

6. Inspect the Pod status and Node assignment again:

```
$ kubectl get pods --namespace default --output=custom-
columns="NAME:.metadata.name,STATUS:.status.phase,NODE:.
spec.nodeName"
```

```
NAME                                                STATUS
NODE
nginx-deployment-example-6b774d7f6c-hthwj           Running
aks-nodepool1-77120516-vmss000001

nginx-deployment-example-6b774d7f6c-jfvqn           Running
aks-nodepool1-77120516-vmss000001

nginx-deployment-example-6b774d7f6c-q94kw           Running
aks-nodepool1-77120516-vmss000002

nginx-deployment-example-6b774d7f6c-wszfn           Running
```

```
aks-nodepool1-77120516-vmss000001
nginx-deployment-example-6b774d7f6c-z8jx2    Running
aks-nodepool1-77120516-vmss000002
```

In the output you can see that the Pods are now distributed between Node 1 and Node 2 – exactly as we wanted.

7. Now, remove *both* taints from the Node 0:

```
$ kubectl taint node aks-nodepool1-77120516-vmss000000
machine-check-exception-
node/aks-nodepool1-77120516-vmss000000 untainted
```

8. Restart the Deployment to reschedule the Pods using the following command:

```
$ kubectl rollout restart deploy nginx-deployment-example
deployment.apps/nginx-deployment-example restarted
```

9. Inspect the Pod status and Node assignment again:

```
$ kubectl get pods --namespace default --output=custom-
columns="NAME:.metadata.name,STATUS:.status.phase,NODE:.
spec.nodeName"
NAME                                                STATUS
NODE
nginx-deployment-example-56f4d4d96d-nf82h    Running
aks-nodepool1-77120516-vmss000002
nginx-deployment-example-56f4d4d96d-v8m9c    Running
aks-nodepool1-77120516-vmss000000
nginx-deployment-example-56f4d4d96d-vzqn4    Running
aks-nodepool1-77120516-vmss000000
nginx-deployment-example-56f4d4d96d-wpv78    Running
aks-nodepool1-77120516-vmss000001
nginx-deployment-example-56f4d4d96d-x7x92    Running
aks-nodepool1-77120516-vmss000001
```

The Pods are again distributed evenly between all three Nodes.

10. And finally, let's see how the combination of `NoExecute` and `NoSchedule` taints work, with `tolerationSeconds` for `NoExecute` set to `60`. Apply two taints to Node 0 again:

```
$ kubectl taint node aks-nodepool1-77120516-vmss000000
machine-check-exception=memory:NoSchedule
node/aks-nodepool1-77120516-vmss000000 tainted
$ kubectl taint node aks-nodepool1-77120516-vmss000000
machine-check-exception=memory:NoExecute
node/aks-nodepool1-77120516-vmss000000 tainted
```

11. Immediately after that, start watching Pods with their Node assignments. Initially, you will see that the Pods are still running on Node 0 for some time. But after 60 seconds, you will see:

```
$ kubectl get pods --namespace default --output=custom-
columns="NAME:.metadata.name,STATUS:.status.phase,NODE:.
spec.nodeName"
```

NAME	STATUS
NODE	
nginx-deployment-example-56f4d4d96d-44zvt	Running
aks-nodepool1-77120516-vmss000002	
nginx-deployment-example-56f4d4d96d-9rg2p	Running
aks-nodepool1-77120516-vmss000001	
nginx-deployment-example-56f4d4d96d-nf82h	Running
aks-nodepool1-77120516-vmss000002	
nginx-deployment-example-56f4d4d96d-wpv78	Running
aks-nodepool1-77120516-vmss000001	
nginx-deployment-example-56f4d4d96d-x7x92	Running
aks-nodepool1-77120516-vmss000001	

As we expected, the Pods have been evicted after 60 seconds and there were no eviction-schedule loops.

This has demonstrated a more advanced use case for taints which you cannot easily substitute with Node affinity rules. In the next section, we will give a short overview of kube-scheduler scheduling policies.

Scheduling policies

kube-scheduler decides for which Node a given Pod should be scheduled, in two phases: **filtering** and **scoring**. To quickly recap, *filtering* is the first phase when kube-scheduler finds a set of Nodes that can be used for the running of a Pod. For example, if a Pod tolerates Node taints. In the second phase, *scoring*, the filtered Nodes are ranked using a scoring system to find the most suitable Node for the Pod.

The way the default kube-scheduler executes these two phases is defined by the **scheduling policy**. This policy is configurable and can be passed to the kube-scheduler process using the additional arguments `--policy-config-file <filename>` or `--policy-configmap <configMap>`.

> **Important note**
>
> In **managed** Kubernetes clusters, such as the managed Azure Kubernetes Service, you will *not* be able to change scheduling policy of kube-scheduler, as you do not have access to Kubernetes master Node.

There are two configuration fields that are most important in scheduling policy:

- **Predicates**: Implement the rules for filtering
- **Priorities**: Implement the scoring system

The full list of currently supported predicates and priorities is available in the official documentation: `https://kubernetes.io/docs/reference/scheduling/policies/`. We will give an overview of a few of the most interesting ones that show how flexible the default kube-scheduler is. Some of the selected predicates are shown in the following list:

- `PodToleratesNodeTaints`: As the name suggests, implements a basic check if a Pod has defined a toleration for current Node taints

- `PodFitsResources`: Implements a check if the Node has enough free resources to meet the requirements specified by a Pod

- `CheckNodePIDPressure`: Implements a check if the Node has enough available process IDs to safely continue running Pods

- `CheckVolumeBinding`: Implements a check if the Node is compatible with PVCs that the Pod requires

Some of the interesting available priorities are as follows:

- `SelectorSpreadPriority`: Ensures that Pods that belong to the same ReplicationController, StatefulSet, and ReplicaSet objects (this includes Deployments) are evenly spread across the Nodes. This ensures better fault tolerance in case of Node failures.

- `NodeAffinityPriority` and `InterPodAffinityPriority`: Implements the soft Node affinity and inter-Pod affinity/anti-affinity.

- `ImageLocalityPriority`: Prioritizes the Nodes that already have the container images required by a Pod in the local cache to reduce start up time and decrease unnecessary network traffic.

- `ServiceSpreadingPriority`: Attempts to spread the Pods by minimizing the number of Pods belonging to the same Service object running on the same Node. This ensures better fault tolerance in case of Node failures.

The preceding examples are just a subset of the available predicates and priorities, but this already gives an overview of how many complex use cases and scenarios are supported out of the box in kube-scheduler.

Summary

This chapter has given an overview of advanced techniques for Pod scheduling in Kubernetes. First, we recapped the theory behind kube-scheduler implementation. We have explained the process of scheduling Pods. Next, we introduced the concept of Node affinity in Pod scheduling. You learned the basic scheduling methods which use Node names and Node selectors, and based on that we have explained how more advanced Node affinity works. We also explained how you can use the affinity concept to achieve anti-affinity, and what inter-Pod affinity/anti-affinity is. After that, we discussed taints for Nodes and tolerations specified by Pods. You learned about some different effects of the taints, and have put the knowledge into practice in an advanced use case involving `NoExecute` and `NoSchedule` taints on a Node. Lastly, we discussed the theory behind scheduler policies that can be used to configure the default kube-scheduler.

In the next chapter, we are going to discuss **autoscaling** of Pods and Nodes in Kubernetes – this will be a topic that will show how flexibly Kubernetes can run workloads in cloud environments.

Further reading

For more information regarding Pod scheduling in Kubernetes, please refer to the following PacktPub books:

- *The Complete Kubernetes Guide*, by *Jonathan Baier, Gigi Sayfan, Jesse White* (`https://www.packtpub.com/virtualization-and-cloud/complete-kubernetes-guide`)

- *Getting Started with Kubernetes – Third Edition*, by *Jonathan Baier, Jesse White* (`https://www.packtpub.com/virtualization-and-cloud/getting-started-kubernetes-third-edition`)

- *Kubernetes for Developers*, by *Joseph Heck* (`https://www.packtpub.com/virtualization-and-cloud/kubernetes-developers`)

You can also refer to official documents:

- Kubernetes documentation (`https://kubernetes.io/docs/home/`), which is always the most up-to-date source of knowledge about Kubernetes in general.

- Node affinity is covered at `https://kubernetes.io/docs/concepts/scheduling-eviction/assign-pod-node/`.

- Taint and tolerations are covered at `https://kubernetes.io/docs/concepts/scheduling-eviction/taint-and-toleration/`.

- Pod priorities and preemption (which we have not covered in this chapter) are described at `https://kubernetes.io/docs/concepts/configuration/pod-priority-preemption/`.

- Advanced kube-scheduler configuration using scheduling profiles (which we have not covered in this chapter) is described at `https://kubernetes.io/docs/reference/scheduling/config`.

20
Autoscaling Kubernetes Pods and Nodes

Needless to say, having **autoscaling** capabilities for your cloud-native application is considered the holy grail of running applications in cloud. In short, by autoscaling, we mean a method to automatically and dynamically adjust the amount of computational resources, such as CPU and RAM memory, available to your application. The goal of it is to cleverly add or remove available resources based on the **activity and demand** of end users. So, for example, the application may require more CPU and RAM memory during daytime hours, when users are most active, but much less during the night. Similarly, if you are running an e-commerce business, you can expect a huge spike in demand during so-called *Black Friday*. In this way, you can not only provide a better, highly available service to users but also reduce your **cost of goods sold** (**COGS**) for the business. The fewer resources you consume in the cloud, the less you pay, and the business can invest the money elsewhere – this is a *win-win* situation. There is, of course, no single rule that fits all use cases, hence good autoscaling needs to be based on critical usage metrics and should have **predictive features** to anticipate the workloads based on history.

Kubernetes, as the most mature container orchestration system available, comes with a variety of built-in autoscaling features. Some of these features are natively supported in every Kubernetes cluster and some require installation or specific type of cluster deployment. There are also multiple *dimensions* of scaling that you can have:

- **Vertical for Pods**: This involves adjusting the amount of CPU and memory resources available to a Pod. Pods can run under limits specified for CPU and memory, to prevent excessive consumption, but these limits may require automatic adjustment rather than a human operator guessing. This is implemented by a **VerticalPodAutoscaler** (**VPA**).

- **Horizontal for Pods**: This involves dynamically changing the number of Pod replicas for your Deployment or StatefulSet. These objects come with nice scaling features out of the box, but adjusting the number of replicas can be automated using a **HorizontalPodAutoscaler** (**HPA**).

- **Horizontal for Nodes**: Another dimension of horizontal scaling (scaling *out*), but this time at the level of a Kubernetes Node. You can scale your whole cluster by adding or removing the Nodes. This requires, of course, a Kubernetes Deployment that runs in an environment that supports the dynamic provisioning of machines, such as a cloud environment. This is implemented by a **Cluster Autoscaler** (**CA**), available for some cloud vendors.

In this chapter, we will cover the following topics:

- Pod resource requests and limits

- Autoscaling Pods vertically using a Vertical Pod Autoscaler

- Autoscaling Pods horizontally using a Horizontal Pod Autoscaler

- Autoscaling Kubernetes Nodes using a Cluster Autoscaler

Technical requirements

For this chapter, you will need the following:

- A Kubernetes cluster deployed. We recommend using a **multi-node**, cloud-based Kubernetes cluster.

- Having a multi-node **Google Kubernetes Engine** (**GKE**) cluster is a recommended prerequisite to follow the second section relating to the **Vertical Pod Autoscaler** (**VPA**). AKS and EKS currently require the manual installation of a VPA, which we are going to demonstrate, but GKE has support for it out of the box.

- Having a multi-node AKS, EKS, or GKE cluster is a prerequisite for following the final section regarding a CA.

- A Kubernetes CLI (`kubectl`) installed on your local machine and configured to manage your Kubernetes cluster.

Basic Kubernetes cluster deployment (local and cloud-based) and `kubectl` installation have been covered in *Chapter 3, Installing Your First Kubernetes Cluster*.

The following chapters can provide you with an overview of how to deploy a fully functional Kubernetes cluster on different cloud platforms and install the requisite CLIs to manage them:

- *Chapter 14, Kubernetes Clusters on Google Kubernetes Engine*

- *Chapter 15, Launching a Kubernetes Cluster on Amazon Web Services with the Amazon Elastic Kubernetes Service*

- *Chapter 16, Kubernetes Clusters on Microsoft Azure with the Azure Kubernetes Service*

You can download the latest code samples for this chapter from the official GitHub repository at `https://github.com/PacktPublishing/The-Kubernetes-Bible/tree/master/Chapter20`.

Pod resource requests and limits

Before we dive into the topics of autoscaling in Kubernetes, we need to explain a bit more about how you can control the CPU and memory resource (known as **compute resources**) usage by Pod containers in Kubernetes. Controlling the use of compute resources is important since, in this way, you can enforce **resource governance** – this allows better planning of the cluster capacity and, most importantly, prevents situations when a single container can consume all compute resources and prevent other Pods from serving the requests.

When you create a Pod, it is possible to specify how much compute resources its containers **require** and what the **limits** are in terms of permitted consumption. The Kubernetes resource model provides an additional distinction between two classes of resources: **compressible** and **incompressible**. In short, a compressible resource can be easily throttled, without severe consequences. A perfect example of such a resource is the CPU – if you need to throttle CPU usage for a given container, the container will operate normally, just slower. On the other hand, we have incompressible resources that cannot be throttled without sever consequences – RAM memory allocation is an example of such a resource. If you do not allow a process running in a container to allocate more memory, the process will crash and result in container restart.

> **Important Note**
>
> If you want to know more about the philosophy and design decisions for the Kubernetes resource governance model, we recommend reading the official design proposal documents. Resource model: `https://github.com/kubernetes/community/blob/master/contributors/design-proposals/scheduling/resources.md`. Resource quality of service: `https://github.com/kubernetes/community/blob/master/contributors/design-proposals/node/resource-qos.md`.

To control the resources for a Pod container, you can specify two values in its specification:

- `requests`: This specifies the **guaranteed** amount of a given resource provided by the system. You can also think the other way round – this is the amount of a given resource that the Pod container **requires** from the system in order to function properly. This is important as Pod scheduling is dependent on the `requests` value (not `limits`), namely, the `PodFitsResources` predicate and the `BalancedResourceAllocation` priority.

- `limits`: This specifies the **maximum** amount of a given resource provided by the system. If specified together with `requests`, this value must be greater than or equal to `requests`. Depending on whether the resource is compressible or incompressible, exceeding the limit has different consequences – compressible resources (CPU) will be throttled, whereas incompressible resources (RAM) *may* result in container kill and restart.

If you use different values for `requests` and `limits`, you can allow for **resource overcommit**. This technique is useful for efficiently handling short bursts of resource usage while allowing better resource usage on average. The reasoning behind this is that you will rarely have all containers on the Node requiring maximum resources, as they specify in `limits`, at the same time. This gives you better bin packing of your Pods for the majority of the time. The concept is similar to overprovisioning for virtual machine hypervisors or, in the real world, overbooking for airplane flights.

If you do not specify `limits` at all, the container can consume as much of the resource on a Node as it wants. This can be controlled by namespace **resource quotas** and **limit ranges** – you can read more about these objects in the official documentation: `https://kubernetes.io/docs/concepts/policy/limit-range/`.

> **Tip**
>
> In more advanced scenarios, you can also control **huge pages** and **ephemeral storage** `requests` and `limits`.

Before we dive into the configuration details, we need to look at what are the units for measuring CPU and memory in Kubernetes. For CPU, the base unit is **Kubernetes CPU (KCU)**, where 1 is equivalent to, for example, 1 vCPU on Azure, 1 core on GCP, or 1 hyperthreaded core on a bare-metal machine. Fractional values are allowed: 0.1 can be also specified as 100m (*milliKCUs*). For memory, the base unit is **byte**; you can, of course, specify standard unit prefixes, such as M, Mi, G, or Gi.

To enable compute resource `requests` and `limits` for Pod containers in our `nginx` Deployment that we used in the previous chapters, you can make the following changes to the YAML manifest, `nginx-deployment.yaml`:

```
apiVersion: apps/v1
kind: Deployment
metadata:
  name: nginx-deployment-example
spec:
  replicas: 5
  selector:
    matchLabels:
      app: nginx
      environment: test
  template:
    metadata:
```

```
      labels:
        app: nginx
        environment: test
    spec:
      containers:
      - name: nginx
        image: nginx:1.17
        ports:
        - containerPort: 80
        resources:
          limits:
            cpu: 200m
            memory: 60Mi
          requests:
            cpu: 100m
            memory: 50Mi
```

For each container that you have in the Pod, you can specify the `.spec.template.spec.containers[*].resources` field. In this case, we have set `limits` at `200m` KCU and `60Mi` for RAM, and `requests` at `100m` KCU and `50Mi` for RAM.

When you apply the manifest to the cluster using `kubectl apply -f ./nginx-deployment.yaml`, you can describe one of the Nodes in the cluster that run Pods for this Deployment and you will see detailed information about compute resources quotas and allocation:

```
$ kubectl describe node aks-nodepool1-77120516-vmss000000
...
Non-terminated Pods:          (5 in total)
  Namespace                   Name
CPU Requests   CPU Limits   Memory Requests   Memory Limits   AGE
  ---------                   ----
------------   ----------   ---------------   -------------   ---
  default                     nginx-deployment-example-
5d8b9979d4-9sd9x    100m (5%)      200m (10%)   50Mi (1%)
60Mi (1%)         8m12s
  default                     nginx-deployment-example-
5d8b9979d4-rbwv2    100m (5%)      200m (10%)   50Mi (1%)
60Mi (1%)         8m10s
```

```
   default                    nginx-deployment-example-
5d8b9979d4-sfzx9      100m (5%)      200m (10%)   50Mi (1%)
60Mi (1%)          8m10s
   kube-system                kube-proxy-q6xdq
100m (5%)        0 (0%)      0 (0%)           0 (0%)         10d
   kube-system                omsagent-czm6q
75m (3%)        500m (26%)  225Mi (4%)       600Mi (13%)    17d
Allocated resources:
   (Total limits may be over 100 percent, i.e., overcommitted.)
   Resource                    Requests      Limits
   --------                    --------      ------
   cpu                         475m (25%)    1100m (57%)
   memory                      375Mi (8%)    780Mi (17%)
   ephemeral-storage           0 (0%)        0 (0%)
   hugepages-1Gi               0 (0%)        0 (0%)
   hugepages-2Mi               0 (0%)        0 (0%)
   attachable-volumes-azure-disk 0           0
```

Now, based on this information, you could experiment, and set `requests` for CPU for the container to a value higher than the capacity of a single Node in the cluster, in our case, `2000m` KCU. When you do that and apply the changes to the Deployment, you will notice that new Pods hang in the `Pending` state because they cannot be scheduled on a matching Node. In such cases, inspecting the Pod will reveal the following:

```
$ kubectl describe pod nginx-deployment-example-56868549b-5n61j
...
Events:
   Type    Reason          Age    From             Message
   ----    ------          ----   ----             -------
   Warning FailedScheduling 25s   default-scheduler 0/3 nodes
are available: 3 Insufficient cpu.
```

There were no Nodes that could accommodate a Pod that has a container requiring `2000m` KCU, and therefore the Pod cannot be scheduled at this moment.

With knowledge of how to manage compute resources, we will move on to autoscaling topics: first, we are going to explain the vertical autoscaling of Pods.

Autoscaling Pods vertically using a Vertical Pod Autoscaler

In the previous section, we have been managing `requests` and `limits` for the compute resources manually. Setting these values correctly requires some accurate human *guessing*, observing metrics, and performing benchmarks to adjust. Using overly high `requests` values will result in a waste of compute resources, whereas setting it too low may result in Pods being packed too densely and having performance issues. Also, in some cases, the only way to scale the Pod workload is to do it **vertically** by increasing the amount of compute resources it can consume. For bare-metal machines, this would mean upgrading the CPU hardware and adding more physical RAM memory. For containers, it is as simple as allowing them more of the compute resource quotas. This works, of course, only up to the capacity of a single Node. You cannot scale vertically beyond that unless you add more powerful Nodes to the cluster.

To help resolve these issues, Kubernetes offers a **Vertical Pod Autoscaler** (**VPA**), which can increase and decrease CPU and memory resource `requests` for Pod containers dynamically. The goal is to better match the *actual* usage rather than rely on hardcoded, predefined values. Controlling `limits` within specified ratios is also supported.

The VPA is created by a **Custom Resource Definition** (**CRD**) object named `VerticalPodAutoscaler`. This means that this object is not part of standard Kubernetes API groups and needs to be installed in the cluster. The VPA is developed as part of an autoscaler project (`https://github.com/kubernetes/autoscaler`) in the Kubernetes ecosystem.

There are three main components of a VPA:

- **Recommender**: Monitors the current and past resource consumption and provides recommended CPU and memory request values for a Pod container.

- **Updater**: Checks for Pods with incorrect resources and **deletes** them, so that the Pods can be recreated with the updated `requests` and `limits` values

- **Admission plugin**: Sets the correct resource `requests` and `limits` on new Pods created or recreated by their controller, for example, a Deployment object, due to changes made by the updater

The reason why the updater needs to terminate Pods and the VPA has to rely on the admission plugin is that Kubernetes does not support dynamic changes to the resource `requests` and `limits`. The only way is to terminate the Pod and create a new one with new values. In-place modifications of values are tracked in KEP1287 (`https://github.com/kubernetes/enhancements/pull/1883`) and, when implemented, will make the design of the VPA much simpler, thereby ensuring improved high availability.

> **Important note**
>
> A VPA can run in recommendation-only mode where you see the suggested values in the VPA object, but the changes are not applied to the Pods. A VPA is currently considered **experimental** and using it in a mode that recreates the Pods may lead to downtimes of your application. This should change when in-place updates of Pod `requests` and `limits` are implemented.

Some Kubernetes offerings come with one-click support for installing a VPA. Two good examples are OpenShift and GKE. We will now quickly explain how you can do that if you are running a GKE cluster.

Enabling a VPA in GKE

Assuming that your GKE cluster is named `k8sforbeginners`, as in *Chapter 14, Kubernetes Clusters on Google Kubernetes Engine*, enabling a VPA is as simple as running the following command:

```
$ gcloud container clusters update k8sforbeginners --enable-
vertical-pod-autoscaling
```

Note that this operation causes a restart to the Kubernetes control plane.

If you want to enable a VPA for a new cluster, you can use the additional argument `--enable-vertical-pod-autoscaling`, for example:

```
$ gcloud container clusters create k8sforbeginners
--num-nodes=2 --zone=us-central1-a --enable-vertical-pod-
autoscaling
```

The GKE cluster will have a VPA CRD available, and you can use it to control the vertical autoscaling of Pods.

Enabling a VPA for other Kubernetes clusters

In the case of different platforms such as AKS or EKS (or even local deployments for testing), you need to install a VPA manually by adding a VPA CRD to the cluster. The exact, most recent steps are documented in the corresponding GitHub repository: `https://github.com/kubernetes/autoscaler/tree/master/vertical-pod-autoscaler#installation`.

To install a VPA in your cluster, please perform the following steps:

1. Clone the Kubernetes autoscaler repository (`https://github.com/kubernetes/autoscaler`):

    ```
    $ git clone https://github.com/kubernetes/autoscaler
    ```

2. Navigate to the VPA component directory:

    ```
    $ cd autoscaler/vertical-pod-autoscaler
    ```

3. Begin installation using the following command. This assumes that your current `kubectl` context is pointing to the desired cluster:

    ```
    $ ./hack/vpa-up.sh
    ```

4. This will create a bunch of Kubernetes objects. You can verify that the main component Pods are started correctly using the following command:

    ```
    $ kubectl get pods -n kube-system
    NAME                                         READY
    STATUS      RESTARTS    AGE
    vpa-admission-controller-688857d5c4-419c2    1/1
    Running     0           10s
    vpa-recommender-74849cc845-qbfpg             1/1
    Running     0           11s
    vpa-updater-6dbd6569d6-9np22                 1/1
    Running     0           12s
    ```

The VPA components are running, and we can now proceed to testing a VPA on real Pods.

Using a VPA

For demonstration purposes, we need a Deployment with Pods that cause actual consumption of CPU. The Kubernetes autoscaler repository has a good, simple example that has **predictable** CPU usage: `https://github.com/kubernetes/autoscaler/blob/master/vertical-pod-autoscaler/examples/hamster.yaml`. We are going to modify this example a bit and do a step-by-step demonstration. Let's prepare the Deployment first:

1. Create the `hamster-deployment.yaml` YAML manifest file:

```
apiVersion: apps/v1
kind: Deployment
metadata:
  name: hamster
spec:
  selector:
    matchLabels:
      app: hamster
  replicas: 5
  template:
    metadata:
      labels:
        app: hamster
    spec:
      containers:
      - name: hamster
        image: ubuntu:20.04
        resources:
          requests:
            cpu: 100m
            memory: 50Mi
        command:
        - /bin/sh
        - -c
        - while true; do timeout 0.5s yes >/dev/null;
sleep 0.5s; done
```

It's a real hamster! The `command` that is used in the Pod's `ubuntu` container consumes the maximum available CPU of 0.5 seconds and does nothing for 0.5 seconds, all the time. This means that the actual CPU usage will stay, on average, at around 500m KCU. However, the `requests` for resources specify that it requires 100m KCU. This means that the Pod will consume more than it declares, but since there are no `limits` set, Kubernetes will not throttle the container CPU. This could potentially lead to incorrect scheduling decisions by Kubernetes Scheduler.

2. Apply the manifest to the cluster using the following command:

```
$ kubectl apply -f ./hamster-deployment.yaml
deployment.apps/hamster created
```

3. Let's verify what the CPU usage of the Pod is. The simplest way is to use the `kubectl top` command:

```
$ kubectl top pod
NAME                       CPU(cores)   MEMORY(bytes)
hamster-779cfd69b4-5bnbf   475m         1Mi
hamster-779cfd69b4-8dt5h   497m         1Mi
hamster-779cfd69b4-mn5p5   492m         1Mi
hamster-779cfd69b4-n7nss   496m         1Mi
hamster-779cfd69b4-rl29j   484m         1Mi
```

As we expected, the CPU consumption for each Pod in the deployment oscillates at around 500m KCU.

With that, we can move on to creating a VPA for our Pods. VPAs can operate in four **modes** that you specify by means of the `.spec.updatePolicy.updateMode` field:

*   `Recreate`: Pod container `limits` and `requests` are assigned on Pod creation and dynamically updated based on calculated recommendations. To update the values, the Pod must be restarted. Please note that this may be disruptive to your application.

*   `Auto`: Currently equivalent to `Recreate`, but when in-place updates for Pod container `requests` and `limits` are implemented, this can automatically switch to the new update mechanism.

*   `Initial`: Pod container `limits` and `requests` are assigned on Pod creation only.

*   `Off`: A VPA runs in recommendation-only mode. The recommended values can be inspected in the VPA object, for example, by using `kubectl`.

We are going to first create a VPA for `hamster` Deployment, which runs in `Off` mode, and later we will enable `Auto` mode. To do this, please perform the following steps:

1. Create a VPA YAML manifest named `hamster-vpa.yaml`:

```yaml
apiVersion: autoscaling.k8s.io/v1
kind: VerticalPodAutoscaler
metadata:
  name: hamster-vpa
spec:
  targetRef:
    apiVersion: apps/v1
    kind: Deployment
    name: hamster
  updatePolicy:
    updateMode: "Off"
  resourcePolicy:
    containerPolicies:
    - containerName: '*'
      minAllowed:
        cpu: 100m
        memory: 50Mi
      maxAllowed:
        cpu: 1
        memory: 500Mi
      controlledResources:
      - cpu
      - memory
```

This VPA is created for a Deployment object with the name `hamster`, as specified in `.spec.targetRef`. The mode is set to `"Off"` in `.spec.updatePolicy.updateMode` (`"Off"` needs to be specified in quotes to avoid being interpreted as a Boolean) and the container resource policy is configured in `.spec.resourcePolicy.containerPolicies`. The policy that we used allows Pod container `requests` for CPU to be adjusted automatically between `100m` KCU and `1000m` KCU, and for memory between `50Mi` and `500Mi`.

2. Apply the manifest file to the cluster:

```
$ kubectl apply -f ./hamster-vpa.yaml
verticalpodautoscaler.autoscaling.k8s.io/hamster-vpa
created
```

3. You need to wait a while for the recommendation to be calculated for the first time. Then, you can check what the recommendation is by describing the VPA:

```
$ kubectl describe vpa hamster-vpa
...
Status:
  Conditions:
    Last Transition Time:  2021-03-28T14:33:33Z
    Status:                True
    Type:                  RecommendationProvided
  Recommendation:
  Container Recommendations:
    Container Name:  hamster
    Lower Bound:
      Cpu:     551m
      Memory:  262144k
    Target:
      Cpu:     587m
      Memory:  262144k
    Uncapped Target:
      Cpu:     587m
      Memory:  262144k
    Upper Bound:
      Cpu:     1
      Memory:  378142066
```

The VPA has recommended allocating a bit more than the expected 500m KCU and 262144k memory. This makes sense, as the Pod should have a safe buffer for CPU consumption.

4. Now we can check the VPA in practice and change its mode to `Auto`. Modify `hamster-vpa.yaml`:

```
apiVersion: autoscaling.k8s.io/v1
kind: VerticalPodAutoscaler
metadata:
  name: hamster-vpa
spec:
...
  updatePolicy:
    updateMode: Auto
...
```

5. Apply the manifest to the cluster:

```
$ kubectl apply -f ./hamster-vpa.yaml
verticalpodautoscaler.autoscaling.k8s.io/hamster-vpa
configured
```

6. After a while, you will notice that the Pods for the Deployment are being restarted by the VPA:

```
$ kubectl get pod
```

NAME AGE	READY	STATUS	RESTARTS
hamster-779cfd69b4-5bnbf 45m	1/1	Running	0
hamster-779cfd69b4-8dt5h 45m	1/1	Terminating	0
hamster-779cfd69b4-9tqfx 60s	1/1	Running	0
hamster-779cfd69b4-n7nss 45m	1/1	Running	0
hamster-779cfd69b4-wdz8t 60s	1/1	Running	0

7. We can inspect one of the restarted Pods to see the current `requests` for resources:

```
$ kubectl describe pod hamster-779cfd69b4-9tqfx
...
```

```
Annotations:   vpaObservedContainers: hamster
               vpaUpdates: Pod resources updated by
hamster-vpa: container 0: cpu request, memory request
...
Containers:
  hamster:
...
     Requests:
       cpu:          587m
       memory:       262144k
...
```

As you can see, the newly started Pod has CPU and memory `requests` set to the values recommended by the VPA!

> **Important note**
>
> A VPA should not be used with an HPA running on CPU/memory metrics at this moment. However, you can use a VPA in conjunction with an HPA running on custom metrics.

Next, we are going to discuss how you can horizontally autoscale Pods using a **Horizontal Pod Autoscaler** (HPA).

Autoscaling Pods horizontally using a Horizontal Pod Autoscaler

While a VPA acts like an optimizer of resource usage, the true scaling of your Deployments and StatefulSets that run multiple Pod replicas can be done using a **Horizontal Pod Autoscaler** (**HPA**). At a high level, the goal of the HPA is to automatically scale the number of replicas in Deployment or StatefulSets depending on the current CPU utilization or other custom metrics (including multiple metrics at once). The details of the algorithm that determines the target number of replicas based on metric values can be found here: `https://kubernetes.io/docs/tasks/run-application/horizontal-Pod-autoscale/#algorithm-details`. HPAs are highly configurable and, in this chapter, we will cover a standard scenario in which we would like to autoscale based on target CPU usage.

> **Important note**
>
> An HPA is represented by a built-in `HorizontalPodAutoscaler`
> API resource in Kubernetes in the `autoscaling` API group. The current
> stable version that supports CPU autoscaling only can be found in the
> `autoscaling/v1` API version. The beta version that supports autoscaling
> based on RAM and custom metrics can be found in the `autoscaling/`
> `v2beta2` API version.

The role of the HPA is to monitor the configured metric for Pods, for example, CPU usage, and determine whether there is a change to the number of replicas needed. Usually, the HPA will calculate the average of the **current** metric value from all Pods and determine whether adding or removing replicas will bring the metric value closer to the specified **target** value. For example, you set the target CPU usage to be 50%. At some point, increased demand for the application causes the Deployment Pods to have 80% CPU usage. The HPA will decide to add more Pod replicas so that the average usage across all replicas will fall and be closer to 50%. And the cycle repeats. In other words, the HPA tries to maintain the average CPU usage to be as close to 50% as possible. This is like a continuous, closed-loop controller – in real life, a thermostat reacting to temperature changes in the building is a good, similar example. HPA additionally uses mechanisms such as a **stabilization window** to prevent the replicas from scaling down too quickly and causing unwanted replica **flapping**.

> **Tip**
>
> GKE has added support for **multidimensional** Pod autoscaling that
> combines horizontal scaling using CPU metrics and vertical scaling based
> on memory usage at the same time. You can read more about this feature
> in the official documentation: `https://cloud.google.com/`
> `kubernetes-engine/docs/how-to/multidimensional-`
> `pod-autoscaling`.

As an HPA is a built-in feature of Kubernetes, there is no need to perform any installation. We just need to prepare a Deployment for testing and create a `HorizontalPodAutoscaler` API object.

Using an HPA

To test an HPA, we are going to rely on the standard CPU usage metric. This means that we need to configure `requests` for CPU on the Deployment Pods, otherwise autoscaling is not possible as there is no absolute number that is needed to calculate the percentage metric. On top of that, we again need a Deployment that can consume a predictable amount of CPU resources. Of course, in real use cases, the varying CPU usage would be coming from actual demand for your application from end users.

Unfortunately, there is no simple way to have predictable and varying CPU usage in a container out of the box, so we have to prepare a Deployment with a Pod template that will do that. We will modify our `hamster` Deployment approach and create an `elastic-hamster` Deployment. The small shell script running continuously in the container will behave slightly differently. We will assign the **total** desired work by hamsters in all Pods together. Each Pod will query the Kubernetes API to check how many replicas there are **currently** running for the Deployment. Then, we will divide the total desired work by the number of replicas to get the amount of work that a **single** hamster needs to do. So, for example, we will say that all hamsters together should do `1.0` of work, which *roughly* maps to the total consumption of KCU in the cluster. Then, if you deploy five replicas for the Deployment, each of the hamsters will do *1.0/5 = 0.2* work, so they will work for `0.2` seconds and sleep for `0.8` seconds. Now, if we scale the Deployment manually to 10 replicas, the amount of work per hamster will fall to `0.1` seconds, and they will sleep for `0.9` seconds. As you can see, they collectively always work for `1.0` second, no matter how many replicas we use. This kind of reflects a real-life scenario where end users cause some amount of traffic to handle, and you distribute it among the Pod replicas. The more Pod replicas you have, the less traffic they have to handle and, in the end, the CPU usage metric will be lower on average.

Querying Deployments via the Kubernetes API will require some additional RBAC setup. You can find more details in *Chapter 18, Authentication and Authorization on Kubernetes*. To create the deployment for the demonstration, please perform the following steps:

1. Create an `elastic-hamster` `ServiceAccount` `manifest` file named `elastic-hamster-serviceaccount.yaml`:

   ```
   apiVersion: v1
   kind: ServiceAccount
   metadata:
     name: elastic-hamster
     namespace: default
   ```

2. Create a `deployment-reader` Role `manifest` file named `deployment-reader-role.yaml`. This role allows Deployments to obtain information from the Kubernetes API:

```
apiVersion: rbac.authorization.k8s.io/v1
kind: Role
metadata:
  namespace: default
  name: deployment-reader
rules:
- apiGroups: ["apps"]
  resources: ["deployments"]
  verbs: ["get", "watch", "list"]
```

3. Create a `read-deployments` RoleBinding manifest file named `read-deployments-rolebinding.yaml`. This RoleBinding associates the ServiceAccount with the role:

```
apiVersion: rbac.authorization.k8s.io/v1
kind: RoleBinding
metadata:
  name: read-deployments
  namespace: default
subjects:
- kind: ServiceAccount
  name: elastic-hamster
  namespace: default
roleRef:
  kind: Role
  name: deployment-reader
  apiGroup: rbac.authorization.k8s.io
```

4. Finally, create an `elastic-hamster` Deployment manifest file named `elastic-hamster-deployment.yaml`, which will have Pods running on the `elastic-hamster` ServiceAccount. Let's take a look at the first part, without the shell command (the full file is also available in the book's GitHub repository: `https://github.com/PacktPublishing/Kubernetes-for-Beginners/blob/master/Chapter20/03_hpa/elastic-hamster-deployment.yaml`):

```
apiVersion: apps/v1
kind: Deployment
metadata:
  name: elastic-hamster
spec:
  selector:
    matchLabels:
      app: elastic-hamster
  replicas: 5
  template:
    metadata:
      labels:
        app: elastic-hamster
    spec:
      serviceAccountName: elastic-hamster
      containers:
      - name: hamster
        image: ubuntu:20.04
        resources:
          requests:
            cpu: 200m
            memory: 50Mi
        env:
        - name: TOTAL_HAMSTER_USAGE
          value: "1.0"
        command:
        - /bin/sh
        - -c
        - |
... shell command available in the next step ...
```

While it is not a good practice to have long shell scripts in the YAML manifest definitions, it is easier for demonstration purposes than creating a dedicated container image, pushing it to the image repository, and consuming it. Let's take a look at what is happening in the `manifest` file. Initially, we need to have five replicas. Each Pod container has `requests` with `cpu` set to `200m` KCU and `memory` set to `50Mi`. We also define an environment variable, `TOTAL_HAMSTER_USAGE`, with an initial value of `"1.0"` for more readability. This variable defines the total collective work that the hamsters are expected to do.

5. Now, let's take a look at the continuation of the file, at the part with the shell script for the container (the indentation has been removed and, in the YAML file, you need to correctly indent the script, as in the GitHub repository):

```
# Install curl and jq
apt-get update && apt-get install -y curl jq || exit 1
SERVICEACCOUNT=/var/run/secrets/kubernetes.io/
serviceaccount
TOKEN=$(cat ${SERVICEACCOUNT}/token)
while true
  # Calculate CPU usage by hamster. This will dynamically
adjust to be 1.0 / num_replicas. So for initial 5
replicas, it will be 0.2
  HAMSTER_USAGE=$(curl -s --cacert $SERVICEACCOUNT/ca.crt
--header "Authorization: Bearer $TOKEN" -X GET https://
kubernetes/apis/apps/v1/namespaces/default/deployments/
elastic-hamster | jq ${TOTAL_HAMSTER_USAGE}/'.spec.
replicas')
  # Hamster sleeps for the rest of the time, with a small
adjustment factor
  HAMSTER_SLEEP=$(jq -n 1.2-$HAMSTER_USAGE)
  echo "Hamster uses $HAMSTER_USAGE and sleeps $HAMSTER_
SLEEP"
  do timeout ${HAMSTER_USAGE}s yes >/dev/null
  sleep ${HAMSTER_SLEEP}s
done
```

The shell script, as the very first step, installs `curl` and `jq` packages from the APT repository. We define `SERVICEACCOUNT` and `TOKEN` variables, which we need to query the Kubernetes API. Then, we retrieve the `elastic-hamster` Deployment from the API using `https://kubernetes/apis/apps/v1/namespaces/default/deployments/elastic-hamster`. The result is parsed using the `jq` command, we extract the `.spec.replicas` field, and use it to divide the total work between all hamsters. Based on this number, we make the hamster *work* for a calculated period of time and then *sleep* for the rest. As you can see, if the number of replicas for the Deployment changes, either by means of a manual action or autoscaling, the amount of work to be done by an individual hamster will change. And therefore, the CPU usage will decrease the more Pod replicas we have.

6. We are now ready to apply all `manifest` files in the directory with the following command:

```
$ kubectl apply -f ./
role.rbac.authorization.k8s.io/deployment-reader created
deployment.apps/elastic-hamster created
serviceaccount/elastic-hamster created
rolebinding.rbac.authorization.k8s.io/read-deployments
created
```

7. When the Pods are fully started, you will be able to see in the logs that the hamster *work and sleep* cycle has begun:

```
$ kubectl logs elastic-hamster-5897858459-26bdd
...
Running hooks in /etc/ca-certificates/update.d...
done.
Hamster uses 0.2 and sleeps 1
Hamster uses 0.2 and sleeps 1
...
```

8. After a while, you will see in the output of the `kubectl top` command that the CPU usage is about the expected `200m` KCU. Of course, this method is **not** precise because there is more CPU usage by the container than just the *work and sleep* cycle:

```
$ kubectl top pods
NAME                                 CPU(cores)
MEMORY(bytes)
elastic-hamster-5897858459-26bdd     229m          40Mi
```

```
elastic-hamster-5897858459-f2856    210m          40Mi
elastic-hamster-5897858459-lmphl    236m          40Mi
elastic-hamster-5897858459-m6j58    225m          40Mi
elastic-hamster-5897858459-qfh76    227m          41Mi
```

9. We can test how it reacts to change in a number of replicas. Scale down the Deployment imperatively to two replicas using the `kubectl scale` command:

```
$ kubectl scale deploy elastic-hamster --replicas=2
deployment.apps/elastic-hamster scaled
```

10. You can inspect the Pod logs again and, after a while, when metrics are processed, you will see the CPU usage change in the `kubectl top` command output, which is, as expected, around 500m KCU per Pod:

```
$ kubectl top pods
NAME                                CPU(cores)
MEMORY(bytes)
elastic-hamster-5897858459-m6j58    462m          40Mi
elastic-hamster-5897858459-qfh76    474m          40Mi
```

With the Deployment ready, we can start using the HPA to automatically adjust the number of replicas, which will target 75% of average CPU utilization across individual Pods. To do that, perform the following steps:

1. Create an `elastic-hamster-hpa.yaml` YAML manifest file for the HPA:

```yaml
apiVersion: autoscaling/v1
kind: HorizontalPodAutoscaler
metadata:
  name: elastic-hamster-hpa
spec:
  minReplicas: 1
  maxReplicas: 10
  targetCPUUtilizationPercentage: 75
  scaleTargetRef:
    apiVersion: apps/v1
    kind: Deployment
    name: elastic-hamster
```

The HPA targets `elastic-hamster` deployment, which we have provided using `.spec.scaleTargetRef`. The configuration that we specified ensures that the HPA will always keep the number of replicas between `minReplicas: 1` and `maxReplicas: 10`. The most important parameter in the HPA targeting the CPU metric is `targetCPUUtilizationPercentage`, which we have set to 75%. This means that the HPA will try to target 75% of the container `requests` value for `cpu`, which we set to be `200m` KCU. As a result, the HPA will try to keep the CPU consumption at around `150m` KCU. Our current Deployment with two replicas only is consuming much more, on average, `500m` KCU.

2. Apply the `manifest` file to the cluster:

```
$ kubectl apply -f ./elastic-hamster-hpa.yaml
horizontalpodautoscaler.autoscaling/elastic-hamster-hpa
created
```

3. After a while, the HPA will start adjusting the number of replicas to match the target CPU usage. Describe the HPA using the `kubectl` command to see the details:

```
$ kubectl describe hpa elastic-hamster-hpa
...
Metrics:                                              (
current / target )
  resource cpu on pods  (as a percentage of request):
79% (159m) / 75%

...

Events:
  Type     Reason            Age    From
Message
  ----     ------            ----   ----
-------
  Normal   SuccessfulRescale 15m    horizontal-pod-
autoscaler  New size: 4; reason: cpu resource utilization
(percentage of request) above target
  Normal   SuccessfulRescale 14m    horizontal-pod-
autoscaler  New size: 6; reason: cpu resource utilization
(percentage of request) above target
  Normal   SuccessfulRescale 13m    horizontal-pod-
autoscaler  New size: 8; reason: cpu resource utilization
(percentage of request) above target
```

```
Normal   SuccessfulRescale   11m   horizontal-pod-
autoscaler   New size: 9; reason: cpu resource utilization
(percentage of request) above target
```

In the output, you can see that the Deployment was gradually scaled up over time as it eventually stabilized at 9 replicas. Note that for you, the numbers may vary slightly. If you hit the maximum number of allowed replicas (10), you may try increasing the number or adjust the `targetCPUUtilizationPercentage` parameter.

> **Tip**
>
> It is possible to use an imperative command to achieve a similar result:
> `kubectl autoscale deploy elastic-hamster --cpu-percent=75 --min=1 --max=10`.

Congratulations! You have successfully configured horizontal autoscaling for your Deployment using an HPA. In the next section, we will take a look at autoscaling Kubernetes Nodes using a CA which gives even more flexibility when combined with an HPA.

Autoscaling Kubernetes Nodes using a Cluster Autoscaler

So far, we have discussed scaling at the level of individual Pods, but this is not the only way in which you can scale your workloads on Kubernetes. It is possible to scale the cluster **itself** to accommodate changes in demand for compute resources – at some point, we will need more Nodes to run more Pods. This is solved by the **CA**, which is part of the Kubernetes autoscaler repository (`https://github.com/kubernetes/autoscaler/tree/master/cluster-autoscaler`). The CA must be able to provision and deprovision Nodes for the Kubernetes cluster, so this means that vendor-specific plugins must be implemented. You can find the list of supported cloud service providers here: `https://github.com/kubernetes/autoscaler/tree/master/cluster-autoscaler#deployment`.

The CA periodically checks the status of Pods and Nodes and decides whether it needs to take action:

- If there are Pods that cannot be scheduled and are in the `Pending` state because of insufficient resources in the cluster, CA will add more Nodes, up to the predefined maximum size.

- If Nodes are under-utilized and all Pods could be scheduled even with a smaller number of Nodes in the cluster, the CA will remove the Nodes from the cluster, unless it has reached the predefined minimum size. Nodes are gracefully drained before they are removed from the cluster.

- For some cloud service providers, the CA can also choose between different SKUs for VMs to better optimize the cost of operating the cluster.

> **Important note**
> Pod containers must specify `requests` for the compute resources to make the CA work properly. Additionally, these values should reflect real usage, otherwise the CA will not be able to take correct decisions for your type of workload.

As you can see, the CA can complement HPA capabilities. If the HPA decides that there should be more Pods for a Deployment or StatefulSet, but no more Pods can be scheduled, then the CA can intervene and increase the cluster size.

Enabling the CA entails different steps depending on your cloud service provider. Additionally, some configuration values are specific for each of them. We will first take a look at GKE.

Enabling the cluster autoscaler in GKE

For GKE, it is easiest to create a cluster with CA enabled from scratch. To do that, you need to run the following command to create a cluster named `k8sforbeginners`:

```
$ gcloud container clusters create k8sforbeginners
--num-nodes=2 --zone=us-central1-a --enable-autoscaling
--min-nodes=2 --max-nodes=10
```

You can control the minimum number of Nodes in autoscaling by using the `--min-nodes` parameter, and the maximum number of Nodes by using the `--max-nodes` parameter.

In the case of an existing cluster, you need to enable the CA on an existing Node pool. For example, if you have a cluster named `k8sforbeginners` with one Node pool named `nodepool1`, then you need to run the following command:

```
$ gcloud container clusters update k8sforbeginners --enable-
autoscaling --min-nodes=2 --max-nodes=10 --zone=us-central1-a
--node-pool=nodepool1
```

The update will take a few minutes.

You can learn more in the official documentation: `https://cloud.google.com/kubernetes-engine/docs/concepts/cluster-autoscaler`.

Once configured, you can move on to *Using the cluster autoscaler*.

Enabling the cluster autoscaler in the Amazon Elastic Kubernetes Service

Setting up the CA in Amazon EKS cannot currently be realized in a one-click or one-command action. You need to create an appropriate IAM policy and role, deploy the CA resources to the Kubernetes cluster, and undertake manual configuration steps. For this reason, we will not cover this in the book and we request that you refer to the official instructions: `https://docs.aws.amazon.com/eks/latest/userguide/cluster-autoscaler.html`.

Once configured, you can move on to *Using the cluster autoscaler*.

Enabling the cluster autoscaler in the Azure Kubernetes Service

AKS provides a similar CA setup experience to GKE – you can use a one-command procedure to either deploy a new cluster with CA enabled or update the existing one to use the CA. To create a new cluster named `k8sforbeginners-aks` from scratch in the `k8sforbeginners-rg` resource group, execute the following command:

```
$ az aks create --resource-group k8sforbeginners-rg --name
k8sforbeginners-aks --node-count 2 --enable-cluster-autoscaler
--min-count 2 --max-count 10
```

You can control the minimum number of Nodes in autoscaling by using the `--min-count` parameter, and the maximum number of Nodes by using the `--max-count` parameter.

To enable the CA on an existing AKS cluster named `k8sforbeginners-aks`, execute the following command:

```
$ az aks update --resource-group k8sforbeginners-rg --name
k8sforbeginners-aks --enable-cluster-autoscaler --min-count 2
--max-count 10
```

The update will take a few minutes.

You can learn more in the official documentation: `https://docs.microsoft.com/en-us/azure/aks/cluster-autoscaler`. Additionally, the CA in AKS has more parameters that you can configure using **autoscaler profile**. Further details are provided in the official documentation at `https://docs.microsoft.com/en-us/azure/aks/cluster-autoscaler#using-the-autoscaler-profile`.

Now, let's take a look at how you can use the CA.

Using the cluster autoscaler

We have just configured the CA for the cluster and now it may take a bit of time until the CA performs its first actions. This depends on the CA configuration, which may be vendor-specific. For example, in the case of AKS, the cluster will be evaluated every 10 seconds (`scan-interval`), whether it needs to be scaled up or down. If scaling down needs to happen after scaling up, there is a 10-minute delay (`scale-down-delay-after-add`). Scaling down will be triggered if the sum of requested resources divided by capacity is below 0.5 (`scale-down-utilization-threshold`).

As a result, the cluster may automatically scale up, scale down, or remain unchanged after the CA was enabled. If you are using exactly the same cluster setup as we did in the examples, you will have the following situation:

- There are three Nodes, each with a capacity of `2000m` KCU, which means that the total KCU in the cluster is `6000m`.

- `elastic-hamster` Deployment is currently automatically scaled by the HPA to 9 replicas, each consuming `200m` KCU, which gives us the total `1800m` KCU requested.

- There is a bit of KCU consumed by the `kube-system` namespace Pods.

- Roughly, the current usage should be around 40%-50% of KCU. You can check the exact number using the `kubectl top nodes` command.

This means that the cluster with the current workload will either scale down by one Node or remain unchanged.

But instead, we can do some modifications to our `elastic-hamster` Deployment to trigger a more *firm* decision from CA. We will increase the total amount of work requested from the `elastic-hamster` Deployment and also increase the `requests` for CPU by its Pods. Additionally, we will allow more replicas to be created by the HPA. This will result in quickly exceeding the cluster capacity of `6000m` KCU and cause the CA to scale the cluster up. To do the demonstration, please perform the following steps:

1. In `elastic-hamster-deployment.yaml`, introduce the following changes. Set the number of `replicas` to `7` and `TOTAL_HAMSTER_USAGE` to `"7.0"` (the second value should be greater than the number of replicas). Set `requests` for cpu to `500m`:

```
apiVersion: apps/v1
kind: Deployment
metadata:
  name: elastic-hamster
spec:
...
  replicas: 7
  template:
...
    spec:
      serviceAccountName: elastic-hamster
      containers:
      - name: hamster
        image: ubuntu:20.04
        resources:
          requests:
            cpu: 500m
            memory: 50Mi
        env:
        - name: TOTAL_HAMSTER_USAGE
          value: "7.0"
...
```

2. In the `elastic-hamster-hpa.yaml` file, change the number of `maxReplicas` to 25:

```
apiVersion: autoscaling/v1
kind: HorizontalPodAutoscaler
metadata:
  name: elastic-hamster-hpa
spec:
  minReplicas: 1
  maxReplicas: 25
...
```

3. Apply all YAML manifests in the directory to the cluster again:

```
$ kubectl apply -f ./
role.rbac.authorization.k8s.io/deployment-reader
unchanged
deployment.apps/elastic-hamster configured
horizontalpodautoscaler.autoscaling/elastic-hamster-hpa
configured
serviceaccount/elastic-hamster unchanged
rolebinding.rbac.authorization.k8s.io/read-deployments
unchanged
```

4. If you soon check the status of the Pods in the cluster, you will see that some of them are pending because of insufficient resources:

```
$ kubectl get pods
NAME                                   READY    STATUS
RESTARTS     AGE
...
elastic-hamster-5854d5f967-cjsmg       0/1      Pending
0            23s
elastic-hamster-5854d5f967-nsnqd       0/1      Pending
0            23s
...
```

5. Check the status of the Nodes in the cluster. The CA should already start provisioning **new** Nodes by that time. In our case, Node 3 has been provisioned successfully, and Node 4 is in the process of provisioning:

```
$ kubectl get node
NAME                                    STATUS     ROLES
AGE       VERSION
aks-nodepool1-77120516-vmss000000       Ready      agent
22d       v1.18.14
aks-nodepool1-77120516-vmss000001       Ready      agent
22d       v1.18.14
aks-nodepool1-77120516-vmss000002       Ready      agent
29h       v1.18.14
aks-nodepool1-77120516-vmss000003       Ready      agent
2m47s     v1.18.14
aks-nodepool1-77120516-vmss000004       NotReady   <none>
5s        v1.18.14
```

6. If you inspect some of the Pods that were in the Pending state, you will see that their events contain information about the CA trigger to create a new Node:

```
$ kubectl describe pod elastic-hamster-5854d5f967-grjbj
...
Events:
  Type       Reason            Age    From
Message
  ----       ------            ----   ----
-------
  Warning    FailedScheduling  5m28s  default-scheduler
0/7 nodes are available: 7 Insufficient cpu.
  Warning    FailedScheduling  3m6s   default-scheduler
0/8 nodes are available: 1 node(s) had taint {node.
kubernetes.io/not-ready: }, that the pod didn't tolerate,
7 Insufficient cpu.
  Normal     Scheduled         2m55s  default-scheduler
Successfully assigned default/elastic-hamster-5854d5f967-
grjbj to aks-nodepool1-77120516-vmss000007
  Normal     TriggeredScaleUp  4m55s  cluster-autoscaler
pod triggered scale-up: [{aks-nodepool1-77120516-vmss
7->8 (max: 10)}]
```

7. Eventually, scaling up using the HPA will be finished, all Pods will become ready, and the CA will not need to autoscale to more Nodes. In our example, we ended at 16 Pod replicas running on 8 Nodes in total, and this resulted in the **stabilization** of average CPU usage at 82%:

```
$ kubectl describe hpa elastic-hamster-hpa
...
Metrics:                                             (
current / target )
   resource cpu on pods  (as a percentage of request):
82% (410m) / 75%
Min replicas:                                        1
Max replicas:                                        25
Deployment pods:                                     16
current / 16 desired
```

8. Node CPU usage is not distributed evenly though – the reason for this is that scaling to new Nodes does not trigger any rescheduling of Pods:

```
$ kubectl top nodes
NAME                                CPU(cores)    CPU%
MEMORY(bytes)    MEMORY%
aks-nodepool1-77120516-vmss000000    981m          51%
2212Mi           48%
aks-nodepool1-77120516-vmss000001    1297m         68%
2121Mi           46%
aks-nodepool1-77120516-vmss000002    486m          25%
883Mi            19%
aks-nodepool1-77120516-vmss000003    475m          25%
933Mi            20%
aks-nodepool1-77120516-vmss000004    507m          26%
945Mi            20%
aks-nodepool1-77120516-vmss000005    902m          47%
987Mi            21%
aks-nodepool1-77120516-vmss000006    1304m         68%
1028Mi           22%
aks-nodepool1-77120516-vmss000007    1263m         66%
1018Mi           22%
```

This shows how the CA has worked together with the HPA to seamlessly scale the Deployment and cluster at the same time to accommodate the workload. We will now show what automatic scaling down looks like. Perform the following steps:

1. To decrease the load in the cluster, we can simply change the value of the TOTAL_ HAMSTER_USAGE environment variable, for example, to "1.0". This will cause a rapid decrease in the load on Pods – if we currently have 16 replicas, the CPU utilization will be roughly 63m KCU per Pod, which gives 13% average CPU usage per Pod. This will cause the HPA to scale down after the **stabilization window** time has passed, which is, by default, 5 minutes. Introduce the changes to the elastic-hamster-deployment.yaml manifest file:

```
apiVersion: apps/v1
kind: Deployment
metadata:
  name: elastic-hamster
spec:
...
  template:
...
    spec:
...
      containers:
      - name: hamster
...
        env:
        - name: TOTAL_HAMSTER_USAGE
          value: "7.0"
...
```

2. Apply the manifest file to the cluster:

```
$ kubectl apply -f ./elastic-hamster-deployment.yaml
deployment.apps/elastic-hamster configured
```

3. Now, you have to wait patiently for a bit. First, the HPA must get past the stabilization window, which can take around 5 minutes, and after the Deployment is scaled down to around 3 replicas, you will still have to wait around 10 minutes for the cluster to scale down following the recent scale-up. It's time for a good cup of coffee!

4. The HPA has eventually scaled down the Deployment to three Pods and stabilized the CPU usage at 66%:

```
$ kubectl describe hpa elastic-hamster-hpa
...
Metrics:                                                      (
current / target )
    resource cpu on pods  (as a percentage of request):
66% (331m) / 75%
Min replicas:                                                 1
Max replicas:                                                25
Deployment pods:                                             3
current / 3 desired
```

5. At some point, you will notice that the Nodes are being deprovisioned:

```
$ kubectl get nodes
NAME                                   STATUS     ROLES
AGE    VERSION
aks-nodepool1-77120516-vmss000000      Ready      agent
22d    v1.18.14
aks-nodepool1-77120516-vmss000001      Ready      agent
22d    v1.18.14
aks-nodepool1-77120516-vmss000003      NotReady   agent
56m    v1.18.14
aks-nodepool1-77120516-vmss000004      Ready      agent
53m    v1.18.14
aks-nodepool1-77120516-vmss000005      NotReady   agent
51m    v1.18.14
aks-nodepool1-77120516-vmss000006      NotReady   agent
47m    v1.18.14
aks-nodepool1-77120516-vmss000007      NotReady   agent
42m    v1.18.14
```

6. And finally, you will end up with a cluster with only two Nodes, which is the minimum number that we preconfigured:

```
$ kubectl get nodes
NAME                                   STATUS    ROLES    AGE
VERSION
aks-nodepool1-77120516-vmss000000      Ready     agent    22d
```

```
v1.18.14
aks-nodepool1-77120516-vmss000001    Ready    agent    22d
v1.18.14
```

This shows how efficiently the CA can react to a decrease in the load in the cluster when the HPA has scaled down the Deployment. Earlier, without any intervention, the cluster scaled to eight Nodes for a short period of time, and then scaled down to just two Nodes. Imagine the cost difference between having an eight-Node cluster running all the time and using the CA to cleverly autoscale on demand!

> **Tip**
> To ensure that you are not charged for any unwanted cloud resources, you need to clean up the cluster or disable cluster autoscaling to be sure that you are not running too many Nodes.

This demonstration concludes our chapter about autoscaling in Kubernetes. Let's summarize what we have learned in this chapter.

Summary

In this chapter, you have learned about autoscaling techniques in Kubernetes clusters. We first explained the basics behind Pod resource requests and limits and why they are crucial for the autoscaling and scheduling of Pods. Next, we introduced the VPA, which can automatically change requests and limits for Pods based on current and past metrics. After that, you learned about the HPA, which can be used to automatically change the number of Deployment or StatefulSet replicas. The changes are done based on CPU, memory, or custom metrics. Lastly, we explained the role of the CA in cloud environments. We also demonstrated how you can efficiently combine using the HPA with the CA to achieve the scaling of your workload together with the scaling of the cluster.

There is much more that can be configured in the VPA, HPA, and CA, so we have just scratched the surface of powerful autoscaling in Kubernetes!

In the last chapter, we will explain how you can use Ingress in Kubernetes for advanced traffic routing.

Further reading

For more information regarding autoscaling in Kubernetes, please refer to the following PacktPub books:

- *The Complete Kubernetes Guide*, by *Jonathan Baier, Gigi Sayfan, Jesse White* (`https://www.packtpub.com/virtualization-and-cloud/complete-kubernetes-guide`)

- *Getting Started with Kubernetes – Third Edition*, by *Jonathan Baier, Jesse White* (`https://www.packtpub.com/virtualization-and-cloud/getting-started-kubernetes-third-edition`)

- *Kubernetes for Developers*, by *Joseph Heck* (`https://www.packtpub.com/virtualization-and-cloud/kubernetes-developers`)

- *Hands-On Kubernetes on Windows*, by *Piotr Tylenda* (`https://www.packtpub.com/product/hands-on-kubernetes-on-windows/9781838821562`)

You can also refer to the official documentation:

- Kubernetes documentation (`https://kubernetes.io/docs/home/`), which is always the most up-to-date source of knowledge regarding Kubernetes in general.

- General installation instructions for the Vertical Pod Autoscaler are available here: `https://github.com/kubernetes/autoscaler/tree/master/vertical-pod-autoscaler#installation`. EKS documentation offers its own version of the instructions: `https://docs.aws.amazon.com/eks/latest/userguide/vertical-pod-autoscaler.html`.

21
Advanced Traffic Routing with Ingress

This last chapter will give an overview of advanced traffic routing in Kubernetes using Ingress resources. In short, Ingress can be used to expose your Pods running behind a Service object to the **external** world using HTTP and HTTPS routes. We have already discussed ways to expose your application using Service objects directly, especially the `LoadBalancer` Service. But this approach works fine only in cloud environments where you have cloud-controller-manager running and byconfiguring external load balancers to be used with this type of Service. And what is more, each `LoadBalancer` Service requires a separate instance of the cloud load balancer, which brings additional costs and maintenance overhead. We are going to introduce Ingress and Ingress Controller, which can be used in any type of environment to provide routing and load-balancing capabilities for your application. You will also learn how to use the nginx web server as Ingress Controller and how you can configure the dedicated Azure Application Gateway Ingress Controller for your AKS cluster.

In this chapter, we will cover the following topics:

- Refresher: Kubernetes services
- Introducing the Ingress object
- Using nginx as an Ingress Controller
- Azure Application Gateway Ingress Controller for AKS

Technical requirements

For this chapter, you will need the following:

- A Kubernetes cluster deployed. We recommend using a **multi-node**, cloud-based Kubernetes cluster. It is possible to use Ingress in `minikube` after enabling the required add-ons.

- An AKS cluster is required to follow the last section about Azure Application Gateway Ingress Controller.

- The Kubernetes CLI (`kubectl`) needs to be installed on your local machine and configured to manage your Kubernetes cluster.

Basic Kubernetes cluster deployment (local and cloud-based) and `kubectl` installation have been covered in *Chapter 3, Installing Your First Kubernetes Cluster*.

The following previous chapters can provide you with an overview of how to deploy a fully functional Kubernetes cluster on different cloud platforms and install the requisite CLIs to manage them:

- *Chapter 14, Kubernetes Clusters on Google Kubernetes Engine.*

- *Chapter 15, Launching a Kubernetes Cluster on Amazon Web Services with the Amazon Elastic Kubernetes Service.*

- *Chapter 16, Kubernetes Clusters on Microsoft Azure with the Azure Kubernetes Service.*

You can download the latest code samples for this chapter from the official GitHub repository at `https://github.com/PacktPublishing/The-Kubernetes-Bible/tree/master/Chapter21`.

Refresher: Kubernetes services

In the previous chapters, you have learned about **Service** objects, which can be used to expose Pods to load-balanced traffic, both internal as well as external. Internally, they are implemented as **virtual IP addresses** managed by kube-proxy at each of the Nodes. We are going to do a quick recap of different types of services:

- `ClusterIP`
- `NodePort`
- `LoadBalancer`

To make it easier to explain, we will assume that we have a Deployment running three replicas of Pods running the `nginx` container, which has the following YAML manifest:

```yaml
apiVersion: apps/v1
kind: Deployment
metadata:
  name: nginx-deployment-example
spec:
  replicas: 3
  selector:
    matchLabels:
      environment: test
  template:
    metadata:
      labels:
        environment: test
    spec:
      containers:
      - name: nginx
        image: nginx:1.17
        ports:
        - containerPort: 80
```

The Pod exposes TCP port `80`, which is used by the `nginx` process to serve the requests, and we will now discuss the details of using the `ClusterIP` Service to expose this Deployment internally.

The ClusterIP Service

Let's now take a look at the `ClusterIP` Service type. This type of Service exposes Pods using internally visible virtual IP addresses managed by kube-proxy on each Node. This means that the Service will be reachable from within the cluster only. Consider the following manifest for the service:

```yaml
apiVersion: v1
kind: Service
metadata:
  name: nginx-deployment-example-clusterip
spec:
```

```
  selector:
    environment: test
  type: ClusterIP
  ports:
  - port: 8080
    protocol: TCP
    targetPort: 80
```

The `ClusterIP` Service is configured in such a way that it will map requests coming from its IP and TCP port `8080` to the container's TCP port `80`. The actual `ClusterIP` address is assigned **dynamically**, unless you specify one explicitly in the specifications. The internal DNS Service in a Kubernetes cluster is responsible for resolving the `nginx-deployment-example` name to the actual `ClusterIP` address as a part of service discovery.

> **Important note**
>
> In the rest of the section, we will provide diagrams that represent how the Service types are implemented **logically**. In fact, under the hood, kube-proxy is responsible for managing the virtual IP addresses on the Nodes and modifying all forwarding rules. So, services exist only as a logical concept inside the cluster. There is no physical process that runs inside the cluster for each Service and does the proxying.

We have visualized the `ClusterIP` Service principles in the following diagram:

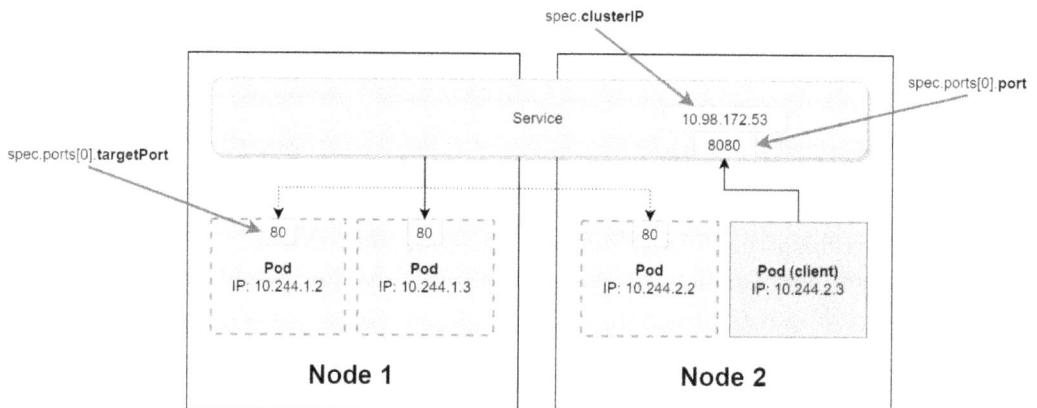

Figure 21.1 – ClusterIP Service

The diagram includes references to the Kubernetes objects specifications to make it easier to understand the connections. ClusterIP Services are the most basic type of Service in Kubernetes and they are part of other Service types that allow Pods to be exposed to external traffic: NodePort and LoadBalancer.

NodePort service

This type of Service is similar to the ClusterIP Service but additionally, it can be reached by **any** cluster node IP address and specified port. To achieve that, kube-proxy exposes the same port on each Node in the range 30000-32767 (which is configurable) and sets up forwarding so that any connections to this port will be forwarded to ClusterIP.

Let's take a look at an example YAML manifest of the NodePort Service:

```
apiVersion: v1
kind: Service
metadata:
  name: nginx-deployment-example-nodeport
spec:
  selector:
    environment: test
  type: NodePort
  ports:
  - port: 8080
    nodePort: 31001
    protocol: TCP
    targetPort: 80
```

In this case, TCP port 31001 is used as the external port on each Node. If you do not specify nodePort, it will be allocated **dynamically** using the range. For internal communication, this Service still behaves like a simple ClusterIP Service, and you can use its ClusterIP address.

These principles have been visualized in the following diagram:

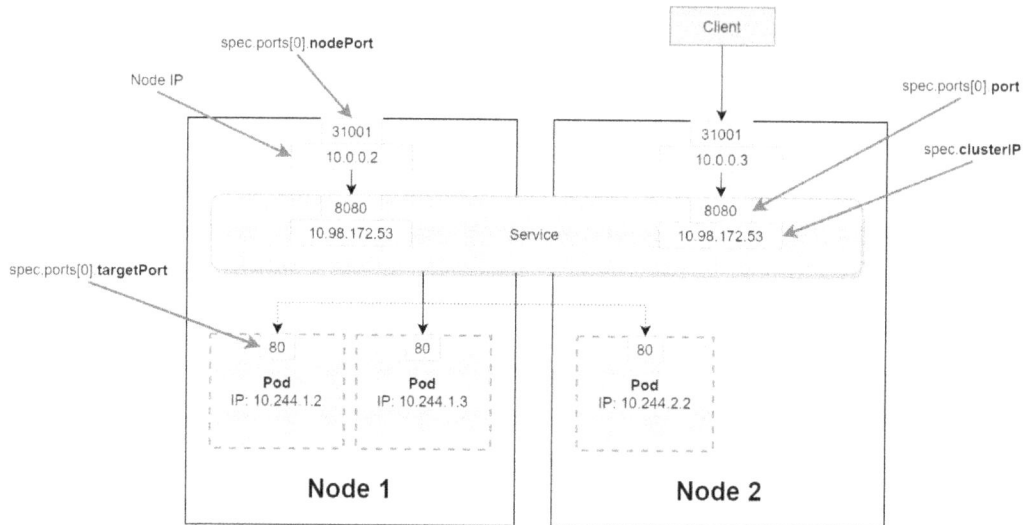

Figure 21.2 – NodePort service

You will need the `NodePort` Service if you want to set up load balancing externally, without using any Kubernetes constructs. They can also be used for communicating directly with the Pods, but exposing the external IP addresses of Nodes is usually not a good idea in terms of security. Please note that many other components rely on `NodePort` Services; for example, they are used in `LoadBalancer` Service implementations or when exposing Ingress Controller, which we are going to discuss in this chapter.

We will now do a quick recap of the `LoadBalancer` Service.

The LoadBalancer service

This is the second type of Service that allows external traffic to the Pods. The `LoadBalancer` Service is usually used in cloud environments where you have **software-defined networking** (**SDN**), and you can configure load balancers on demand that redirect traffic to your cluster. The automatic provisioning of load balancers in the cloud is done by vendor-specific plugins in cloud-controller-manager. This type of service combines the approach of the `NodePort` Service with an additional external load balancer in front of it, which routes traffic to `NodePorts`.

Let's take a look at an example YAML manifest of the LoadBalancer Service:

```
apiVersion: v1
kind: Service
metadata:
  name: nginx-deployment-example-lb
spec:
  selector:
    environment: test
  type: LoadBalancer
  ports:
  - port: 8080
    protocol: TCP
    targetPort: 80
```

When you create such a Service, it will behave as a NodePort Service with a randomly assigned port and a cloud load balancer that serves requests at TCP port 8080 and forwards them to NodePorts. The **external** IP address of the service is provided by the load balancer and is also available in the Service object in .status.loadBalancer. ingress[0].ip. These principles have been visualized in the following diagram:

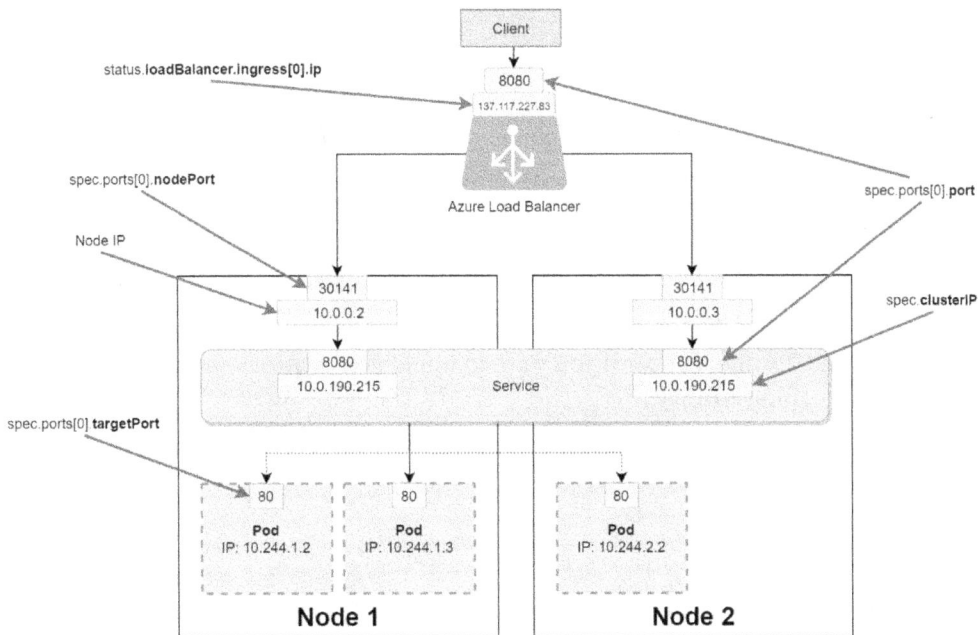

Figure 21.3 – LoadBalancer service

You can still, of course, use the service internally via its ClusterIP.

It may seem appealing to always use Kubernetes services for allowing external traffic to the cluster, but there are a few disadvantages of using them all the time. We will now introduce the Ingress object and discuss why it is needed and when it should be used instead of Services to manage external traffic.

Introducing the Ingress object

In the previous section, we did a short recap of Service objects in Kubernetes and their role in routing traffic. From the perspective of external traffic, the most important are the `NodePort` Service and the `LoadBalancer` Service. In general, the `NodePort` Service can only be used in conjunction with a different routing and load balancing component, as exposing multiple external endpoints on all Kubernetes Nodes is not secure. This leaves us with the `LoadBalancer` Service, which, under the hood, relies on `NodePort`. There are a few problems with this type of Service in some use cases:

- The `LoadBalancer` Service is used for **L4 load balancing**, which means it is done at OSI layer 4 (transport). The load balancer can make the decisions based on the TCP/UDP protocol. Applications that use HTTP or HTTPS protocols often require **L7 load balancing**, which is done at OSI layer 7 (application).

- The L4 load balancer cannot do **HTTPS traffic termination and offloading**.

- You cannot implement **name-based virtual hosting** using the same L4 load balancer for multiple domain names.

- You need an L7 load balancer to implement **path-based routing**. For example, configuring requests to `https://<loadBalancerIp>/service1` to be redirected to the Kubernetes Service named `service1`, and requests to `https://<loadBalancerIp>/service2` to be redirected to the Kubernetes Service named `service2` is not possible with the L4 load balancer – it is not aware of the HTTP(S) protocol.

- You need an L7 load balancer if you want to implement features such as **sticky sessions** or **cookie affinity**.

In Kubernetes, you can solve these problems using an **Ingress** object, which can be used for implementing and modeling L7 load balancing. The Ingress object is used for defining the routing and balancing rules only, for example, which path should be routed to which Kubernetes Service. Let's take a look at an example YAML manifest file, `example-ingress.yaml`, for Ingress:

```
apiVersion: networking.k8s.io/v1beta1
kind: Ingress
metadata:
  name: example-ingress
  annotations:
    nginx.ingress.kubernetes.io/rewrite-target: /
spec:
  rules:
  - http:
      paths:
      - path: /service1
        pathType: Prefix
        backend:
          serviceName: example-service1
          servicePort: 80
      - path: /service2
        pathType: Prefix
        backend:
          serviceName: example-service2
          servicePort: 80
```

Simply put, Ingress is an abstract definition of routing rules for your Services. Alone, it is not doing anything; it requires **Ingress Controller** to actually process and implement these rules – you can apply the `manifest` file, but at this point, it will have no effect. But first, we will explain how the Ingress HTTP routing rules are built. Each of these `rules` in the specification contains the following:

- **Optional host**: In the example, we are not using this field, so the rule that we defined is applied to all incoming traffic. If the field value is provided, then the rule applies only to requests that have this host as the destination – you can have multiple hostnames resolving to the same IP address. The `host` field supports wildcards.

- **List of path routings**: Each of the paths has an associated Ingress backend that you define by providing `serviceName` and `servicePort`. In the preceding example, all requests arriving at the path with the prefix `/service1` will be routed to Pods of the `example-service1` Service, and all requests arriving at the path with the prefix `/service2` will be routed to Pods of the `example-service2` Service. The `path` fields support prefixes and exact matching, and it is also possible to use implementation-specific matching, which is carried out by the underlying Ingress Controller.

In this way, you can configure complex routing rules that involve multiple Services in the cluster, but externally they will be visible as a **single** endpoint with multiple paths available. This is especially useful when you create API gateways in microservice architecture for frontend or client applications.

> **Important note**
>
> In Kubernetes 1.19, Ingress has become a `networking.k8s.io/v1` resource. There are a few changes compared to `networking.k8s.io/v1beta1`; for example, the backend is defined differently: `https://v1-19.docs.kubernetes.io/docs/reference/generated/kubernetes-api/v1.19/#ingressbackend-v1-networking-k8s-io`.

To materialize Ingress objects, we need to have an Ingress Controller installed in the cluster.

Using nginx as an Ingress Controller

An **Ingress Controller** is a Kubernetes controller that is deployed manually to the cluster, most often as a DaemonSet or a Deployment object that runs dedicated Pods for handling incoming traffic load balancing and smart routing. It is responsible for processing the Ingress objects (which specify that they especially want to use the Ingress Controller) and dynamically configuring real routing rules. A commonly used Ingress controller for Kubernetes is `ingress-nginx` (`https://www.nginx.com/products/nginx/kubernetes-ingress-controller`), which is installed in the cluster as a Deployment of an `nginx` web host with a set of rules for handling Ingress API objects. The Ingress Controller is exposed as a Service with a type that depends on the installation – in cloud environments, this will be `LoadBalancer`.

> **Important note**
>
> In cloud environments, you will often see dedicated Ingress Controllers that leverage vendor-specific features that allow **direct** communication from the external load balancer to the Pods. In such cases, there are no additional Pods involved and there may even be no need for `NodePort` Services. The routing is handled at SDN and CNI levels and the load balancer can use private IPs of the Pods. We will show an example of such an approach in the next section when we discuss the Application Gateway ingress controller for AKS.

The installation of `ingress-nginx` is described for different environments in the official documentation: `https://kubernetes.github.io/ingress-nginx/deploy/`. Note that it is also possible to use Helm to install this Ingress Controller, which makes management and upgrades a bit easier. For cloud environments, the installation is usually very simple and involves applying a single `YAML manifest` file, which creates multiple Kubernetes objects. Let's demonstrate this in AKS. Please execute the following command:

```
$ kubectl apply -f https://raw.githubusercontent.com/
kubernetes/ingress-nginx/controller-v0.44.0/deploy/static/
provider/cloud/deploy.yaml

namespace/ingress-nginx created

serviceaccount/ingress-nginx created

configmap/ingress-nginx-controller created

...
```

Now, we can also create our example Services together with the Ingress object that defines routings for them. First, we will create the Service objects and Deployment objects. There is no need to explain the YAML manifests as they are simple web servers printing out a welcome message with information on which Service you have reached:

```
$ kubectl apply -f https://raw.githubusercontent.com/
PacktPublishing/Kubernetes-for-Beginners/master/Chapter21/02_
ingress/example-services.yaml
deployment.apps/example-service1 created
deployment.apps/example-service2 created
service/example-service1 created
service/example-service2 created
```

Next, we can apply the Ingress object to the cluster that we created earlier:

```
$ kubectl apply -f ./example-ingress.yaml
ingress.networking.k8s.io/example-ingress created
```

By this time, the LoadBalancer Service that is running as part of Ingress Controller should already be functional and have an external IP address available. You can get it using the following command:

```
$ kubectl describe svc -n ingress-nginx ingress-nginx-
controller
...
LoadBalancer Ingress:      137.117.227.83
...
```

At this point, we can visualize what is happening behind Ingress Controller in the following diagram:

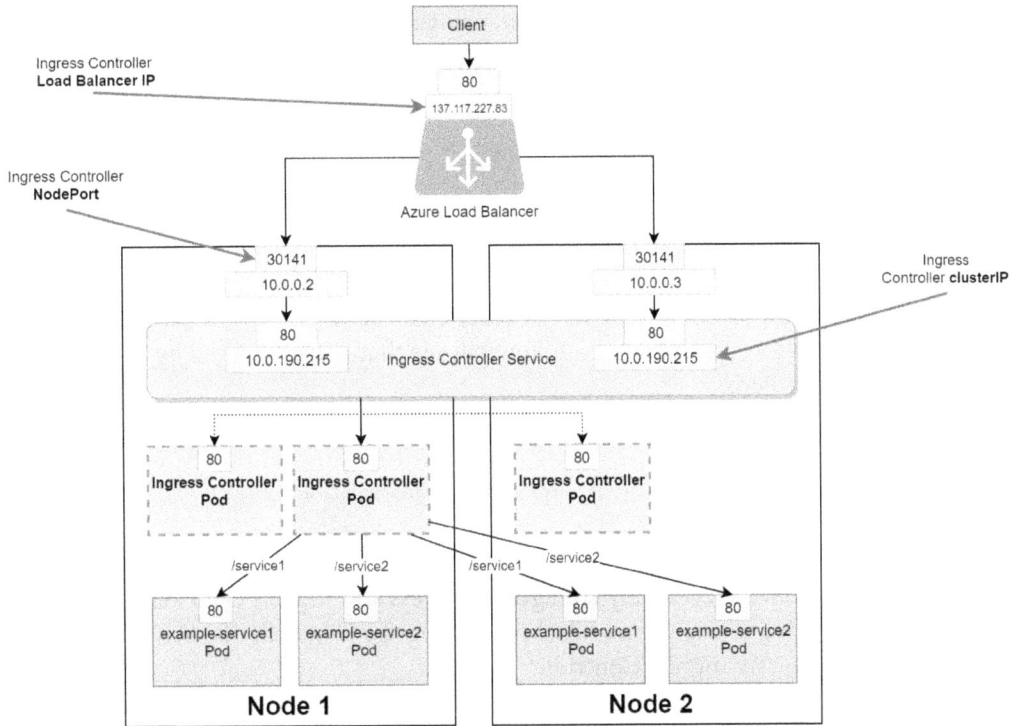

Figure 21.4 – Using nginx as Ingress Controller in a cloud environment

When you perform an HTTP request to `http://<ingressServiceLoadBalan-cerIp>/service1`, the traffic will be routed by `nginx` to `example-service1`. Similarly, when you use the `/service2` path, the traffic will be routed to `example-ser-vice2`. Note that you are using only **one** cloud load balancer for this operation, and that the actual routing to Kubernetes Services is performed by the Ingress Controller Pods using path-based routing.

Important note

In practice, you need to set up SSL certificates for your HTTP endpoints to ensure proper security. In our examples, we are not doing that for simplicity and to make the demonstrations clearer.

Let's verify this in practice. In your web browser, navigate to the /service1 path. In our case, this will be http://137.117.227.83/service1. You will see that you are served by example-service1 Pods:

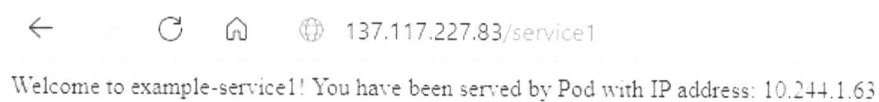

←   C   ⌂   ⊕ 137.117.227.83/service1

Welcome to example-service1! You have been served by Pod with IP address: 10.244.1.63

Figure 21.5 – Routing to example-service1 via the nginx Ingress Controller

Now, navigate to the /service2 path. In our case, this will be http://137.117.227.83/service2. You will see that you are served by example-service2 Pods:

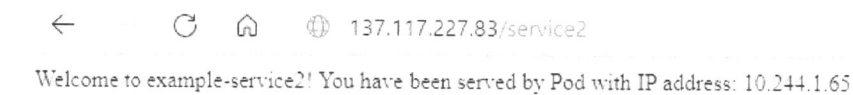

←   C   ⌂   ⊕ 137.117.227.83/service2

Welcome to example-service2! You have been served by Pod with IP address: 10.244.1.65

Figure 21.6 – Routing to example-service2 via the nginx Ingress Controller

Congratulations! You have successfully configured Ingress and Ingress Controller in your cluster. We are now going to explore a special type of Ingress Controller for AKS named Azure Application Gateway Ingress Controller.

Azure Application Gateway Ingress Controller for AKS

The Ingress Controller based on the nginx web server that we showed in the last section is a type of **generic** Ingress Controller that can be used in almost any environment. It relies on standard Kubernetes objects such as Deployments, Pods, and Services, and does not require any external components. If any external components are provisioned, this is done by cloud-controller-manager, and not the Ingress Controller itself.

This approach has a few drawbacks if you use it in a cloud environment such as Azure Kubernetes Service:

- You have an Azure load balancer just to proxy the requests to nginx Ingress Controller Pods via NodePorts. Then, there is *another* level of load balancing after the request reaches the Node, performed by kube-proxy as part of the Service object for Ingress Controller. Request routing based on paths is done by Ingress Controller Pods. And eventually, the last level of kube-proxy load balancing is at the target Service (in our demo, example-service1 or example-service2).

- There are more points of failure, especially if you scale cluster Nodes, drain Nodes, and so on.

Instead, it is possible to leverage a native L7 load balancer service in Azure named **Application Gateway** (`https://docs.microsoft.com/en-us/azure/application-gateway/overview`). AKS offers **Application Gateway Ingress Controller** (**AGIC**), which uses Application Gateway to directly communicate with Pods using their private IP addresses. This is achieved by AGIC Pods monitoring the Kubernetes API and instructing Azure Resource Manager to make changes to Application Gateway depending on Pods and services changes. Application Gateway can communicate directly with Pods thanks to Azure SDN features such as **VNet peering** (`https://docs.microsoft.com/en-us/azure/virtual-network/virtual-network-peering-overview`). This design is shown in the following diagram:

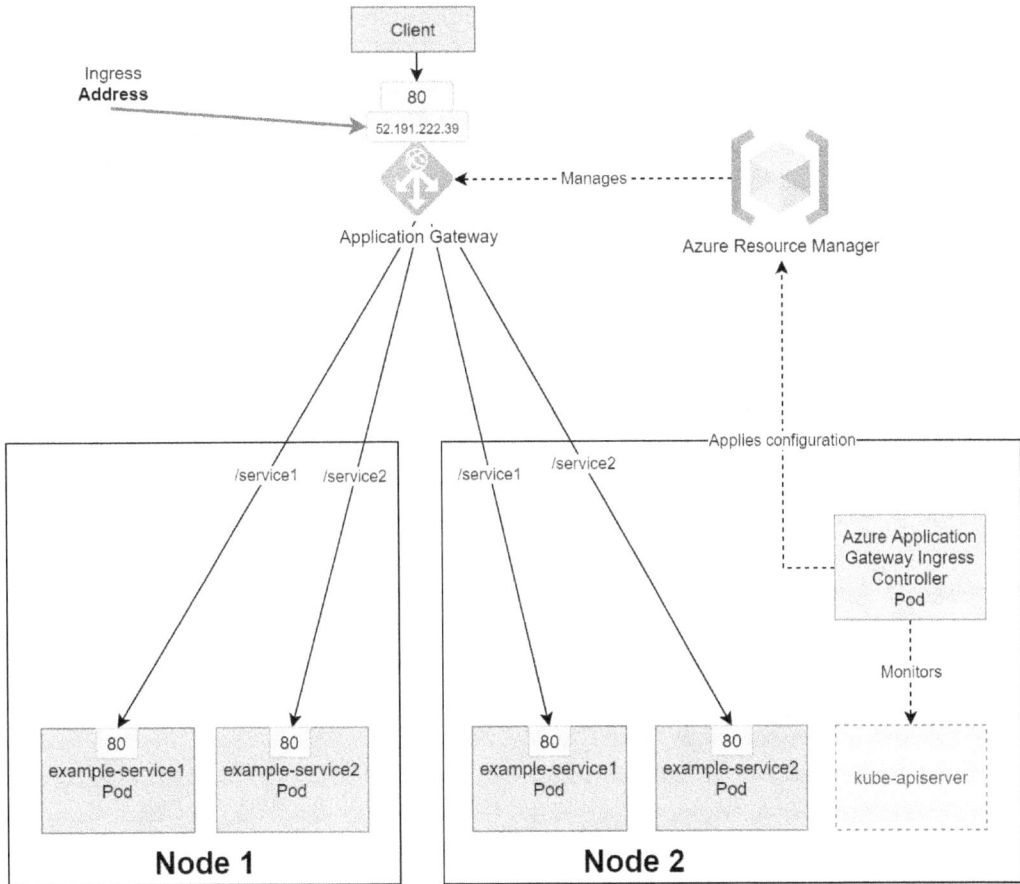

Figure 21.7 – Application Gateway ingress controller in AKS

It is possible to configure AGIC on an existing AKS cluster, and this is described in the official documentation: `https://docs.microsoft.com/en-us/azure/application-gateway/tutorial-ingress-controller-add-on-existing`. For simplicity, we will create a new AKS cluster with AGIC enabled – this is a one-command action. To deploy the two-node cluster named `k8sforbeginners-aks-agic` in the `k8sforbeginners-rg` resource group, execute the following command:

```
$ az aks create --resource-group k8sforbeginners-rg --name
k8sforbeginners-aks-agic --node-count 2 --network-plugin
azure --enable-managed-identity -a ingress-appgw --appgw-
name AksApplicationGateway --appgw-subnet-cidr "10.2.0.0/16"
--generate-ssh-keys
```

This will create an Azure Application Gateway named `AksApplicationGateway` with the subnet CIDR `10.2.0.0/16`.

When the cluster finishes deploying, we need to generate `kubeconfig` to use it with `kubectl`. Run the following command (it will switch to a new context so you will still have the old context available later):

```
$ az aks get-credentials --resource-group k8sforbeginners-rg
--name k8sforbeginners-aks-agic
Merged "k8sforbeginners-aks-agic" as current context in .kube/
config
```

Now, we can apply the same YAML manifest for Deployments and Services as in the previous section:

```
$ kubectl apply -f https://raw.githubusercontent.com/
PacktPublishing/Kubernetes-for-Beginners/master/Chapter21/03_
aks-agic/example-services.yaml
deployment.apps/example-service1 created
deployment.apps/example-service2 created
service/example-service1 created
service/example-service2 created
```

The final step will involve creating an Ingress object. We need to modify the YAML manifest slightly – the reason for this is that AGIC 1.4 does not yet support apiVersion: networking.k8s.io/v1beta1, and we need to use apiVersion: extensions/v1beta1, which can use kubernetes.io/ingress.class annotation instead of the ingressClassName field in the specification. Create an example-ingress.yaml file with the following content:

```
apiVersion: extensions/v1beta1
kind: Ingress
metadata:
  name: example-ingress
  annotations:
    kubernetes.io/ingress.class: azure/application-gateway
    appgw.ingress.kubernetes.io/backend-path-prefix: "/"
spec:
  rules:
  - http:
      paths:
      - path: /service1
        backend:
          serviceName: example-service1
          servicePort: 80
      - path: /service2
        backend:
          serviceName: example-service2
          servicePort: 80
```

For AGIC to pick up this Ingress object, it must define the kubernetes.io/ingress.class annotation with the value azure/application-gateway. We additionally need to ensure that the prefix for requests to the backend service is simply "/". After that, we can apply the manifest to the cluster:

```
$ kubectl apply -f ./example-ingress.yaml
ingress.extensions/example-ingress configured
```

You will need to wait a few minutes for Application Gateway to reconfigure. To get the external IP address of the Ingress, execute the following command:

```
$ kubectl get ingress
NAME              CLASS      HOSTS   ADDRESS          PORTS   AGE
example-ingress   <none>     *       52.191.222.39    80      36m
```

In our case, the IP address is 52.191.222.39.

Let's verify whether AGIC works correctly. In your web browser, navigate to the /service1 path. In our case, this will be http://52.191.222.39/service1. You will see that you are served by example-service1 Pods:

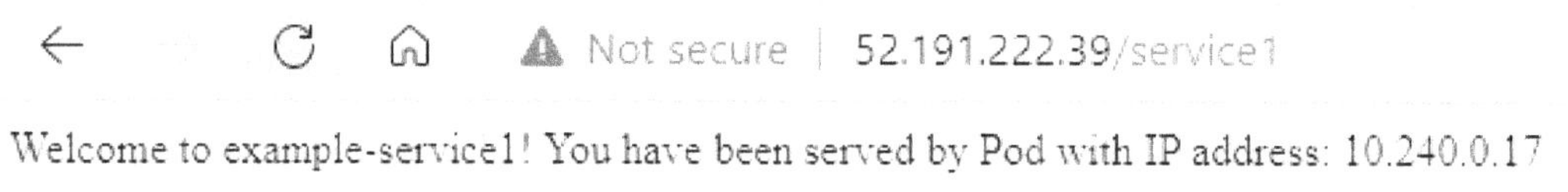

←   C   ⌂   ⚠ Not secure | 52.191.222.39/service1

Welcome to example-service1! You have been served by Pod with IP address: 10.240.0.17

Figure 21.8 – Routing to example-service1 via Application Gateway Ingress Controller in AKS

Now, navigate to the /service2 path. In our case, this will be http://52.191.222.39/service2. You will see that you are served by example-service2 Pods:

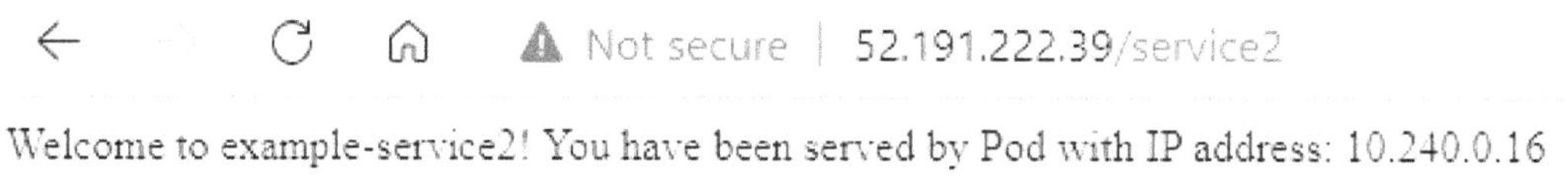

←   C   ⌂   ⚠ Not secure | 52.191.222.39/service2

Welcome to example-service2! You have been served by Pod with IP address: 10.240.0.16

Figure 21.9 – Routing to example-service2 via Application Gateway Ingress Controller in AKS

Congratulations! You have successfully configured and tested Application Gateway Ingress Controller in AKS! This was the last practical demonstration in this book, so let's now summarize what you have learned.

Summary

In this last chapter, we have explained advanced traffic routing approaches in Kubernetes using Ingress objects and Ingress Controllers. At the beginning, we did a brief recap of Kubernetes Service types. We refreshed our knowledge regarding `ClusterIP`, `NodePort`, and `LoadBalancer` Service objects. Based on that, we introduced Ingress objects and Ingress Controller and explained how they fit into the landscape of traffic routing in Kubernetes. Now, you know that simple Services are commonly used when L4 load balancing is required, but if you have HTTP or HTTPS endpoints in your applications, it is better to use L7 load balancing offered by Ingress and Ingress Controllers. You learned how to deploy the nginx web server as Ingress Controller and we tested this on example Deployments. Lastly, we explained how you can approach Ingress and Ingress Controllers in cloud environments where you have native support for L7 load balancing outside of the Kubernetes cluster. As a demonstration, we deployed an AKS cluster with **Application Gateway Ingress Controller** (**AGIC**) to handle Ingress objects.

Congratulations! This has been a long journey into the exciting territory of Kubernetes and container orchestration. Good luck with your further Kubernetes journey and thanks for reading.

Further reading

For more information regarding autoscaling in Kubernetes, please refer to the following Packt books:

- *The Complete Kubernetes Guide*, by *Jonathan Baier, Gigi Sayfan, Jesse White* (`https://www.packtpub.com/virtualization-and-cloud/complete-kubernetes-guide`)

- *Getting Started with Kubernetes – Third Edition*, by *Jonathan Baier, Jesse White* (`https://www.packtpub.com/virtualization-and-cloud/getting-started-kubernetes-third-edition`)

- *Kubernetes for Developers*, by *Joseph Heck* (`https://www.packtpub.com/virtualization-and-cloud/kubernetes-developers`)

- *Hands-On Kubernetes on Windows*, by *Piotr Tylenda* (`https://www.packtpub.com/product/hands-on-kubernetes-on-windows/9781838821562`)

You can also refer to the following official documentation:

- Kubernetes documentation (`https://kubernetes.io/docs/home/`), which is always the most up-to-date source of knowledge regarding Kubernetes in general.

- A list of many available Ingress Controllers can be found at the following link: `https://kubernetes.io/docs/concepts/services-networking/ingress-controllers/`.

- Similar to AKS, GKE offers a built-in, managed Ingress Controller called **GKE Ingress**. You can learn more in the official documentation: `https://cloud.google.com/kubernetes-engine/docs/concepts/ingress`. You can also check the Ingress features that are implemented in GKE here: `https://cloud.google.com/kubernetes-engine/docs/how-to/ingress-features`.

- For Amazon EKS, there is **AWS Load Balancer Controller**. You can find more information in the official documentation: `https://docs.aws.amazon.com/eks/latest/userguide/alb-ingress.html`.

Index

Symbols

--expose flag
  avoiding 199

A

adapter design pattern 155, 156
adapter multi-container Pod
  creating 156, 157
AKS cluster
  accessing, with AAD integration
    enabled 537-539
  deleting 479, 480
  launching 465-468
  workload, interacting with 468
AKS cluster, conditional access
  reference link 539
AKS cluster, JIT access
  reference link 539
Amazon EBS PersistentVolume
  YAML 266, 267
Amazon EKS cluster
  deleting 454, 455
  interacting with 447
  launching 444-447
  workload, deploying 447-449

Amazon Elastic Block Storage (EBS) 438
Amazon Elastic Kubernetes Service
  about 436, 438, 547
  Cluster Autoscaler (CA), enabling 601
  eksctl, installing 443
  Kubernetes cluster, deleting on 88
  multi-node Kubernetes cluster,
    launching on 81-88
  used, for installing Kubernetes
    cluster 80, 81
Amazon Machine Image (AMI) 444
Amazon Route 53 438
Amazon Web Services (AWS) 8,
    17, 21, 190, 412, 436, 437
ambassador design pattern 152, 153
ambassador multi-container Pod
  example 153, 154
annotations
  about 109
  adding, to Pod 113
  versus labels 109
anonymous requests 513
anti-affinity 558
application
  decoupling 161

N

R

X

Y

Z

Packt‹t›

Packt.com

Subscribe to our online digital library for full access to over 7,000 books and videos, as well as industry leading tools to help you plan your personal development and advance your career. For more information, please visit our website.

Why subscribe?

- Spend less time learning and more time coding with practical eBooks and Videos from over 4,000 industry professionals

- Improve your learning with Skill Plans built especially for you

- Get a free eBook or video every month

- Fully searchable for easy access to vital information

- Copy and paste, print, and bookmark content

Did you know that Packt offers eBook versions of every book published, with PDF and ePub files available? You can upgrade to the eBook version at packt.com and as a print book customer, you are entitled to a discount on the eBook copy. Get in touch with us at customercare@packtpub.com for more details.

At www.packt.com, you can also read a collection of free technical articles, sign up for a range of free newsletters, and receive exclusive discounts and offers on Packt books and eBooks.

Other Books You May Enjoy

If you enjoyed this book, you may be interested in these other books by Packt:

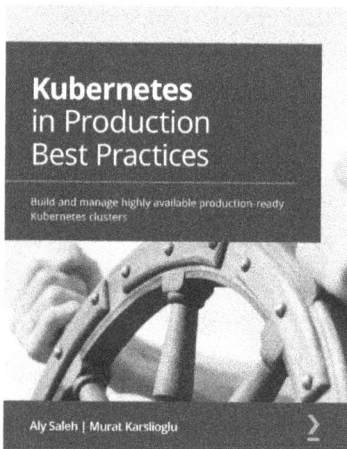

Kubernetes in Production Best Practices

Aly Saleh, Murat Karslioglu

ISBN: 9781800202450

- Explore different infrastructure architectures for Kubernetes deployment
- Implement optimal open source and commercial storage management solutions
- Apply best practices for provisioning and configuring Kubernetes clusters, including **infrastructure as code** (**IaC**) and **configuration as code** (**CAC**)
- Configure the cluster networking plugin and core networking components to get the best out of them
- Secure your Kubernetes environment using the latest tools and best practices
- Deploy core observability stacks, such as monitoring and logging, to fine-tune your infrastructure

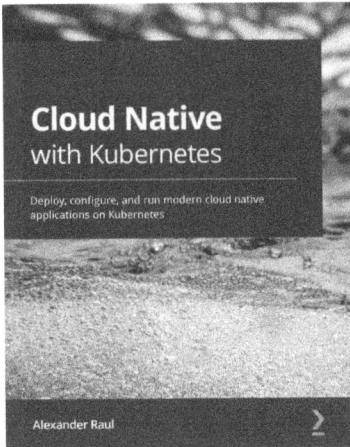

Cloud Native with Kubernetes

Alexander Raul

ISBN: 9781838823078

- Set up Kubernetes and configure its authentication
- Deploy your applications to Kubernetes
- Configure and provide storage to Kubernetes applications
- Expose Kubernetes applications outside the cluster
- Control where and how applications are run on Kubernetes
- Set up observability for Kubernetes
- Build a **continuous integration and continuous deployment (CI/CD)** pipeline for Kubernetes
- Extend Kubernetes with service meshes, serverless, and more

Packt is searching for authors like you

If you're interested in becoming an author for Packt, please visit `authors.packtpub.com` and apply today. We have worked with thousands of developers and tech professionals, just like you, to help them share their insight with the global tech community. You can make a general application, apply for a specific hot topic that we are recruiting an author for, or submit your own idea.

Share Your Thoughts

Now you've finished *The Kubernetes Bible*, we'd love to hear your thoughts! Scan the QR code below to go straight to the Amazon review page for this book and share your feedback or leave a review on the site that you purchased it from.

`https://packt.link/r/1838827692`

Your review is important to us and the tech community and will help us make sure we're delivering excellent quality content.

www.ingramcontent.com/pod-product-compliance
Lightning Source LLC
Chambersburg PA
CBHW082102220326
41598CB00066BA/4609